T0291807

Harmonious Triads

Transformations: Studies in the History of Science and Technology
Jed Z. Buchwald, general editor

Jesuit Science and the Republic of Letters
Mordechai Feingold, editor

Ships and Science: The Birth of Naval Architecture in the Scientific Revolution, 1600–1800
Larrie D. Ferreiro

Wireless: From Marconi's Black-Box to the Audion
Sungook Hong

The Path Not Taken: French Industrialization in the Age of Revolution, 1750–1830
Jeff Horn

Harmonious Triads: Physicists, Musicians, and Instrument Makers in Nineteenth-Century Germany
Myles W. Jackson

Spectrum of Belief: Joseph von Fraunhofer and the Craft of Precision Optics
Myles W. Jackson

Affinity, That Elusive Dream: A Genealogy of the Chemical Revolution
Mi Gyung Kim

American Hegemony and the Postwar Reconstruction of Science in Europe
John Krige

Conserving the Enlightenment: French Military Engineering from Vauban to the Revolution
Janis Langins

Picturing Machines 1400–1700
Wolfgang Lefèvre, editor

Secrets of Nature: Astrology and Alchemy in Early Modern Europe
William R. Newman and Anthony Grafton, editors

Historia: Empiricism and Erudition in Early Modern Europe
Gianna Pomata and Nancy Siraisi, editors

Nationalizing Science: Adolphe Wurtz and the Battle for French Chemistry
Alan J. Rocke

Harmonious Triads

Physicists, Musicians, and Instrument Makers in Nineteenth-Century Germany

Myles W. Jackson

The MIT Press
Cambridge, Massachusetts
London, England

First MIT Press paperback edition, 2008

This book was set in stone sans and stone serif by SNP Best-set Typesetter Ltd., Hong Kong.

Library of Congress Cataloging-in-Publication Data

Jackson, Myles W.
Harmonious triads : physicists, musicians, and instrument makers in nineteenth-century Germany / Myles W. Jackson.
p. cm.—(Transformations: Studies in the history of science and technology)
Includes bibliographic references and index.
ISBN-13: 978-0-262-10116-5 (hc: alk. paper)—978-0-262-60075-0 (pb: alk. paper)
1. Music—Acoustics and physics—19th century. 2. Music—Germany—19th century—Philosophy and aesthetics. 3. Germany—Intellectual life—19th century. I. Title. II. Series: Transformations (MIT Press).

ML3805.J33 2006
784.1943'09034—dc22

2006044991

The MIT Press is pleased to keep this title available in print by manufacturing single copies, on demand, via digital printing technology.

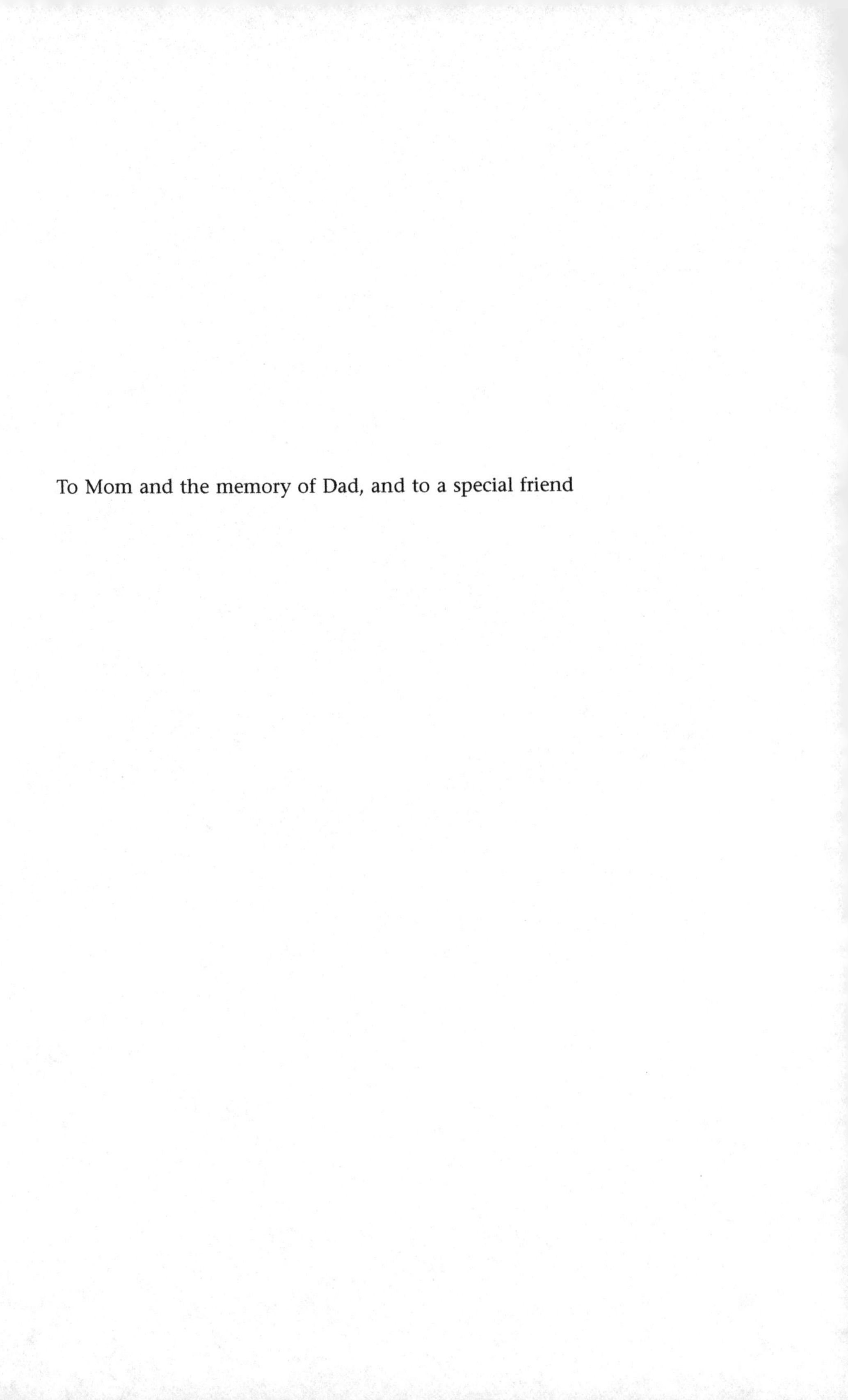

To Mom and the memory of Dad, and to a special friend

Contents

Acknowledgments

This book is the culmination of eight years of labor. During that time my work has benefited greatly from various scholarships and fellowships as well as the comments from a number of colleagues. An Alexander-von-Humboldt Fellowship in 2000–2001 at the Max-Planck-Institut für Wissenschaftsgeschichte in Berlin, a Senior Fellowship at the Dibner Institute for the History of Science and Technology at MIT in 2003–2004, a Willamette University Atkinson Grant in 2002, and an Oregon Council of the Humanities Research Fellowship in 2002 enabled me to carry out and complete this project. And my colleagues at Willamette will attest to the university's generosity in permitting me to take time off in order to scour the globe for archives.

Lorraine Daston served as my Humboldt Fellowship advisor and read manuscripts from various stages of the book. I thank her for her time and comments. Rob Iliffe, Rob Kohler, Michael Gordin, and David Kaiser all read and made insightful comments to various chapters at earlier stages. While at the Dibner Institute, Elfrieda and Erwin Hiebert, David Pantalony and I created an informal reading group on the history of acoustics. They read all the chapters of this book. I thank them for sharing their wisdom, collegiality, and friendship. I also benefited from numerous discussions with George Smith. John Peel, Michael Nord, and Gaetano DeLeonibus kindly offered me their expertise. Finally, I owe much gratitude to Jed Z. Buchwald, who painstakingly and thoughtfully read through an earlier version of this book. His insightful critiques greatly improved this work. Cambridge University Press and Science History Publications Ltd. kindly granted me permission to draw upon earlier arguments of mine, newer versions of which comprise portions of chapters 3 and 8.

I would also like to thank the librarians and archivists at various institutions, who helped make this book possible: Geheimes Staatsarchiv—Preußischer Kulturbesitz in Dahlem (Berlin); the Max-Planck-Institut für Wissenschaftsgeschichte, Berlin; Willamette University; the Deutsches Museum, Munich; the Niedersächsiche Staats- und Universitätsbibliothek Göttingen; the Staatsbibliothek Preußischer Kulturbesitz Berlin, Houses I and II; Staatliches Institut für Musikforschung in Berlin; the Dibner Library at MIT; and the Loeb Music Library and Houghton Library at Harvard University.

1 Introduction

Historically, music has been classified both as a field of natural philosophy and as a performing art. It was, of course, a part of mathematics within the quadrivium. Throughout the Scientific Revolution, savants interested in music's association with natural phenomena offered elaborate schemes in order to account for its physical basis. In his seminal article, Claude V. Palisca has argued that theoretical music, or the aspects of music dealing with, among other things, the mathematics of tuning or the music of the celestial spheres, was a branch of the exact sciences until the seventeenth century. He adds, "The separation of musical art from science was an event of considerable importance for the future development of composition and musical practice. . . ."[1] Many concerned themselves, for example, with providing physical accounts that could distinguish dissonance from consonance. They also delved into such theoretical problems as mathematically dividing up the octave in different ways, establishing various musical temperaments.

Debates among musical theorists (so-called because they were interested in the theoretical qualities of music) concerning the appropriate temperament for pieces raged from the sixteenth through eighteenth centuries. Meanwhile, natural philosophers such as Johannes Kepler, Galileo Galilei, René Descartes, Marin Mersenne, and Christiaan Huygens were busying themselves with similar problems. A number of them attempted to explain consonance and dissonance in terms of physical, physiological, and at times psychological processes.

Far removed from the world of musical theoreticians, musical performers were practicing their art in public places throughout Europe. Their virtuosity was based not on any intellectual contribution to the body of scientific knowledge, but rather in their manual and vocal skills. And musical-instrument makers plied their craft in the darkened enclaves of artisanal shops, enveloped in guild secrecy. Both of these groups cared little for music's alliance with natural philosophy. To them, music was their livelihood, their source of income.

Similarly, although science clearly possesses theoretical components, it is also a craft, as sociologists, philosophers, and historians have argued from quite some time now.[2] Scientists not only contemplate, they also tinker. Early historical analyses of the relationship between science and music have tended to emphasize both as theoretical bodies of knowledge, embodying universal principles.[3] More recent studies, however, are beginning to consider both to be cultural enterprises, comprising similar sets of skills and practices.[4]

Both disciplines have undoubtedly been inextricably linked throughout history. And given their epistemological natures, many different types of histories can be written describing their historically contingent relationship. Histories that concentrate on the logic of musical theory—as crucial as they are to the discipline—have seemingly exhausted the themes of temperament, dissonance and consonance, and the music of the heavens. Given the bravado of recent cultural histories of science, which argue for the relevance of the cultural context to the content and conduct of science, it is surely time to construct a history of science and music that will improve upon the disciplinary histories of both fields. By searching out new historical venues where science and music intersect—whether in the analysis of the virtuoso, in the instrument makers' shops, in the architectural design of concert halls, or in the construction of tuning forks and metronomes, one can begin to see how each discipline helped shape the other. Music provided fruitful material for the investigations of natural philosophers as well as important opportunities for social interactions and (later on) the formation of a national identity for scientists. Physicists helped musical-instrument makers improve on the design of instruments by sharing their acoustical expertise, while these craftsmen discussed the various designs and components of their instruments with natural scientists. Musicians turned to physicists for ways to quantify and standardize musical pitch and beat. And the link between music and science did not cease after the Scientific Revolution. Quite the contrary, nineteenth-century physicists, much less interested in the music of the spheres or in musical temperament than natural philosophers centuries earlier, were nevertheless committed to understanding the world of performing musicians and their instruments. Indeed, given the importance of music to the physical and physiological studies of Ernst Florens Friedrich Chladni, Felix Savart, Pierre Louis Dulong, Ernst Heinrich and Wilhelm Eduard Weber, Georg Simon Ohm, Hermann von Helmholtz, Lord Rayleigh, and John Tyndall, to name just a few, histories of music and science promise to open up new methodological vistas and provide rich tapestries woven together by fascinating interdisciplinary narratives.

The goal of this book is to provide such a history. My historical account of German physics and music will discuss how critical experiments dealing with adiabatic phenomena and theories of acoustical resonance, for example, were carried out by physicists on musical instruments, such as organ pipes. Musical instruments provided physicists with experimental systems, and in turn, their work directly contributed to the improvement of musical-instrument manufacture. Debates on the measurement of musical qualities, such as beat, pitch, and virtuosity, became the venue for physicists to apply physical principles in hopes of assisting performing musicians in their craft. And, conversely, music provided investigators of nature with a cultural resource, which in part forged acquaintances and future collaborations.

German *Naturforscher* (investigators of nature), such as Alexander von Humboldt and E. F. F. Chladni, and physicists, including Wilhelm Weber and Helmholtz, were interested in aesthetic issues. But why?[5] As Tim Lenoir has already done, one can trace the personal connections among the members of the German *Bildungsbürgertum*, the educated upper-middle class.[6] Such a prosopography—a study that identifies and draws relationships between various individuals within a specific historical context—reveals that artists and natural scientists, including physicists, physiologists, chemists, and mathematicians, all belonged to the same social circles (and, in some cases, even the same families). They were educated in the same schools and often made their careers in the same institutions, such as the Berlin academies of art and science. They participated in Sunday afternoon salonlike gatherings of friends and family, for concerts or lectures on art, literature, music, or popular science. Many were members of the same associations (Vereine). And, they collaborated on common projects that shaped German *Kultur*.[7] These same social circles also included musicians. For example, the Mendelssohn-Bartholdys and

Humboldts were family friends. And the composer Carl Maria von Weber was the father-in-law of Berlin's professor of zoology and baritone, Hinrich Lichtenstein. Wilhelm Eduard and Ernst Heinrich Weber, who as children lived in the same house as "the father of experimental acoustics" E. F. F. Chladni, published acoustical treatises in the leading musical journals of the early nineteenth century. Some savants, including the lawyer and magistrate (and amateur cellist and flautist) Gottfried Weber, were talented enough to offer theories of musical composition and acoustical theory. And many natural scientists, such as Helmholtz, Chladni, and Lichtenstein, were gifted amateur musicians. These members of the *Bildungsbürgertum* busied themselves with defining the roles aesthetics and science would play in shaping a liberal, united Germany.

Aesthetic issues, views of nature, and the political circumstances of the French Revolution and subsequent Wars of Liberation were all intricately interwoven during the late eighteenth and early nineteenth centuries. Michael Dettelbach shows that Alexander von Humboldt's enterprise, which Humboldt himself described as a "terresterial physics," was simultaneously a physical investigation owing much to the Laplacian school of precision measurement and instrumentation *and* an aesthetic view of nature influenced by Friedrich Schiller, particularly *Ueber die aesthetische Erziehung der Menschen* (*On the Aesthetic Education of Man*).[8] Dettelbach demonstrates that precision measurement and aesthetics existed harmoniously in Humboldt's view of nature. Humboldt informed us that he went to great pains to render the reader "sensitive to nature's beauty," and impress upon him or her how the landscape "determines the imagination and aesthetic sense of peoples."[9] And he was constantly striving to conquer the challenges posed by both "precision" and "picturesqueness" in his representation of vegetation.[10] Humboldt's *Naturanschauung* united the world of mechanical, instrumental mastery that pervaded Paris with the aesthetics of an organic nature emerging in the fledgling German states. Many of the intellectuals in those German states desperately sought the freedom from tyranny for all of humanity, which they hoped the French Revolution would bring about, only to have their hopes dashed by Napoleon's invading forces.

Schiller was one such intellectual: "If man is ever to solve the problem of politics in practice he will have to approach it through the problem of aesthetics, because it is only through Beauty that man makes his way to

Freedom."[11] Aesthetics was for him a political issue.[12] In the last letter of his *On the Aesthetic Education of Man*, Schiller outlined his "*aesthetisches Reich*," whereby, as Dettelbach concludes, Schiller demonstrated that "[t]he domain of the aesthetic is where lasting social order is achieved." The historical context is, of course, the moment of crucial transition in the postrevolutionary German territories.[13]

In a similar fashion, Norton Wise argues that after the defeat of Napoleon, Prussia began to celebrate its victory by distancing itself from French cultural influence and turning to France's perennial nemesis, Great Britain.[14] This transition was most notable in the numerous English-style gardens cropping up throughout Berlin and Potsdam from 1815 to 1850. These Prussian gardens nevertheless differed from their English prototypes, which had been renowned for their depictions of a wild nature left to run its course, unfettered by human intervention, in the former's inclusion of steam engines.

The Glienicke park in particular, with its steam engine lifting water from the Havel River to the previously sandy lands, evoked the feelings of aesthetic beauty and technological innovation. While visiting Glienicke in 1841, the future field marshal Helmuth von Moltke wrote to his wife, "I wish I could escort you around in this exquisite park. For as far as the eye can see, the grass is of the freshest green, the hills are crowned with beautiful deciduous trees; and the stream and the lakes weave their blue ribbon through a landscape in which castles and villas, gardens and vineyards lie scattered. Certainly the Glienicke park is one of the most beautiful in Germany." He continued by commenting on how technology enabled nature to flourish in areas that were previously barren. "It is unbelievable what art has known how to make out of this barren earth. A steam engine works from morning until night to lift the water out of the river Havel up to the sandy heights and to create lush meadows where without the engine only weeds of the heath [heather] would survive. A powerful cascade roars over cliffs under the arch of a bridge, half washed away, seemingly from its violence, and abruptly rages fifty feet down to the Havel onto a terrain where prudent Mother Nature would not have thought to let a pail of water flow, because the parched sand would have had to stand forty years to achieve this mightiness."[15]

Such emotions resonated rather well with the policies of Glienicke's previous owner, Prince Hardenberg, the Prussian chancellor from 1810 to

1822.[16] The chief architect of Glienicke, the landscape artist Peter Josef Lenné strove to unite utility and beauty in his works. His city design of Berlin with the renowned Prussian architect Karl Friedrich Schinkel included lush parks interspersed among factories and workers' dwellings.[17] A variation of this intriguing juxtaposition of the mechanical and organic represents a critical tension addressed in this book, tying the work together. Rather than being an austere contradiction, the steam engine and garden existed harmoniously: each component blending with the other to produce a pleasing effect to the senses.

Emulation of Britain's economic power was an important goal of the liberal P. C. W. Beuth, who founded Berlin's Gewerbeschule (Technical Institute) in 1821. The Gewerbeschule sought to combine basic knowledge of building construction and the physical sciences with metalworking and metallurgical techniques. To do so, it hired a number of professors of the natural sciences who simultaneously taught at the University of Berlin, including the chemists Eilhard Mitscherlich and Rudolf Magnus, the physicists K. D. Turte and Heinrich Dove, and the botanist H. F. Link.[18] Housed in the same building was another of Beuth's inventions, the Verein zur Beförderung des Gewerbefleisses (Society for the Advancement of Industry), which acted as a type of scientific society providing a venue where the most current technological advances were discussed.[19] This society also had natural scientists as active members seeking to use science to bolster the economic strength of the state. And, as Wise suggests, both the Gewerbeschule and the Verein stressed the cultivation of aesthetics.[20] It was in this milieu of science, technology, aesthetics, and reformed economics that the young Helmholtz thrived, seeking "to join through a dedicated pursuit of medicine and science coupled with a developed taste and talent for art and music."[21]

Indeed, a number of German natural scientists saw themselves as *Kulturträger,* "bearers of culture," a word coined by the physician Rudolf Virchow, one of the leaders of the Vorschrittspartei (the Progressive Party) opposed to Otto von Bismarck. Du-Bois Reymond shared Helmholtz's commitment to progressive liberalism as evidenced by his understanding of the relationship between natural science and culture articulated in his lecture of March 1877, entitled "Natural Science and the History of Culture."[22] Helmholtz, too, argued for the importance of the natural sciences to economic transformation and unification.[23] As a *Kulturträger,* Jurkowitz asserts, he was dedicated to depicting the role of science within the "broader development of society and 'civilization'—a keyword for progressive liberals."[24]

Perhaps the most revealing account of Helmholtz's commitment to the relationship between science and art was his 1892 discussion of Goethe, which was far more sympathetic to the titan of German *Kultur* than had been his remarks of 1853. In that early account, Helmholtz praised Goethe's poetic and literary skills, which sharpened his observational abilities, but criticized the poet's misguided diatribes against Newton, which ironically were a result of being blinded by those very same poetic skills. In the later essay, the aged Helmholtz decided that he and Goethe shared the fundamental belief that natural scientists and artists needed to possess both a thorough knowledge of the law-governed behavior of the object being represented and an ability to depict those natural laws.[25]

Helmholtz limited his scientific study of music to the experience of musical notes in their most basic forms, since the comprehension and explanation of the extraordinarily complex effects of music on the listener required a detailed knowledge of historical circumstances and national characteristics.[26] Yet, he did feel that it certainly was within the purview of a physiologist to proffer a general overview of aesthetics, as he himself did in *Tonempfindung*, since "aesthetic intuition depends on the same sort of unconscious mental operations as occur in ordinary sense perception."[27] Hence, as Hatfield points out, Helmholtz felt that the natural sciences could indeed explain the immediate sensational effects of music (and works of art) but were unable to derive "laws" of analysis of the aesthetic effect of entire compositions.[28] This was the domain of the musicologist, artist, and critic. In short, natural scientists such as Helmholtz and Du Bois-Reymond, much like Goethe and Schiller had a half-century before, appreciated the importance of aesthetics to both the development of science and the identity of German culture.[29]

The second chapter of this book analyzes the acoustical works of E. F. F. Chladni. Despite his importance to the history of acoustics little has been written (particularly in English) on his contributions to both experimental acoustics and musical-instrument making. His life offers insight into the fascinating and complex acoustical landscape of late-eighteenth- and early-nineteenth-century German territories; his labors led to the invention of two musical instruments, the euphone and clavicylinder, which were found with increasing frequency in bourgeois households. He was also a rather astute music critic. His instruments were built, in part, to stem the rise of performances that featured the digital gymnastics (or *Augenmusik*, literally "music for the eyes") of crass virtuosi wishing to impress fee-paying audiences.

The third chapter details the importance of the *Liedertafel* (choral society) songs to the creation of an identity for the newly established Versammlung deutscher Naturforscher und Aerzte (Association of German Investigators of Nature and Physicians). Music was able to provide German *Naturforscher* and physicians a vehicle for socializing, male camaraderie, and cultural unity that inspired future intellectual collaborations in the absence of a political unity. Popular folk songs, whose words were rewritten to reflect the scientific interests of these savants, were certainly not viewed as a mere pastime or hobby. Singing these tunes partly defined what a *Naturforscher* or physician did.

Whereas choral society songs represent the organic and social aspect of music, music's mechanical aspects, namely mechanical musical instruments, are discussed in the fourth chapter taking us into the artisan workshops of the early nineteenth century. In these secretive places, artisans tried to build machines that would rival musicians' performances in precision, speed, and expression. This mechanical, avowedly inorganic reproduction of music was diametrically opposed to the ideologies of the members of the *Sturm und Drang* movement and the German Romantics, some of whom parodied the belief that any sound produced by a mechanical contraption could be considered musical, and all of whom felt that the very notion of the mechanical was anathema to the organic essence of music. The mechanical and the organic views of music gave rise to two different types of aesthetic: one of precision and flawlessness, the other the aesthetic of sentimentality. This chapter addresses two aspects of the mechanical. The first was the mechanical world of French Cartesian rationalism, the ideology of the invaders who reduced biological phenomena to lifeless levers and pulleys. German literary savants, most notably Goethe and Schiller, celebrated the human spirit and were fascinated by nature and its forces. They stressed J. F. Blumenbach's *Bildungstrieb* (formative force), which differentiated living from nonliving entities, and limited the ability of Cartesian reduction to explain natural phenomena. Second, "mechanical" also possessed a musical resonance in the form of mechanical musical instruments.

Chapter 5 explores the collaboration of two mechanicians with the experimental physicist, Wilhelm Eduard Weber. Weber, far removed from the world of the Romantics, was fundamentally committed to applying physical principles to areas of manufacturing with the aim of improving Germany's economy. A good example was his work on adiabatic phenomena, wherein he developed a method for the construction of reed pipes that was subsequently deployed by organ builders. The reed pipe was also a scientific instrument, and Weber had a mechanician construct an extremely sensitive one to test the velocity of sound through various gases and determine the ratio of the increase in pressure and density of a sound wave, a critical value in the study of adiabatic processes. The term mechanical here connoted industrialization, which was spreading throughout the German territories (particularly Prussia) from the 1840s onward. An important characteristic of industrialization was indeed the cooperation between physicists and mechanicians. The construction of tuning forks, sirens, and tonometers—direct results of those collaborations—was relevant to the practice of nineteenth-century music and acoustical experimentation. Calibration, precision measurement, and standardization were mechanical processes crucial to both music and physics.

Weber's compensated reed pipe was not the only example of the application of precision measurement to music. Chapter 6 tells the story of the silk manufacturer Johann Heinrich Scheibler, who hoped that the universal principles of mathematics and mechanics would replace weavers as well as keyboard tuners. As is well known, the mechanization of the Industrial Revolution resulted in the replacement of some laborers with machines, continually shifting what counted as a skill. Being the first in Krefeld to power his silk factory with a steam engine, Scheibler developed a technique for tuning keyboard instruments in equal temperament by means of a tonometer built by a mechanician. Performers, with his technique and tuning forks in hand, could tune their organs and pianos to equal temperament with unprecedented accuracy. His technique drew upon the use of combination tones, a phenomenon that supported the beat theory of the superposition of waves. Indeed, physicists praised Scheibler's technique for generating so clearly these physical phenomena.

Precision is also the theme of chapter 7, which uncovers the nineteenth-century attempts to standardize musical beat and pitch by means of the chronometer, metronome, tuning fork, and siren. Debates erupted among musicians, musical-instrument makers, and physicists both as to whether standardizing such aesthetic qualities was anathema to the creative enterprise of music making, and as to what the roles of physics and mathematics should be in determining such values. The procedures involved in standardization were relevant not only to the practice of music, but also to musical-instrument manufacture, national pride, and even improved precision measurement for experimental physicists.

The final chapter teases out yet another fundamental tension created by the interplay between the mechanical and the organic. The early nineteenth century marked a period of unprecedented mechanization. Machines were at the heart of Britain's new political economy, in which the economic cycle became dependent upon mechanization, thereby reducing the skill base of the artisanal trades.[30] Machines were used to discipline, and in some cases replace, the human body.[31] The beauty of specialized handicraft was succumbing to the aesthetic of mass production in many spheres of culture, including music.

Over the past decade, historians of science have discussed the attempts by savants to mechanize certain processes relevant to natural philosophy. For example, Charles Babbage spent over twenty years having skilled artisans attempt to build his difference engine, which was intended to replace the labor of human calculators.[32] Both Babbage and John Herschel were committed to applying the efficiency of industrial manufacture to the human mind.[33] Critical to these historical studies is the role of skill, a topic that has received considerable attention from historians of science.[34] Not coincidentally, all of these studies have focused on Britain, the most mechanized nation of the early nineteenth century. Although the German territories were not nearly as mechanized, what constituted scientific and musical skill and how such skill should be recognized and coordinated were nevertheless hotly contested issues.[35]

This final chapter concerns itself with the mechanization of music and how such mechanization lent itself to physical analysis. It had long been the desire of artisans to craft a mechanical device that could serve as a surrogate for an orchestra. By the mid-nineteenth century, several music critics claimed that the mechanicians had achieved their goal of mimicking the skills of performers. And a number of keyboard instructors turned to mechanical contraptions to inculcate the requisite skills for playing the piano. In the first instance, the mechanical replaced the human; in the second, the human became part of the mechanical.[36] Could machines somehow convey the correct playing technique for young girls and boys learning to play the piano? Could physicists, using the universal principles of mechanics, such as force, momentum, and mass, explain the skilled manipulations of a Paganini or a Liszt with a view to pass on that information to future generations of musicians? Was virtuosity somehow quantifiable? A historical analysis of these questions, with which both physicists and musicians occupied themselves throughout the nineteenth century, not only begins to shed light on the cultural histories of music and physics, but also forces us to admit that a history of one is incomplete without a history of the other.

2 E. F. F. Chladni: The Nodal Point between Acoustician and Musical-Instrument Maker

Labeled the founder of experimental acoustics by his nineteenth-century biographers,[1] Ernst Florens Friedrich Chladni (1756–1827) (figure 2.1) moved in several interesting spheres of late-eighteenth- and early-nineteenth-century German culture, including physics, music, and musical-instrument making. Famous for revealing the patterns that vibrations produce when a metal or glass plate sprinkled with sand is bowed, Chladni tested the acoustical theories of Leonhard Euler, Gottfried Wilhelm Leibniz, Count Giordano Riccati, and Daniel Bernoulli with an impressive range of experiments and, in the process, uncovered new physical phenomena relevant to acoustical, musical, and adiabatic phenomena. Chladni never held a secured position, and his peripatetic existence—which he himself described as a "*nomadische Lebensreise*"[2]—his flexibility, and his poverty encouraged him to travel throughout Europe, sharing his knowledge of acoustics with and gleaning information for his musical-instrument making from leading savants and instrument makers of the period. Remarking on his lifestyle, Chladni argued in 1802 that as an artist (*Künstler*), who was not constrained to one particular place, he had a distinct advantage over the scholar, who generally needed to live a stationary life.[3] Fifteen years later he elaborated: "I find such an existence moreover neither a shame for science [*Wissenschaft*], nor for me. If I were employed in a single place (with the exception of Göttingen, where one can make use of a rich library), I would not be able to work on either acoustics or meteors, because it would be difficult for me to have learned all of the earlier observations and experiments of others, . . . and I would have had most likely not invented both of my [musical] instruments."[4]

Because Chladni moved among several spheres, his life can be used to cast light on the relationships among the physical theory of acoustics,

Figure 2.1
Portrait of E. F. F. Chladni (1756–1827). Source: Guillemin 1881, p. 694.

aesthetical taste in music, and the practices of musical-instrument making. Despite the high level of his education, his career mirrored those of artists and mechanicians aiming to sell their wares. And his ability to intervene in arenas that seem today to be so far apart was the rule, rather than the exception. Many acousticians of the period found themselves offering their sought-after advice to musicians and instrument makers. Indeed, numerous savants, who wrote essays on tonal theory, also made their views known on acoustical theories and instrument design. Chladni, himself one of those cherished acousticians, periodically penned reviews of newly invented musical instruments in the pages of the *Allgemeine musikalische Zeitung*, the leading musical journal of the period.[5]

After attending the Fürstenschule of Grimma near Wittenberg, Chladni informed his father, the first Professor of Law at the University of Wittenberg, that he wanted to study medicine, as the natural sciences fascinated him the most.[6] His stern father objected, informing his son that a career in law would guarantee him a handsome salary. The disappointed lad departed for Leipzig to embark upon his legal studies. Completing his degree in 1782, he returned to Wittenberg, where his father had secured his employment. Later that same year, Chladni's father died. This loss shaped the development of the career of the bereaved Chladni in two critical ways. First, Chladni was now free to pursue his interest in the natural sciences. Second, the twenty-five-year-old was financially responsible for both himself and his stepmother. He offered lectures on geometry, physical and mathematical geography, and natural history at the University of Wittenberg.[7] Unfortunately, his lectures were not very well attended; hence, his salary was barely sufficient to support himself, let alone his stepmother. Chladni needed to pursue a different course, and acoustics was the path he chose.

While a student in Grimma, Chladni learned to play the piano. From that point on, music was to play a major role in his life. During the 1780s, he began to study the mathematical works of D. Bernoulli, Riccati, and Euler on the vibrations of strings and other resonating bodies. He, however, wished to approach acoustics via experimentation, rather than mathematics. His acoustical investigations owed much to the work of Georg Christian von Lichtenberg on electricity. In 1771 the renowned Göttingen physicist had discovered that an electric spark leaves behind characteristic traces when it jumps over a nonconducting material. These traces can be made visible by sprinkling the area with a powder, such as resin, or dust. If the spark derives from a positively charged conductor, the powder assumes a shape characteristic of a star or tree. If the spark is negative, the powder takes on the form of a cloud.[8] Chladni reckoned that this rendering visible of the invisible could be applied to the vibrations of resonating bodies. He set a metal plate on a stand and sprinkled its surface with grains of sand. He then placed the fingers of his left hand on various parts of the plate while bowing its edges perpendicular to its surface with his right hand (figure 2.2). To his amazement, a ten-sided figure resulted (see number 8 in figure 2.3). Varying where the plate was bowed and held generated numerous figures with differing geometrical shapes. These shapes, later to be called "Chladni figures," served as the basis of his first book, *Entdeckungen über die Theorie des Klanges* (Discoveries on the Theory of Sound), in 1787.

Chladni commenced this work, which was presented to the Royal Academy of Sciences in St. Petersburg, with the observation that, while much research had already been undertaken on the vibrations of strings and rods, the properties of elastic curved surfaces with numerous

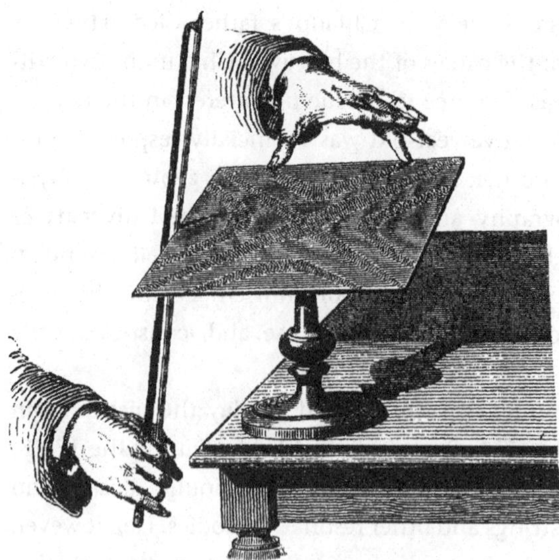

Figure 2.2
Generating the Chladni figures. Source: Guillemin 1881, p. 695.

dimensions were still generally unknown. Every sounding body generates different tones and assumes a type of vibrating motion, in which the resulting curve either transects the axis (or the form of the body when at rest) or not (figure 2.4). Chladni called the points of the vibrating body that pass through the axis nodal points (*Schwingungsknoten*). They remain still while other points of the resonating body vibrate.[9]

Chladni was the first savant to carry out detailed experimental investigations of elastic materials that were sent into motion by being bowed. He reviewed the available literature provided by Daniel Bernoulli, Euler, and Riccati on the acoustical properties of rods, strings, and metal strips.[10] In the 1730s, Bernoulli had shown mathematically that the movements of strings comprise an infinite number of superimposed harmonic vibrations. During the 1770s, Euler had investigated the properties of one-dimensional vibrating rods, considering the case of a horizontal rod experiencing small vertical displacements. He had argued, based on mathematical grounds, that the number of the rod's vibrations per second: $s = (m^2/L^2)\ \sqrt{(Dr/g)}$, where m is the variable depending on the rod's degrees of freedom, D is the diameter of the rod, L its length, r is the coefficient of elasticity, and g is the specific gravity. Shortly thereafter, Count Giordano Riccati

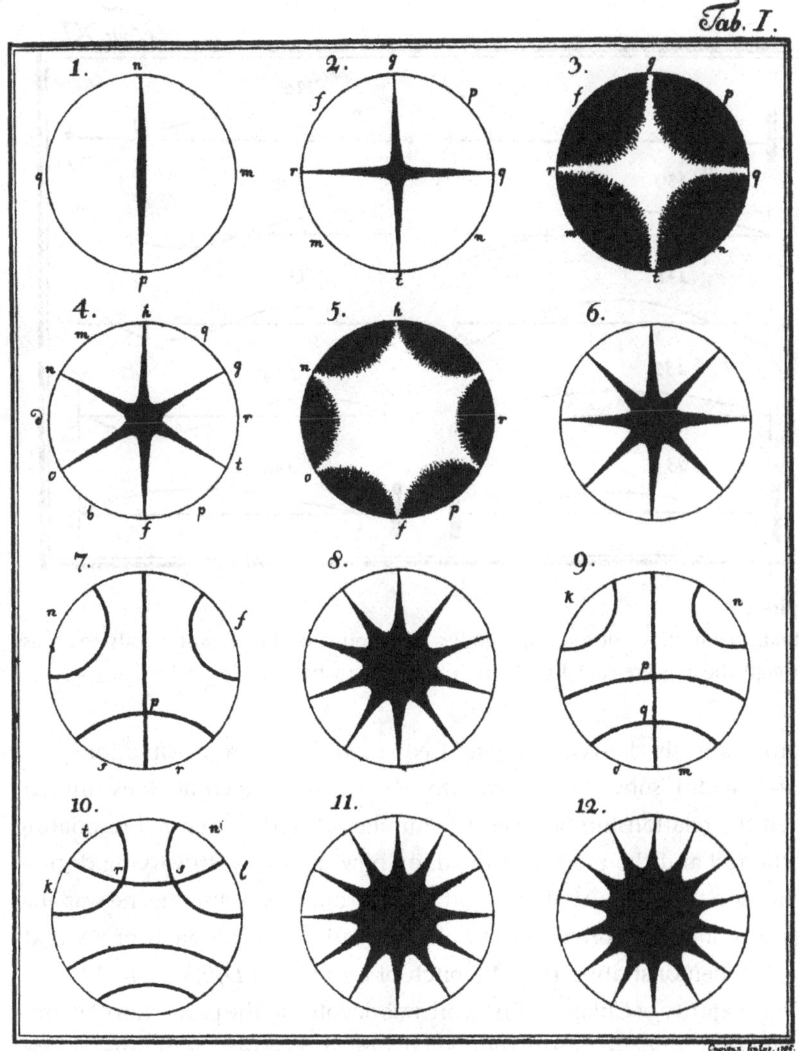

Figure 2.3
Chladni figures. Source: Chladni 1787, table I.

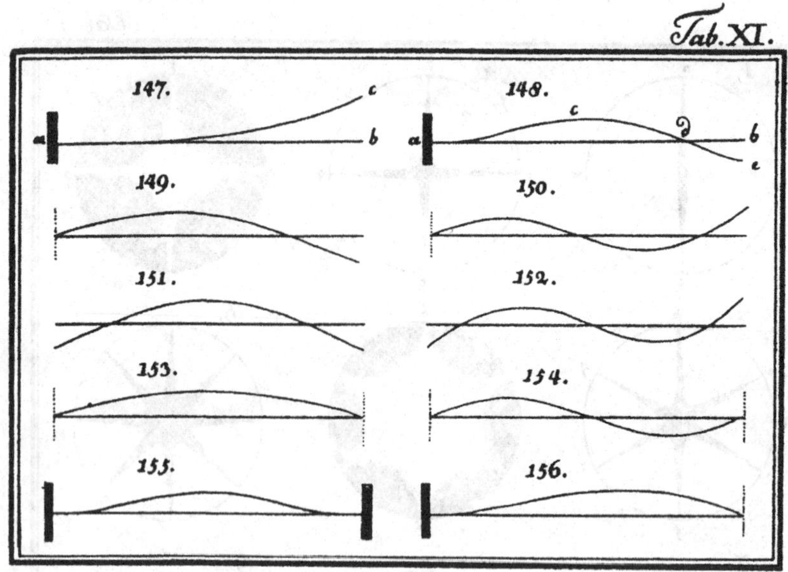

Figure 2.4
Vibrational motions of sounding bodies. The points of the vibrating body that pass through the axis are called nodal points. Source: Chladni 1787, table XI.

mathematically derived a different equation for the rod's vibrations, one that Chladni subsequently confirmed. Chladni experimentally investigated the relationship between the number of nodal lines of a resonating elastic rod and the pitch generated, by bowing rods with varying degrees of freedom.[11] By means of his rigorous experimental technique, he was able to verify Riccati's correction of Euler's equation for the pitch of a vibrating rod, demonstrating that the pitch of a rod is $(m^2D)/L^2\ (\sqrt{(r/g)})$.[12]

The majority of Chladni's first work was devoted to the properties of vibrating motions of bells, bowls, rounded plates, and vessels, or in other words, entities with elastic curved surfaces.[13] Before Chladni's work there had been very little written on the topic, as previous scholars had merely assumed (wrongly as it turns out) that the bell, bowl, rounded plate, and vessel could all be analyzed as though they were composed of an infinite number of elastic rings. Hence, it had been argued, the acoustical properties of solids with these shapes were the same as those of elastic rings, which are essentially one-dimensional curves. Chladni, however, experimentally demonstrated that the acoustical properties of these sorts of objects could best be understood by analyzing the properties of elastic curved surfaces, which then remained by and large unknown. To demonstrate this claim, he again resorted to the visualization of vibrational patterns, in particular for resonating bells. He placed a thin piece of paper on the bottom of an inverted bell and sprinkled the paper with grains of sand or iron filings and bowed the bell at a chosen point (see figure 2.3, no. 2). When the bow was applied perpendicular to the bell's surface at either *f*, *p*, *n*, or *m*, the grains of sand formed a cross pattern on the paper. The bell could be touched at any point lying on the cross-bars *gt* or *rq* without disturbing its vibrating parts; hence, the bell's tone was not altered. As we shall see below, these experiments were crucial to Chladni's invention of two musical instruments, the euphone and the clavicylinder. His novel technique for determining pitches was a marked improvement over the common use of a monochord (a string instrument used to determine pitches of other musical instruments).[14]

The Glass Harmonica

The acoustics of elastic curved surfaces was of particular interest to Chladni, since it was applicable to a range of popular instruments of the period, namely the glass harmonica and several related instruments. And, as we shall see, the glass harmonica served as a prototype for his euphone. Although its roots date back to the seventeenth century, the glass harmonica was the instrument of choice for bourgeois households during the late eighteenth and early nineteenth centuries.[15] Composed of glass bowls, which were connected to a revolving axis (figure 2.5), its tones were elicited by rubbing a finger lubricated with water on the revolving bowls.[16] In the early 1760s, Benjamin Franklin had constructed such a device using thirty to forty glass bowls of decreasing size, each having a hole in the center, through which a metal spindle was placed. This spindle, which was attached to a desklike apparatus containing the glass, could be set into motion by depressing a pedal with one's foot. By gliding a finger along the wetted moving glass rims, full chords could be produced. The *Musikalische Almanach für Deutschland auf das Jahr 1782* (The Musical Almanac for Germany for the Year 1782) proclaimed, "Of all musical inventions, the one of Mr. Franklin has created perhaps the greatest excitement."[17] German Romantic circles, in particular, cherished the harmonica, which appeared in a number of novels, where its evocative harmonies often caused the

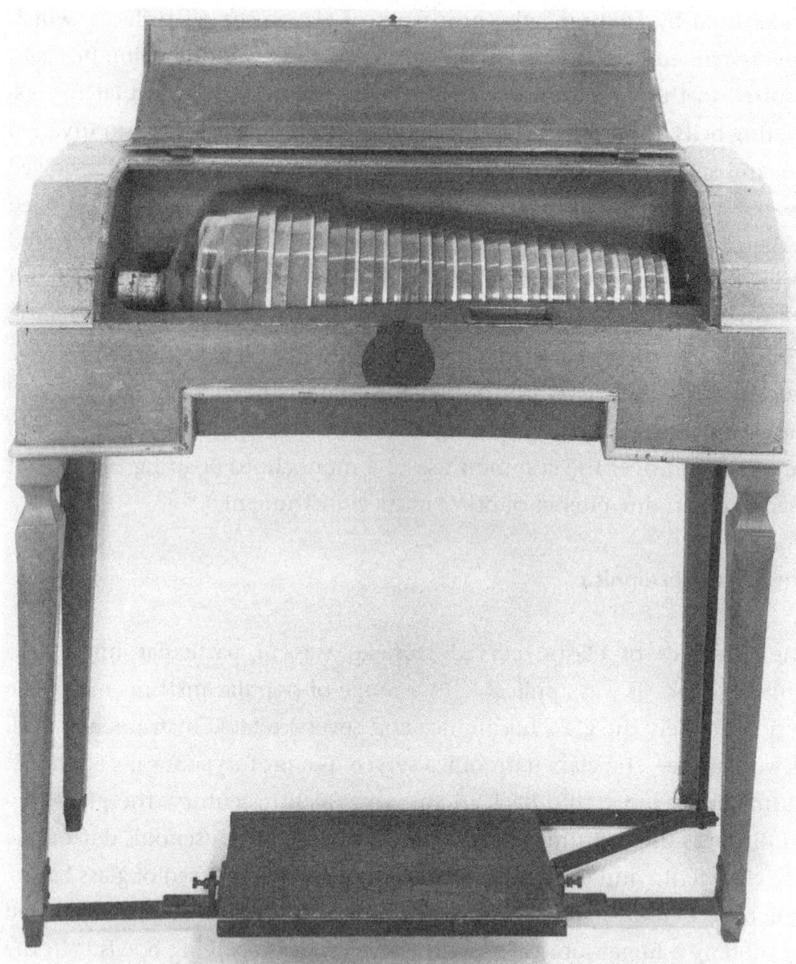

Figure 2.5
Glass harmonica of the late eighteenth century, based on the design by Benjamin Franklin. Photo: Deutsches Museum, Munich.

heroine to faint.[18] Its tones were thought to be ethereal and spiritual, associated with *Bewunderung* ("wonder") and *Herzensempfindung* ("sensitivity of the heart")[19] and sounded as though they came "from another world," thereby offering the perfect musical resource for Romantic authors interested in the relationship between natural and supernatural entities.[20] As the early-twentieth-century musicologist Curt Sachs remarks, "one must picture that we are dealing with the heyday of the glass harmonica, the clavicylinder, and the aeolharp. Music aficionados searched 'for a deep, far-reaching effect,' the romantic-supernatural of the diminishing tone from nothingness, everything ethereal, which neither the organ nor the piano could provide."[21] Sachs astutely points to the turn of the eighteenth into the nineteenth century as a time when musical and aesthetic tastes were changing. This type of ethereal, household music was precisely what Chladni's instruments elicited.

Throughout the 1770s many argued that playing the harmonica posed serious threats to the health of the players, many of whom were women.[22] Not only would the constant stimulation through the eardrum pose dire consequences, it was argued, but also the constant vibrations of the glass would travel through the fingertips and adversely affect the performer's nervous system. These claims were backed by numerous examples of virtuosi who needed to stop playing after a decade or so due to nervous breakdowns.[23]

Several musical-instrument makers, whose works were familiar to Chladni, fought to counter what they claimed was a ridiculous, unjustifiable attack against their musical devices. Most notable of these was Johann Christian Müller.[24] Müller admitted that the harmonica could indeed summon forth unusual sensations from individuals, depending on their mood at the time, but continued, "that it should be a disadvantage to one's health has still not been proven."[25] "If the harmonica brings its players closer to death," Müller added, "or at least cause certain illnesses that would be admittedly very bad. But where are the facts? Demoiselle Bause, the renowned glass harmonica player, died from natural causes."[26] He then listed six harmonica players who had been playing for over a decade with no physical complaints whatsoever.

Müller's protests notwithstanding, several musicians and instrument makers reformed the construction of the harmonica in order to avoid any perceived health complications. Joseph Philip Frick, former Court Organist of the Margrave of Baden-Baden who became a leading London musician, constructed a keyboard that controlled a moist material, similar to human skin, that would rub against the glass. He returned to the German territories in 1769 for a one-year tour with his harmonica.[27] His invention was, however, generally ignored. Abbot Mazzuchi employed a violin bow dipped in turpentine, wax, or soap to replace human fingers.[28] He also used bowls made from metal and wood, rather than glass, as these materials tended not to produce unwanted additional tones (*Beitöne*).[29] By the 1770s, many mechanicians had begun to offer their own improvements on the instrument, which, despite its hefty price and alleged harmful effects, was immensely popular.[30] Joseph Aloys Schmittbauer, *Kapellmeister* to the Baron of Baden, believing that keyboards damaged the tones more so than the performer's health, decided that his harmonicas would have to be played with the fingers. However, he decided to substitute glass with crystal in order to generate more melodious tones. His harmonicas enjoyed the impressive range of over two octaves, from e to f^2 (the E below middle C [c′] on the piano, 256 Hz in just temperament, to the F above middle C), the widest of the day (see the appendix). Schmittbauer was the teacher of one of the most famous virtuosi of the period, the blind Frau Marie Kirchgessern.[31]

In 1785 the mechanician Hessel from St. Petersburg invented in Berlin a *Clavier-Harmonika*, or a harmonica with a keyboard; it was a greatly improved version of Frick's earlier attempt.[32] Drawing upon his artisanal expertise, Hessel constructed his instrument using three rods that rubbed against glass bowls, producing pitches between G and g^3 (the G key more than octave below middle C to the G two octaves above)—a range more than an octave greater than Schmittbauer's earlier harmonica. The glass bowls were set into motion by a foot pedal. The keyboard was located on the left side of the desk-shaped instrument. By means of the keys, the desired pressure could be transferred to the rotating bells.[33] Hessel's instrument enjoyed immense popularity.

Berlin quickly became a major site of harmonica manufacture. The musician and composer Karl L. Röllig of Berlin sought to improve on Hessel's design, a feat that he accomplished in 1786. In his instrument the larger bowls generating the bass pitches were hung by silk cords. These bass vessels were set into motion by either a keyboard or finger. For the lighter, higher tones, Röllig used white glass only, whereas most harmonicas had kept Franklin's original color scheme: C was red, D orange, E yellow, F green, G blue, A indigo, and B violet, with the half tones being composed of white glass.[34] Arguing for the importance of the quality of the glass to the tone of the instrument, Röllig avoided the use of colored glass and scoured glass huts throughout Europe to obtain the highest quality glass. He had a gold lining added to the rims of the glass vessels of the half tones, to distinguish them from the whole tones. Not only did Röllig build his own *Tastenharmonika* (keyboard harmonica), he was also one of its leading virtuosi[35] and composed numerous pieces for his instrument, which he published in 1789 as *Kleine Tonstücke für Harmonika oder das Pianoforte* (Little Musical Pieces for Harmonica or Pianoforte) and which were generally very well received.[36] Because of his extraordinary ability to play the harmonica as well as the various improvements he made on the instrument, Röllig historically (if not quite accurately) has been credited as the inventor of the keyboard harmonica.[37]

Heinrich Klein, professor of music at the National-schule of Pressburg (now called Bratislava), further improved on the keyboard harmonica.[38] He reckoned that in order to improve the tones across the musical range of his instrument, the smaller bowls (corresponding to the higher pitches) needed to revolve around their axis more often than the larger bowls with lower tones. Hence, Klein decided to build three shafts, with the lowest-range vessels on the first, the medium-range bowls on the second, and the highest-range bowls on the third. The first shaft made one revolution in the time it took the second to make four, and the third to make three.[39] To generate the tones, Klein employed small pieces of a bathing sponge that were fitted on a pad of either felt or horsehair. This sponge was kept moist during the performance of the instrument, which possessed a range of four octaves from F to f^3, the F two octaves below middle C to the F two octaves above it.[40]

In short, the glass harmonica, as originally designed by Franklin and then modified by a host of musicians and instrument makers, was a worthy enough instrument to warrant the attention of Wolfgang Amadeus Mozart and Ludwig van Beethoven, both of whom composed pieces for the instrument.[41] Although it never became a standard instrument in music halls, it was popular in the bourgeois German household.

The Euphone and Clavicylinder

Chladni's acoustical research fits within this context of musical-instrument design. He familiarized himself with the plethora of new instruments appearing during the last quarter of the eighteenth century, such as the harmonica, the harmonichord, Röllig and Matthias Müller's xanorphica, Thomas Kunzen's orchestrion, the triphone, and the iron violin.[42] He discussed the glass harmonica and its various permutations in his *Entdeckungen über die Theorie des Klanges*[43] and commented on the problem of using glass bowls for the harmonica. Since the elasticity of the glass may not be uniform throughout the bowl, some glass bowls could produce more than one tone when rubbed, thereby rendering the bowl useless. Chladni detailed the types of tones a glass bowl or plate could produce (other than the fundamental tone), by touching certain spots on the bowl while it was revolving.[44] This analysis was followed by similar investigations of sounding quadratic disks.[45] He concluded the section by arguing that "the preceding remarks on the sound of disks can be applied to the construction of instruments made of glass or metal disks, which can be played with two or more violin bows or perhaps via another method. Its tones should resemble those of the usual harmonica, but perhaps with a greater range; it should be quite strong, particularly in the higher tones."[46] Chladni's use of nodal lines was applicable not only to the study of acoustical pitch, but also to the design of musical instruments, particularly his euphone and clavicylinder. Chladni decided to increase his meager wages as university lecturer by attempting to sell musical instruments that he invented. Critical to the design of these instruments, detailed below, was the determination of the precise location of the nodal lines of glass and metal rods and strips, so that one could attach a rubbing rod without affecting the vibrating portions of the sounding material, and therefore the pitch. Hence, he reemployed his method for rendering the invisible visible by sprinkling dust (and iron filings for the rods and strips generating the lower notes) and bowing the rods or strips at specific places. Once the nodal lines were determined, Chladni marked these points, which would be the precise spots where the rubbing rods needed to be placed.

In March of 1790, or three years after publishing his *Entdeckungen über die Theorie des Klanges*, Chladni finally announced the invention of his first musical instrument, the euphone, in the *Journal von und für Deutschland* (Journal of and for Germany) and the *Journal des Luxus und der Moden* (Journal of Luxury and Fashions) the following October.[47] He had originally thought of constructing a harmonica with a keyboard, but since the design had already been realized, he decided to create something new. He reckoned that one could rub glass rods in a straight line, along their lengths, rather than the edges of glass dishes, as was the case with the harmonica. By rubbing these rods, either cylindrical or parallelepipal in shape, with wet fingers, one produced notes whose tonal color was quite distinct from that produced by the harmonica.[48] His euphone was to resemble both an organ and piano. Organ tones could be sustained for an extended period of time, but their volume could not be altered. Piano tones, on the other hand, could be played loudly or softly, but could not be sustained as long as an organ's. Chladni wanted the best of both worlds: the expression of the piano and the stamina of the organ.[49]

The euphone was a desklike structure composed of glass rods or strips (rather than bowls or vessels as was the case with the harmonica) and rubbing rods (*Streichstäbe*) (figure 2.6).[50] The mechanism for generating the tone was straightforward (figure 2.7). The glass rod *ss'* was attached to the instrument's sounding board *rr'* by two small metallic rods *ed* and *e'd'*. In the middle of the glass rod, Chladni placed a smaller glass rod, *ml*, which was rubbed by the moistened fingers of the performer. The vibrations resulting from the rubbing were transferred to *ss'* and gave rise to the transverse vibrations producing the tone. Chladni placed (horizontally) forty glass rods (*ss'*), originally thermometer and barometer rods, in a box the size of a desktop.[51] Each rod had a different length, thereby generating a different tone. The relationship between the lengths of rods and the tones generated was discussed in his *Entdeckungen über die Theorie des Klanges*. The range of the instrument was from f to f^3, the F below middle C on the piano to the F two octaves above it. Half tones were painted black so as to be easily distinguishable from whole tones.

After making various improvements, Chladni soon toured with his instrument in order to increase its sales, as well as officially claim priority over mechanicians, who were constructing similarly designed instruments. In 1792 he embarked on a journey throughout northeastern Europe. His first destination was one of Europe's leading cities for musicians and instrument makers, Dresden, where the prince presented him with a box made of gold in appreciation for his performance on the new instrument. He continued on to Berlin later that same year. By then Chladni had made various improvements to the instrument. It was now composed of forty-seven glass cylinder rods, which were no longer made from thermometer rods but from basalt and cobalt milky-white glass.[52] Those rods that produced half tones were painted black. But since the paint often cracked

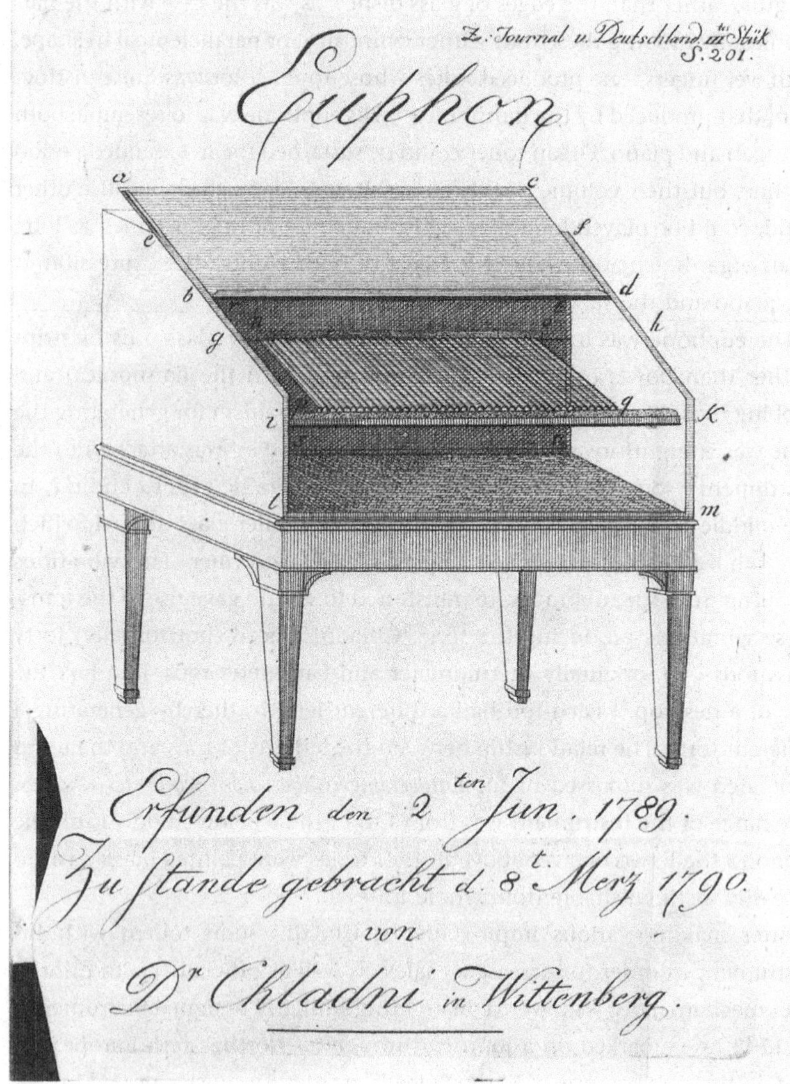

Figure 2.6
Chladni's euphone. By permission of the Houghton Library of Harvard University.

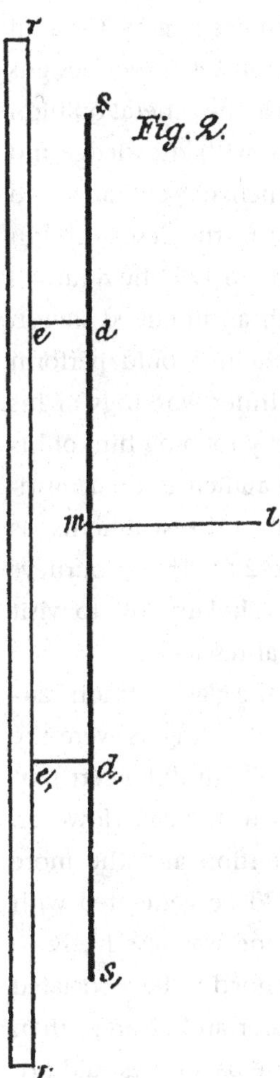

Figure 2.7
The mechanism for generating the tones of the euphone. Source: Melde 1866/1888 (second edition), p. 17.

because of the moisture used to lubricate the performer's fingers, Chladni decided to use colored glass (black and milky-white) in his newer designs of the instrument. One Berlin reviewer remarked: "It is the general opinion of the public that Dr. Chladni handled his instrument with knowledge and skill, and one can only accuse the instrument, which in general found much support, of possessing a weak bass in relation to the descant."[53] In 1793 he journeyed to Hamburg and Copenhagen, and in 1794 he returned to Thuringia and finished the two-year sojourn with an intended stay in Russia later that year. Throughout his travels, Chladni would perform pieces on his euphone, but he never revealed the inner workings of his device, as he feared that pirates would copy it, thereby robbing him of his major source of income.[54] He also lectured to paying audiences on acoustical theory and its application to the building of musical instruments, as well as demonstrated his Chladni figures (see figure 2.8). These journeys enabled Chladni to exchange ideas with leading scholars and to visit numerous libraries where he continued his acoustical research.

The euphone's tones resembled those produced by the glass harmonica—"soft and ethereal,"[55] not surprisingly, since the player's fingers were the source of friction for both instruments, although Chladni did assert that the lower tones of the euphone were more pleasant to the ear. However, he did wish to differentiate between his new invention and the more popular glass harmonica. The euphone's tones could be generated with much less effort than the harmonica's; therefore, one was less likely to suffer from the nervous disorders controversially claimed to be associated with the harmonica. The euphone's tones were "purer and clearer" than those of its predecessor, because they (particularly the bass tones) did not continue to sound after striking the rod, as was the case with the harmonica. Also, each rod produced the same pitch when touched at various points, whereas the bowls of a harmonica often generated different pitches when touched at different spots, rendering its playing much more complex. The euphone was also about half the price of a harmonica, and one could purchase replacement parts much more cheaply. Finally, the euphone's architecture was more symmetric and therefore, he argued, more aesthetically pleasing to the eye. To be fair, Chladni did admit two disadvantages of his euphone. First, the tones of the harmonica could be played for a longer duration with much greater ease. Second, if extreme heat caused a performer's hands to perspire, then more rods than desired might

Figure 2.8
Chladni demonstrating his figures to a fee-paying audience. Photo: Deutsches Museum, Munich.

accidentally be activated; and this was, it seems, a greater danger with the euphone than the harmonica.[56]

The announcement of Chladni's completion of his second invention, the clavicylinder, appeared in Voigt's *Magazin für den neuesten Zustand der Naturkunde* (Magazine for the Newest State of Natural History) in 1800. It was slightly larger than the euphone,[57] but was still compact enough so that it could travel well. The instrument's sound was produced by a keyboard, which brought a cylinder 2 1/2 inches in diameter in contact with glass rods and metal strips. The cylinder would spin by means of a metal pedal that was depressed by the player's foot. The spinning cylinder would rub against the rods and strips and produce a tone. Referring to figure 2.9, *TT′* is a clavicylinder key connected to the sounding board *RR′* by peg *p* and through a cushion atop peg *q*. A metal strip *AB* is connected to the key at *e* and *e′*. The two strips *e* and *e′* touch *AB* at the nodal points *n* and *n′*; therefore, they do not affect the generated pitch. Chladni here used his technique of generating Chladni figures in order to determine the precise location of the nodal points. When the key *TT′* is depressed, the metal strip *AB* approaches a rotating cylinder *c*. *AB* contacts the cylinder by means of a cloth at *v* (near *A*), creating a ringing tone. A spring connected at *T* then hits against a small piece of wood *K*, causing the key to return to its original position.

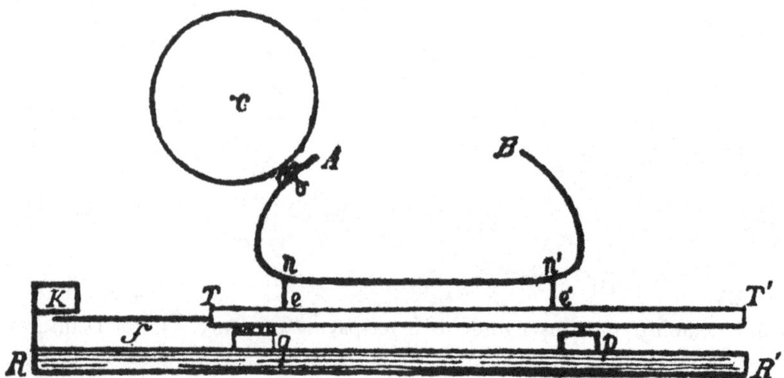

Figure 2.9
The mechanism for generating the tones of the clavicylinder. Source: Melde 1866/1888 (second edition), p. 26.

Chladni's clavicylinder originally had a range of three octaves and a major sixth, from g to e^3, the G below middle C to the E two octaves above it.[59] It was comparable to an organ in that its tones could be sustained for a longer period of time than the piano, and it, too, was better suited for the slower, more melodic musical pieces "with a nice harmony" than the more rapid, "jumpy passages" that are better left for the piano.[60] Although the clavicylinder could not rival the organ's power and richness of tone, it did have the distinct advantage of being able to change the volume of its notes by the performer altering the pressure of his or her fingers on the keyboard.[61] Chladni described the tones of his new instrument as being different from either those of the euphone or the harmonica. Some of the tones, he argued, resembled the softer register of the organ, while lower tones sounded like a bassoon, and the higher tones like an oboe.[62] And, the clavicylinder's tones could not fall out of tune.

When contrasting his own two instruments, Chladni emphasized that the clavicylinder, unlike the euphone, was different from the harmonica. He argued that that the clavicylinder's sound was between a harmonica and a well-played wind instrument.[63] He himself offered the following contrast between his two instruments: "It appears to me that the playing and listening to the clavicylinder is to be compared more to a hearty and nourishing meal, which can fill one's appetite. Playing and listening to the euphone, however, must be compared to something delicious, enjoyed in small amounts and only rarely, something like a dessert."[64] The euphone was the instrument of choice (as was the harmonica) for light andantes and adagios.[65] The clavicylinder could perform the slurred, legato movements composed for the organ, since both had keyboards.[66] It could perform faster passages than the euphone, but slower "full and slurred movements" were preferred.[67]

A year after his invention of the clavicylinder, Chladni's stepmother passed away. He quickly returned to Wittenberg and started to write his second major work, *Die Akustik*. This first comprehensive work on acoustics to appear in German (indeed, one of the first to appear in any language) was divided into four parts. The first part dealt with the general theory of tones and the arithmetic part of acoustics, consonance and dissonance, musical intervals, the number of vibrations per second for pitches, and equal and unequal temperaments. The second part, labeled the mechanical part of acoustics, reviewed the properties of the vibrations of resonating bodies already researched by Euler, D. Bernoulli, and Riccati. He also detailed the three types of vibrating motions of resonating bodies: longitudinal, transverse, and spinning (now called torsional vibrations), and repeated portions of his *Entdeckungen über die Theorie des Klanges* dealing with the vibrations of curved rods, plates (triangular, quadratic, round, and elliptic), and bowls, as well as the tones that these shapes elicit. He concluded the second part with a discussion of the combinations of different types of vibrations. The third part dealt with the propagation of sound through fluids and rigid bodies, while the fourth and final section, which was primarily experimental and explanatory, reviewed the physiology of the ear in humans and animals.

After Chladni completed his work on acoustics, he decided it was once again time to embark on several journeys in order to publicize the clavicylinder. From 1802 to 1805 he traveled south throughout Saxony, Saxony-Anhalt, and Bavaria.[68] After spending a year and a half at home, he headed westward in 1807. He performed pieces by Handel on the clavicylinder in Brunswick, Cassel, and Heidelberg in the spring of 1807. A review of his Heidelberg performances commented on the improvements made to the bass notes of the instrument, which now sounded oboe-like.[69] The clavicylinder's overall tone "unified the beauty of a harmonica with that of a French horn."[70] After a lengthy stay in Holland, he traveled to Brussels and then to Paris in late 1808, where he stayed until April of 1810.[71] Of all of his journeys, the only one for which we have a detailed account is his stay in Paris, the world's leading scientific city of the period.[72]

While in Paris Chladni was able to discuss both his acoustical research and musical-instrument design with leading French savants. He spoke to the members of the physics and mathematics section and the arts section of the French Institute,[73] where Pierre-Simon Laplace encouraged him to write a treatise on acoustics in French. As described below, Laplace was particularly interested in Chladni's research as it applied to his own work on the speed of sound in various substances. Chladni informed Laplace that he would fulfill the request on two conditions: first, he needed to be paid for his effort, and second, several French savants were required to assist him with the translation. Laplace immediately agreed to the second request: Jean Baptiste Biot, Siméon-Denis Poisson, and Georges Cuvier were put at Chladni's disposal for his translation of *Die Akustik*.[74] As to the first, both Laplace and his friend the chemist Claude Louis Berthollet needed to appeal to a higher authority, Napoleon. After their consultation, the emperor, who was certainly no stranger to science and mathematics, agreed to pay Chladni 6,000 francs for the translation.[75] As a gesture of appreciation for this financial intervention, Chladni entertained Napoleon with a performance on his clavicylinder in February 1809. The new instrument intrigued Europe's most powerful man, who quietly asked Laplace whether Chladni would reveal the inner workings of the instrument. Laplace replied that Chladni wished to keep the details of the instrument secret for fear that someone might copy his design.[76] Napoleon understood and withdrew his request.[77] Chladni then demonstrated his acoustical figures to Napoleon, who wanted a detailed mathematical account of his guest's experiments (figure 2.10).[78] Chladni, however, was at a loss to provide such an account. In 1809 Chladni's *Traité d'acoustique* was finally made available to French scholars by the Parisian publisher of mathematical texts, Courcier. Chladni dedicated his book, not surprisingly, to his new patron, Napoleon: "Napoléon le Grand a daigné agreér la dédicace de cet ourage, après en avoir vu les expériences fondamentales."[79]

A group comprising three members of the Class of Mathematical and Physical Sciences and the Class of Beaux-Arts praised Chladni's work and called for those concerned with the advancement of mathematical physics to devote themselves to exploring the phenomenon of plate vibrations. On 13 February 1809, the First Class of the Institute, composed of René-Just Haüy, Laplace, Halle, Count Rumford, and Joseph Louis Gay-Lussac, recommended "that the 3,000 francs that were not rewarded this year for a discovery in galvanism be employed to encourage the mathematical analysis of the experiments made by M. Chladni on the vibrating of resonating plates."[80] The Parisian savants were particularly interested in the curves and lines where no motion could be detected. "M. Chladni has discovered a means as simple as it is ingenious for making these curves visible. He first covers the plate with powder. When the plate is excited, the powder abandons all oscillating regions of the body, collects and settles down motionless at the boundaries of these regions. The curved lines of equilibrium so formed take on very different but regular patterns."[81] Within a week, Napoleon approved the change in the competition. In early April 1809, the *prix extraordinaire* was announced, calling for "a rigorous theory explaining all phenomena rendered sensible by these experiments."[82] The

Figure 2.10
Chladni demonstrates his figures to Napoleon. Photo: Deutsches Museum, Munich.

French Institute set a two-year limit on the competition, ending on 1 October 1811.[83]

The French Institute was not the first to host such a competition. The Royal Academy of Sciences in St. Petersburg and the Batavia Academy of Sciences in Haarlem had offered similar rewards for providing a mathematical explanation of Chladni's empirical research, but the contests of both Academies had failed to produce a winner. Indeed, the task was so difficult that even the French Institute needed to reopen its competition before someone could come up with an attempt at a mathematical account. That someone was Sophie Germain. The aged A. M. Legendre, one of the judges of the competition, argued that her principal equation was incorrect, and she needed to consult Joseph Louis Lagrange's 1811 edition of *Mécanique analytique*. Lagrange himself argued that the solution to Chladni's vibrating plactes would require a new type of mathematical analysis. The judges were Lagrange, Laplace, Lacroix, and Malus.[84] The contest was reopened and the deadline extended to 1 October 1813. Once again, Germain's essay was the only entry. Her essay argued that Lagrange's equation did indeed yield Chladni's figures in several instances, but she could not "satisfactorily" derive Lagrange's equation from physical principles. The contest was reopened a third time, with the deadline extended to 1 October 1815.[85] This time her submission, which was subsequently published, *Memoir on the Vibrations of Elastic Bodies,* earned her the prize medal of one kilogram of gold. Some of the judges, however, felt that her mathematical analysis lacked a certain rigor in some places. She was indeed able to provide the correct fourth-order equation; however, her choice of boundary conditions was incorrect.[86] Chladni's experimental observations were now bolstered by the impressive mathematics of the French.

Longitudinal Vibrations

Although Chladni never possessed the impressive range of mathematical skills as his French colleagues, it would be a mistake to see him as merely offering experimental techniques to confirm or correct previous acoustical theories. As a result of his investigation, he uncovered a novel physical phenomenon. Introduced at the conclusion to Chladni's *Entdeckungen über die Theorie des Klanges,* Wilhelm Eduard Weber later claimed the phenomenon to be the most important of all Chladni's discoveries. Chladni had discovered longitudinal vibrations in solid bodies.[87] The following passage of Chladni's seems almost to be an afterthought. He does not expound upon the point until several years after the publication of his first major work.

If one bows a long and thin string [along its length] with an acute angle, a type of vibration occurs (which as far as I know, no one has noticed before). When the bowing occurs on the middle of the string (or near its vicinity), one hears a sound that is approximately 3 to 5 octaves higher than the normal fundamental tone. If the middle of the string is gently touched by a finger (or is similarly muted), and the middle of one of the two halves is bowed in the aforementioned manner, one hears a pitch an octave higher than the tone previously produced. Likewise, one can produce still more tones by touching the nodal points of the string that divide it into 3, 4, or more parts and bowing the middle of the vibrating part (as long as the length permits). These tones have the same relationship to the first tone mentioned above, just as the harmonic tones produced by the same partitioning of the string through the perpendicular bowing stand in relation to the fundamental tone. All these tones are uncomfortable to the ear, and therefore cannot be put to any practical use. However, they do merit a closer investigation because of their deviation from all other types of vibration. They do not have a definitive relationship to the tones produced by perpendicular bowing. They also vary only slightly with the increased tension of the string, such that when the usual tones are raised about an octave by strongly tightening the string, the pitch of these newly observed tones only increase by about a halftone.[88]

In June 1792 he penned a brief article published in *Musikalische Monatschrift* (Musical Monthly Journal)[89] entitled "Ueber die Längentöne einer Saite" (On the Longitudinal Tones of a String), in which he stretched strings made of various materials (such as brass, steel, and gut) between two pegs, which were approximately four feet apart, creating a monochord. He attached weights to one of the ends of the string to see how its pitch would be affected. He noted that an increase in the tension of the string by additional weight, while greatly affecting the pitch of the fundamental tone, barely changed the pitches of the longitudinal tones. Chladni also noticed that these longitudinal tones were more dependent on the string's material than was the fundamental tone.[90]

This essay was Chladni's response to a paper published by the professor of mathematics and physics in Dessau (and later Freiberg), Friedrich Gottlieb von Busse.[91] Noting Chladni's *Entdeckungen über die Theorie des Klanges*, Busse claimed that he had heard such tones some twenty years earlier when listening to a lecture and demonstration of the violinist Rust.[92]

Busse called these unique tones "*Vogeltöne,*" as their sounds were reminiscent of birdcalls. Such tones were not to be confused with the well-known harmonic tones (or harmonics). They required a unique bowing technique and grasp, and they did not possess the "soft and unforced" sound as the harmonic tones.[93] Busse then listed the properties of his *Vogeltöne*, which (not surprisingly) were identical to Chladni's longitudinal tones. Chladni demonstrated in his 1792 essay that Busse's *Vogeltöne* were indeed his own longitudinal tones. Chladni chose the term "*Längentöne*" since the tones were generated by a string vibrating in the same direction as its length, that is, longitudinally.[94]

By 1796 Chladni had thoroughly researched the properties of this phenomenon.[95] These tones were a product of longitudinal vibrations, which were found only in rigid, solid bodies and were generated by bowing a string in such a fashion that the bow crossed the string at an acute angle, rather than along a perpendicular (as is usually the case), or by rubbing a string with violin rosin and then gliding a finger or cloth covered in rosin dust along the string's length.[96] Transverse and longitudinal vibrations possess different physical properties. With transverse vibrations the pitch of the tone is a function of the string's elasticity, length, thickness, and weight, whereas with longitudinal vibrations, these parameters play a minimal role, raising the pitch no more than a third of a full tone.[97] The string's composition, however, plays a critical role in determining the pitch of the longitudinal vibrations. For example, the pitch of a brass string is approximately a sixth or minor seventh higher than a string of gut of the same dimensions. And a steel string produces a pitch about a fourth to a fifth higher than a brass string of equal dimensions.[98]

Chladni also discovered that the vibrating string is not the only shape that produces longitudinal vibrations; an elastic vibrating rod can produce them as well. And longitudinal vibrations of rods behave differently than transverse vibrations generated by those same rods. The rod's surface had to be smooth, and the investigator needed to hold it with two fingers at a nodal point and stroke the vibrating portion along its length with a small, dampened cloth held in the other hand. For glass rods, Chladni recommended the use of thermometer tubes. Metal and wooden rods could also produce longitudinal vibrations if rubbed with rosin.[99] He investigated the properties of three types of rods: rods that could vibrate freely, rods with one free end and one secured, and rods with both ends secured.[100] And he concluded his essay by underscoring the fact that an analogous set of vibrations governs the tones of pipes. As will be discussed in chapter 5, during the 1820s and '30s, Wilhelm Eduard Weber drew upon and expanded upon Chladni's work on longitudinal vibrations in his investigation of reed pipes in organs.

Although longitudinal vibrations were discovered amidst Chladni's acoustical investigations, they were a powerful scientific assay for determining the velocity of sound in solid bodies. To do so, Chladni asserted that the longitudinal vibrations of rods are sounds in precisely the same sense that longitudinal vibrations in air are sounds. If one could imagine immersing oneself in a solid, then the ear would hear the longitudinal vibrations. The difference between the sounds produced by transverse and longitudinally vibrating rods is that in the transverse case the oscillations of the rods themselves do not sound. Rather, they produce longitudinal oscillations, and so sound in air. By contrast, a longitudinally vibrating rod does itself sound, and its "sounds" are directly transmitted to the neighboring air without conversion between transverse and longitudinal forms. In both cases the frequency of the oscillation is unchanged between the rod and the air. Chladni, however, demonstrated that the length of the vibration of a longitudinally vibrating rod (we would now call it a wavelength) is very different from the length of the vibrations of an air column (again, what we would call wavelength) of the sounding air that it generates, which at once implied that the speeds of "sound" in air and in the rod had to be very different from one another. To find the speed of the longitudinal waves within the rod, Chladni first fixed the wavelength in the rod by holding it in the middle, or as he himself said, halving the length of the vibrating rod, which meant that the rod's length had to be half the wavelength of oscillation. Chladni then compared the pitch of the sound generated in air with that of an open-ended pipe. Since the frequency, or as he called it "the number of vibrations of air in a pipe," was the same in the rod and air, this gave him the frequency corresponding to the wavelength in the rod as well.[101] Multiplying the two together at once gave the velocity in the rod.[102] As a direct consequence, oscillations in the air and the rod that had the same wavelength would necessarily have different frequencies. On this basis Chladni argued that his measurements further indicated that the velocity of sound in a given solid material increases with frequency, which was not, it was thought, the case in air.

Chladni's method could be used to examine the propagation of sounds through many different substances—solids, liquids, and gases. For example, the pitches of tones emanating from a tin rod were two octaves and a major seventh higher than the tonal pitches of an air column of an open-ended pipe of the same length (i.e., for equal wavelengths the rod's oscillations were that much greater in frequency than air oscillations); the pitch of a silver rod was three octaves and a full tone higher, the pitch of a copper rod nearly three octaves and a fourth, the pitch of an iron and glass rod approximately four octaves and a half tone higher.[103] Hence, Chladni continued, the velocity of sound through tin should be about 7,800 Parisian feet per second (approximately 2,500 m/sec), through silver, 9,300 Pf/s (3000 m/s), through copper 12,500 Pf/s (4,000 m/s), and through glass and iron 17,500 Pf/s (5,700 m/s).[104]

In 1798 he turned his attention to using vibrations[105] for determining the velocity of sound in various gases. He commenced his essay on the topic by claiming that "[t]he labours undertaken by Newton, Euler, Daniel Bernoulli, Lambert, Giordano Riccati, and others, to determine, from general mechanical principles, the velocity with which sound is conducted, have given no other results than those taught by experience."[106] The mathematically determined distance that sound travels through air, or any expansible fluid substance, is $\sqrt{(2ga)/b}$, where g is the distance a heavy body travels in a second, a is the elasticity of the fluid, and b is the fluid's density. Chladni noticed, however, that whereas the theoretical value of the velocity of sound through air yielded a result of 900 Parisian feet per second, the experimental value was 1,038 Pf/s. Chladni concluded that "[f]rom the present experiments it follows, in my opinion, that the velocity of the vibrations of an expansible fluid substance cannot be determined by the mechanical principles hitherto admitted alone. . . ."[107] He detailed his procedure for ascertaining the pitch of various gases.[108] He took an open-ended organ pipe made of tin six inches long and connected its upper end into the neck of a bell glass that could be shut by a stopcock. The bell and pipe were immersed in water so that the air would not contaminate the experiment. A bladder, which was also furnished with a stopcock, was screwed to the bell's neck. It was squeezed, ensuring that it was void of air. The bell and bladder were filled with various gases such that the water stood at an equal height inside and out (i.e., the gas pressure was atmospheric).[109] The gas was made to travel through the pipe, causing it to sound, by pressing on the bladder. Since the physical configuration of the system, particularly the temperature, always remained the same, so did the resulting length of the vibrations in the air column (wavelength) no matter what was sent through, which meant that the ratio of speeds in various gases could be found directly from the ratios of the corresponding pitches (frequencies) heard in air.

Chladni determined the pitches produced by various gases by tuning them to two strings suspended on a monochord.[110] He found that oxygen gas liberated from amber sounded with a pitch between a half tone and a whole tone lower than air, while nitrogen gas was a half tone lower, and hydrogen gas elicited "much higher tones" than air. And carbon dioxide, liberated when chalk and sulfuric acid are mixed, was nearly a minor third lower than the pitch generated by air.[111] His extensive experimental investigations revealed a number of instances in which the pitch of the pipe differed greatly from early theoretical claims that the pipe's pitch is inversely proportional to the square root of the weight of the gas compared to air.[112] Chladni argued that if the tone of a pipe of a given length filled with air resonates at c^2 (the C above middle C on the piano), that same pipe filled with oxygen will elicit a pitch between b^2 and $b\flat^2$, with nitrogen b′ (just below middle C), with a mixture of nitrogen and oxygen c^2 (same as atmospheric air), with hydrogen between c^3 (two octaves above middle C) and e^3, with carbon dioxide approximately $g^{\#\prime}$, and nitrous oxide gas b′.

He could thereby determine the velocity of the propagation of sound through each of the aforementioned gases: air was 1038 Parisian feet/sec (338 m/s) with a specific gravity of 887; oxygen gas, 950–960 Pf/s (309–312 m/s) with a specific gravity of 844.57; nitrogen gas, 990 Pf/s (320 m/s) with a specific gravity of 893.7; hydrogen gas, 2100–2500 Pf/s (680–810 m/s) with a specific gravity of 3060; carbon dioxide, 840 Pf/s (270 m/s) with a specific gravity of 724.2; and nitric acid gas, 980 Pf/s (320 m/s) with a specific gravity of 811.4.[113] He concluded from these trials that the velocity of the vibrations of an elastic fluid was determined not merely by its specific gravity, but also by its chemical properties, which warranted further investigation.[114]

Chladni's work on the speed of sound in various gases and solid bodies caught the attention of both Jean-Baptiste Biot and Laplace. In November of 1808 Biot read a paper to the French Institute on the experiments on the propagation of sound through solid bodies and very long cylinders.[115] He drew upon Chladni's method for determining the velocity of sound through solids by using the pitches generated by longitudinally vibrating bodies.[116] Working with Mr. Martin, "a very skilled and eager artiste"[117] and maker of sea clocks, Biot used hollow cylinders made of cast iron, constructed during the first decade of the nineteenth century for carrying water to various gardens throughout Paris.[118] These large cylinders, which averaged 2.515 meters in length, enabled the two men to conduct extensive experiments on the velocity of sound in air and solids. A lead ring, with a thickness of 142.56 millimeters, connected the ends of two pipes. In their first set of experiments Biot and Martin placed 78 cylinders together, totaling 196.17 meters in length, plus 77 lead rings of 1.1 meters in length, culminating in a length of 197.27 meters (or 607.25 Parisian feet). The last cylinder contained an iron ring of equal circumference to which a hammer and bell were attached. Once the hammer struck the bell, two distinct tones could be detected: one traveling through the cast iron, the other through the air enclosed by the cylinders.[119] With Martin's chronometer in hand, they calculated the time interval between the arrival of the two generated tones. After fifty-three trials, they determined the average length of the interval to be 0.542 seconds. Biot and Martin then drew upon the earlier calculations of the Paris Academy, determining that at 11°C and an atmospheric pressure of 0.76 meters, the sound would take 0.579 seconds to traverse the 197.27 meters.[120] Hence, the sound traveled through the cast iron in 0.579 minus 0.542, or 0.037, seconds. After their calculations, however, they quickly added that they did not wish to put too much faith in that result, as the slightest observational error would yield a very different answer.[121]

Bouvard and Étienne Louis Malus conducted similar experiments using twice as many cylinders, totaling 394.55 meters (1214.7 Parisian feet). Again, drawing upon the calculations of the Paris Academy, the sound waves would take 1.158 seconds to travel such a distance.[122] After sixty-four trials, they calculated the time interval between the two tones (generated by the air and cast iron) to be 0.81 seconds, meaning that the sound traveled through the solid body in 0.348 seconds, nearly an order of magnitude slower than Biot and Martin's calculation. Initially Biot thought that this time was simply too slow. Perhaps the instant of Malus and Bouvard's observation did not correspond precisely to the beat of the seconds pendulum. Or, Biot postulated, perhaps not all of the iron cylinders were at the same temperature. Differences in temperature, he reckoned, could create air currents in the tubes, impeding the propagation of the sound waves.

In any event, Malus and Bouvard's results forced Biot and Martin to conduct another series of trials. This time they attached 376 cylinders with lead rings, measuring 951.25 meters (2928 1/3 Parisian feet) in toto. Biot then performed the experiment more than two hundred times on his own. Martin repeated the experiment, and obtained the same result without having prior knowledge of Biot's value. Both men independently determined that the interval between the time the sound propagated through the cast iron and the air was 2.50 seconds. Now at 11°C, the time of travel of the sound through the air is 2.79 seconds; hence the sound traveled through the iron in 0.29 seconds, a value much closer to Malus and Bouvard's calculation than their original. Biot, however, still feared that any temperature difference in the tubes could result in air currents, which would interfere with the sound waves and affect their velocity. He decided to devise a second experiment that would verify their latest results. Armed with synchronized clocks that could count half-seconds, both Biot and Martin positioned themselves at the opposite ends of the cylinders. Martin would hit the end of the cylinder with a hammer when the clock read 0 and 30 seconds, while Biot hit the end of his cylinder with a hammer when his clock read 15 and 45 seconds. They both noted precisely the time of the arrival of the tones. After nineteen trials, they noted that the average time of the propagation of sound through the iron cylinders was only 0.03 seconds off the earlier Paris Academy calculations.[123] They were confident enough now to conclude that sound travels in iron roughly ten times faster than in air.

Eight years later, Laplace argued that the phenomenon of the increase of the speed of sound in air with an increase in temperature must also occur in solids and liquids, but it had been difficult to determine experimentally the precise influence of heat on these substances.[124] He praised "M. Chladni's observations on sounding plates and rods. This learned physicist has developed a method based on these [sounding plates and rods] to determine the speed of sound in various solids."[125] On 25 November 1816 Laplace offered a lecture to the French Institute arguing that it had been well known that the speed of sound in air is increased by one-sixth due to heating. He continued, "the same cause must also undoubtedly change the velocity of sound in all other bodies; however, it is difficult to determine its influence. One can nevertheless succeed in doing so, if one compares the velocity calculated theoretically with the results M. Chladni has obtained with longitudinally vibrating and sounding rods, which provide the actual velocity of sound in these bodies."[126] One now had experimental confirmation of the effect of increase and decrease in temperature on the propagation of sounds through solids and liquids.

Here is a case where a phenomenon was first detected by Chladni and Busse in a musical context, illustrating that the boundaries between musical and physical investigations were rather porous during the late eighteenth and early nineteenth centuries. Busse, recall, first heard the *Vogeltöne* while listening to a performance and lecture by the violinist Rust. Chladni stumbled upon his longitudinal tones during his acoustical investigations connected with his invention of the euphone. Because longitudinal tones were unpleasant to the ear, Chladni did not incorporate them into his musical-instrument designs. However, he did recognize their merit as a scientific tool for studying the velocity of sound in various substances.

In summary, Chladni's success was predicated, in part, on his ability to move in and among different spheres of the late-eighteenth- and early-nineteenth-century European savant and instrument maker. His life mirrored, as it were, the nodal lines that brought him fame, for it lay at the points of intersection of so many vibrant cultural and scientific activities. Part of his success was nevertheless due to the particular circumstances of the period, for during the late eighteenth and early nineteenth centuries it was not uncommon for musicians and savants to construct musical instruments. By the second quarter of the nineteenth century, as we shall see in the ensuing chapters, this became increasingly rare.[127] Professionalization thwarted the attempts of the dilettante to intervene in physics and musical-instrument making.

Chladni's work, however, marks as well the rise of the role of science within the world of music, an impact that extended from the design of instruments to the very heart of musical aesthetics itself. Acoustical experimentation would now be applied to the manufacture of musical instruments—not just to the euphone and the clavicylinder, but, as we shall see, even to organ pipes and reed instruments such as clarinets. Questions were soon raised as to whether aesthetics itself could be analyzed and even understood in physical terms. These queries resonated throughout the nineteenth century and formed the core of Wilhelm Weber's research of the 1820s and early 1830s as well as Hermann von Helmholtz's work during the 1860s, discussed in chapters 5 and 8.

Chladni's instruments affected musical performance as well, for they were meant in part to stem the rise in performances that emphasized the playing of rapid passages.[128] His instruments, meant to be played by many, undoubtedly required skill, but these skills were well distributed throughout society. Music, for Chladni, was a crucial component of the persona of the *Bildungsbürgertum.* It was to be enjoyed not merely by passive audiences, but by all of its members, engaged in active participation and performance. The performers were not intended to dazzle or bewilder. His desire to appeal to the musical interest of the general public fits squarely into attempts of musicians in early nineteenth-century Prussia to improve the level of music education of the *Bildungsbürgertum.* The role of this anti-elitist aspect of music was perhaps best epitomized by the folk-song tradition of the late eighteenth and early nineteenth centuries.

3 Singing Savants: Music for the *Volk*

Music was not only important to the physical investigations of vibrations and the construction of musical instruments; it also played a critical role in the identity of the newly formed Association of German Investigators of Nature and Physicians (Versammlung deutscher Naturforscher und Aerzte, henceforth the Versammlung). The Versammlung has been the subject of numerous studies.[1] By and large these studies have aimed either at offering histories of scientific disciplines and biographies of renowned German scientists and physicians, or at tracing the development of scientific and medical societies in Germany. This chapter, however, will focus on one critical aspect of the annual meetings of the Versammlung relevant to a sociocultural history of science, namely the singing of choral-society (*Liedertafel*) songs.[2] German *Naturforscher* and physicians drew upon these choral-society songs in order to construct their persona.

The Versammlung was certainly not the only organization that drew upon the singing of folk songs as a form of identity construction and camaraderie.[3] The songs of the student corporations, recorded in the Tübingen *Kommersbuch* of 1815, served a similar function. As physicians and *Naturforscher* saw their enterprise as simultaneously cultural and scientific, they drew upon a rather ubiquitous cultural resource, folk songs. German *Gebildete* (educated elite) felt compelled to be well versed not only in the natural or medical sciences but in other aspects of cultural as well, particularly music. As will be argued below, German scholars—including physicians and investigators of nature—were trying to identify and salvage the elements of German science and culture from the rubble wrought by the Wars of Liberation. German scientific and medical communities were tired of looking westward to follow the cues of Parisian savants. The Versammlung underscored Germanic contributions to science, as Humboldt was to argue in 1828. Choral-society songs were perfect in this respect, as they originated with King Friedrich Wilhelm III's return to Berlin in December of 1810 after the French occupation. This rise of scientific *Bildungsbürgertum* rather famously culminated in the persona of Hermann von Helmholtz, physicist, physiologist, physician, amateur pianist, and *Kulturträger*.

The early use of songs as a genre for both all-male camaraderie and the construction of a national, cultural, and scientific identity served as a model for later German scientific and medical societies well into the twentieth century. Although there had been associations of Naturforschende Freunde (Friends Investigating Nature) throughout the German territories during the last quarter of the eighteenth century, the Versammlung was the first to use singing as a cultural resource. As te Heesen has demonstrated, patriotism had certainly played a critical role in the formation of the Naturforschende Freunde of Berlin in 1773.[4] This group's identity stemmed from the collection of natural objects for their natural-history cabinet.[5] As the number of members soared, however, the importance of these objects began to wane, and other forms of camaraderie became necessary. Clearly the practice of science in the late eighteenth century had become increasingly identified with the bourgeoisie, and the numbers of participants had risen tremendously. Similarly, as musicologists have known for quite some time, music had moved away from the court and into the bourgeois "public sphere" during that same period. Ludwig van Beethoven's *bürgerliche* sympathies were preceded by Wolfgang Amadeus Mozart's separation from the constraints of the Habsburg Court. And the *Freischütz* of the less flamboyant, but nevertheless important, Carl Maria von Weber brought the aspirations of the *Bürgertum* to the German operatic stage.

During the 1760s Immanuel Breitkopf of Leipzig had invented a method of typesetting that simplified the publication of keyboard music. He specialized in the sale of straightforward pieces for piano and simple songs in large volume at very low prices.[6] This invention enabled members of the educated public to purchase and perform music at home. And by the 1770s and '80s a plethora of piano pieces composed for four hands had been included in the repertoire of amateur musicians. Indeed, by the end of the century those interested in playing privately needed to search for new pieces not composed for four hands.[7] These works were typically very easy to play, their purpose being to encourage everyone to practice with friends and loved ones in the privacy and *Gemütlichkeit* of the household. Likewise, the singing of folk songs, necessarily easy to perform, became a very popular pastime by the end of the century. As musicologist Charles Rosen aptly points out, "folk music . . . evok[es] not an individual but a communal personality, expressive of the soil."[8] From the mid-eighteenth century through the 1820s, then, music in the German territories served the important purpose of creating an identity for the middle class, and choral-society songs quickly replaced the collection of natural objects as the prevailing form of sociability uniting German *Naturforscher* and physicians.

The Association of German Investigators of Nature and Physicians

On 18 September 1822, the first ever meeting of the Versammlung gathered in Leipzig. The purpose of this meeting was clear to Lorenz Oken, the organizer of the gathering, Romantic biologist, physician, and editor of the new journal dedicated to the study of nature, art, and politics, *Isis*:

> One has often asked: why is it that in France and England so many important works through communal collaboration [*gemeinschaftliches Zusammenwirken*] of top-notched scholars have appeared? There are encyclopedias of the entire natural sciences, medicine, arts, commerce, innumerable dictionaries of this kind, clever and uninterrupted society proceedings, periodicals, etc. Why, in contrast, is there practically nothing of the sort in Germany? Why aren't there three men who can communally collaborate on such works? And why is it that when such an association comes together somewhere [here in Germany], it falls apart shortly thereafter?
>
> One continues by asking: why is it that in France and England such clever reviews, reports, and excellent works appear? Why is it that in these critical pages such a fine and educated tone rules, that, even when people are overwhelmed with reproach, there is hardly a reader who could notice it? Why then, in contrast, in Germany are only the simplistic books reviewed, and when it comes to the clever ones, those reviews are spilling over with bile and are handled with severity? Precisely these clever works are where the most interesting things appear.[9]

Oken proceeded to offer answers to those questions (figure 3.1). Paris was, of course, a cosmopolitan city, where most scholars lived, exchanged ideas, and struck up collaborations. Similarly, London performed the same function for the British. These powerful, unified nations offered sites where experimental natural philosophers could congregate both publicly and privately. Oken concluded, in a large typescript, that "the reason, therefore, for these large advantages of the French and English literature [in the sciences]

Figure 3.1
Portrait of Lorenz Oken (1779–1851). Photo: Deutsches Museum, Munich.

lies in the personal acquaintance of the scholars [*in der persönlichen Bekanntschaft der Gelehrten*]."[10] To remedy this German predicament, Oken drafted the statutes of the Versammlung, one of which declared that "the main purpose of this Versammlung is to provide an opportunity for investigators of nature and physicians to get to know one another personally."[11] Camaraderie was, in Oken's view, the crucial missing ingredient among German circles of the intelligentsia. As the German territories were not united, Oken lamented that "[w]e have in Germany no Paris and no London; we have no place where hundreds of investigators of nature and physicians live together, where those otherwise scattered scholars would stream, in order to make the necessary acquaintances [*die nöthigen Bekanntschaften*] . . . to study the large collections, and to use the odd apparatus."[12]

Oken wished to introduce a form of sociability, which involved the new genre of *Liedertafel* songs, into the ever-increasing circles of German *Naturforscher.* Detailed correspondence among all practitioners of natural philosophy, that treasured method of sociability among the members of the Republic of Letters, was now rather impractical. J. S. C. Schweigger, professor of physics at Halle and Wilhelm Weber's mentor, argued in 1823, "Experience in our Fatherland has taught us that, as large as the amount of correspondents of a learned society may be, all forms of writing in the natural sciences remain only an imperfect surrogate to personal reports."[13] For Schweigger, *persönliche Unterhaltung* ("personal discussions") played a critical role in the *geistiges Leben* ("spiritual life") of the investigator of nature.[14] Even Goethe, who kept a healthy distance from the Versammlung, complained to his confidant, Eckermann, that "[w]e Germans are from yesterday, as we all basically lead an isolated, impoverished life. . . . One sits in Vienna, another in Berlin, another in Königsberg, and another still in Düsseldorf or Bonn. All of us are separated by 50 to 100 miles, so that personal contacts and a personal exchange of ideas are rare occurrences."[15] Nearly three years later, he brought up the importance of personal contacts to the Versammlung with Eckermann: "I know quite well that very little scientific worth comes about at these gatherings. But they are indeed excellent, since one can get to know each other and possibly even learn to love one another, the result of which is that a new doctrine of an important person will be accepted, and this doctrine will be suitable for us to accept and encourage our directions in other disciplines."[16]

Providing a common social experience for companionship among German investigators of nature and physicians was the main purpose of Oken in forming the Versammlung. He hoped that the Versammlung meetings would bring about "the preparation of all-encompassing works on nature, natural history, and medicine, from which will emerge: what German scholars would be able to produce in the highest form, which rank the natural sciences presently could have in Germany—contrary to what foreign scholars wish to claim, and what could be seen as a codex and monument for our times in the future."[17]

By the time Oken established the Versammlung, he was a rather notorious political activist, known for his efforts to assist in the unification of the German territories. His *Zur Kriegskunst* (On the Art of War) of 1811 and *Neue Bewaffnung, neues Frankreich, und neues Deutschland* (New Arming, New France, and New Germany) of 1814, written in the aftermath of the Napoleonic occupation, both argued that a liberal constitutional monarchy could singularly meet the demands of German unity.[18] And his involvement in the *Burschenfest* (festival of fraternities) at Wartburg in October of 1817 commemorating the anniversary of the Battle of Leipzig aroused the suspicions of the Prussian Ministry of Police.[19] The importance of the Versammlung to the formation of a unified, political nation was underscored by a participant at the Hamburg Meeting of 1830: "We never either spoke or wrote about it, but the idea of a single Fatherland, which was interwoven with our gatherings, blossomed independently, and our work was and is, as a result, an involuntary induction. We happily ascribed to the recognition . . . of the ever-increasing triumphant power of science. But the glances and handshakes by our coming and going always informed us that still another power had blazed the trail to the hearts of our people. The extinct German spirit is this secret power. . . . "[20]

For Oken, unity was key to practicing both good science and powerful politics. One major cultural activity that played a critical role in the construction and expression of the *Naturforscher*'s identity, predicated on unity and male camaraderie within the Versammlung, was music. Although different types of music were present at the various annual meetings of the Versammlung, by far the most prevalent were the *Liedertafel* songs. The *Liedertafel* tradition was the invention of Carl Friedrich Zelter, director of Berlin's renowned Singakademie, professor of music of the Akademie der Künste (Academy of Arts) and later of the University of Berlin (figure 3.2). These *Liedertafeln* reflected the fervent rise of German patriotism during the French occupation. An observant critic of the period linked the freedom from French tyranny with both the freedom embodied by the investigation of knowledge in all fields and the rise of the *Singvereine* (singing societies): "One of the numerous blessings of freedom . . . is the awakening of a youthful life in the arts and sciences. The deep and serious, yet immensely cheerful, German mind transforms with new power and excitement, which has carried us from the battle against despotic power, toward the investigation of truth in all branches of knowledge."[21] The critic continued by remarking on the sociability among good friends generated by their collective singing.[22] Zelter wanted to establish a society in which

Figure 3.2
Portrait of Carl Friedrich Zelter (1758–1832). Source: *Allgemeine musikalische Zeitung*, vol. 32 (1830), frontispiece.

singing was an extension of bourgeois domestication. His *Liedertafel* signaled the rise of the importance of music to the Prussian bourgeoisie during the early nineteenth century. He described his newly initiated *Liedertafel* to his dear friend, Johann Wolfgang von Goethe on 26 December 1808:

In celebration of the return of the King [of Prussia, Friedrich Wilhelm III], I have founded a *Liedertafel*: a society of 25 men, the twenty-fifth of whom is the chosen *Meister*, who gather monthly at dinner of two dishes and enjoy themselves by singing their favorite German songs. The members must be poets, singers, or composers. Whoever writes the words to a new song, or composes a new tune, reads or sings it at the table, or has it sung for him. If there is an opportunity, a box is put on the table, whereby everyone who has enjoyed the song will place a *groschen*—or more. The box will be emptied and the contents counted on the table. If the pennies should add up to a *thaler*, this sum of money will be paid to the winner, and he will receive a silver medal. The health of the poet or composer is then toasted and then the beauty of the song will be discussed. If a member can produce twelve silver medals, he will be entertained at the expense of the society, and a wreath will be placed around his neck. He may also choose the wine that he wishes to drink, and will receive a gold metal worth twenty-five *thaler.*[23]

Zelter continued by adding that "should a composer produce something which the members do not enjoy, he will be fined."[24] The *Liedertafel* was quintessentially patriotic. Zelter wrote in the first statue of the *Liedertafel* of "the need of such particularly German multi-voiced songs, which breathe German spirit, seriousness and gaiety and thus no mere mimicking of foreign forms. . . ."[25] These *Liedertafeln* were to serve "the King, the Fatherland, the general welfare, the German spirit, and German loyalty."[26] Zelter argued that music's "communal purpose" was to further *Bildung* (education and self-cultivation), which he defined as the "activity of inner or spiritual force, to the end that man realizes his complete existence and becomes nobler."[27] He reckoned that no other art had been the hallmark of being German, distinguishing Germany from other European countries. As music was a critical element in identifying "Germanness," he felt that it must be revived.[28] It possessed the unique gift of forging "a natural bond between the lower and higher classes of the nation," permitting "all members of the nation without the accidental distinctions of society" to unite.[29] With this view clearly in mind, Michael Traugott Pfeiffer and Hans-Georg Nägeli published their influential text in 1810, *Gesangsbildungs-Lehre nach Pestalozzischen Grundsätzen* (The Study of the Education of Singing According to Pestalozzi's Principles), which detailed the proper instruction of hymn singing by school children.[30] Zelter was not interested in forms of music favoring those few who were highly skilled (such as operatic music or pieces that only a handful of virtuosi could perform), as they could not be used to unite the German people.

Zelter and his predecessor, Karl Friedrich Christian Fasch, wanted to revive the tradition of choral singing, which in stark contrast to solo instrumental music was on the decline, since it was being neglected in both the church and in schools. In 1791 Fasch had gathered twenty-seven friends of music from both sexes in the home of the widow of general surgeon Prof. Voitus, Unter den Linden 59.[31] The gathering was first referred to as the Singakademie two years later.[32] This society dedicated itself to the singing of choral songs with Fasch as its leader, and it was to serve as the model for all German *Singvereine*.[33] According to W. C. Müller, those *Singvereine* were critical to the rejuvenation of church music throughout the German territories. They "raised themselves in dismembered Germany . . . to the noblest emulation, to a national art, to the true blossoming of taste—by means of hospitality, to grandiose performances of beautiful works. . . ."[34]

Sociability played a major role in the formation of these musical associations.[35] Zelter's *Liedertafel* had as its goal the fostering of sociability by each individual member.[36] In contrast to the other form of choral society, namely the *Liederkranz* formed by the Swissman Hans-Georg Nägeli in Stuttgart in 1824, the *Liedertafel* was smaller, generally limiting itself to twenty or thirty members. And unlike other associations of the period, Zelter's *Liedertafel* wished to foster a type of "*Gebildetsein*" ("being cultured"), whereby each member was equal and contributed to the organic whole of the Verein.[37] Although the singing of these songs was meant to foster and underscore inclusion, it was also used to exclude. Women did attend the Versammlung meetings; however, they were not permitted to attend the sessions. Day trips were arranged at these annual meetings for wives and children to enjoy the surrounds. No female voices rang out during the singing of the *Liedertafel* songs at the meetings. The parts were strictly for men, despite the fact there were women in other *Liedertafeln* by the 1820s.

Liedertafeln owed much to the old Berlin genre of *Lieder*, or songs. As Christian Gottlieb Krause had argued back in 1753 in his *Von der musikalischen Poesie* (Of Musical Poetry), the Berlin *Lied* was purposely straightforward, easy to sing and contained a simple story to which everyone could relate.[38] This sentiment was echoed by the music theorist Heinrich Christoph Koch's definition of a *Lied* nearly a half-century later: "a lyrical poem of several stanzas, intended to be sung, and united with a melody that is repeated for each stanza and that also is of such a nature that it can be sung by anyone who has a normal and reasonably flexible voice, whether he has any training in the art or not."[39] By the end of the 1820s, *Liedertafeln* and *Liederkränze* were contributing to Germany's national identity.[40] In a very real sense, these *Gesangvereine* sought to unite the forces of sociability and politics with the harmony of music.[41]

The first instance of a *Liedertafel* evening of the Versammlung occurred at the fifth annual meeting at Dresden in 1826. Letters from the participants spoke of the good mood, wine and song that followed the daily lectures. Oken reprinted in *Isis* "a song to be sung at a friendly meal," written by Tiedge to the melody of "Es kann ja nicht immer so bleiben."[42] A musical round, penned to the tune of "Frisch auf Cameraden," welcomed the *Naturforscher* to Dresden at another festive meal.[43] Oken was toasted at this particular dinner, which concluded with a song whose lyrics were written by the zoologist F. J. M. A. Forster and sung to the melody of "Bekränzt mit Laub."[44] From that meeting on, the *Liedertafel* songs were a beloved feature of the Versammlung. A year later, for example, at the Versammlung meeting in Munich, the *Kammerrat* Waitz of Altenburg sang a greeting penned by Dr. Nürnberger of Sorau at the dinner table to the members of the meeting.[45] At that same meeting, all members sang a festive song to the melody of "Im Kreise froher kluger Zecher," written in honor of the director of the Görlitz Natural Society, J. T. Schneider.[46] It must be stressed that these songs were not simply ornamental. German investigators of nature and physicians considered them to be important events during their meetings. Indeed, singing these songs was part of what a *Naturforscher* or physician did. It belonged to his identity.

Often, the choral societies were composed of members of the Versammlung themselves, *Naturforscher* and physicians, who cherished the opportunity to sing publicly with colleagues. So popular were these *Liedertafel* evenings, in fact, that as the years went by, German natural scientists and physicians began writing and singing more and more songs, and one joked at the Brunswick Meeting of 1841 that "the public at first did not know what to expect from the Versammlung. One named it '*Naturforscherfest*'[47] [or 'festival of investigators of nature'] as well as '*Musikfest.*'"[48] The fruits of these singing savants culminated in the publication in 1867 of a 173-paged songbook, providing the words of the favorite songs sung at the annual gatherings (figure 3.3). Generally, these songs were popular folk songs with the words changed to reflect the interests of the Versammlung. Crucially, there were no solo parts.

Some of these tunes emphasized the purpose of music to the Versammlung. For example, "Zur Weihe des Tages," sung to the tune of "Vom hohen Olymp," reminded the audience that

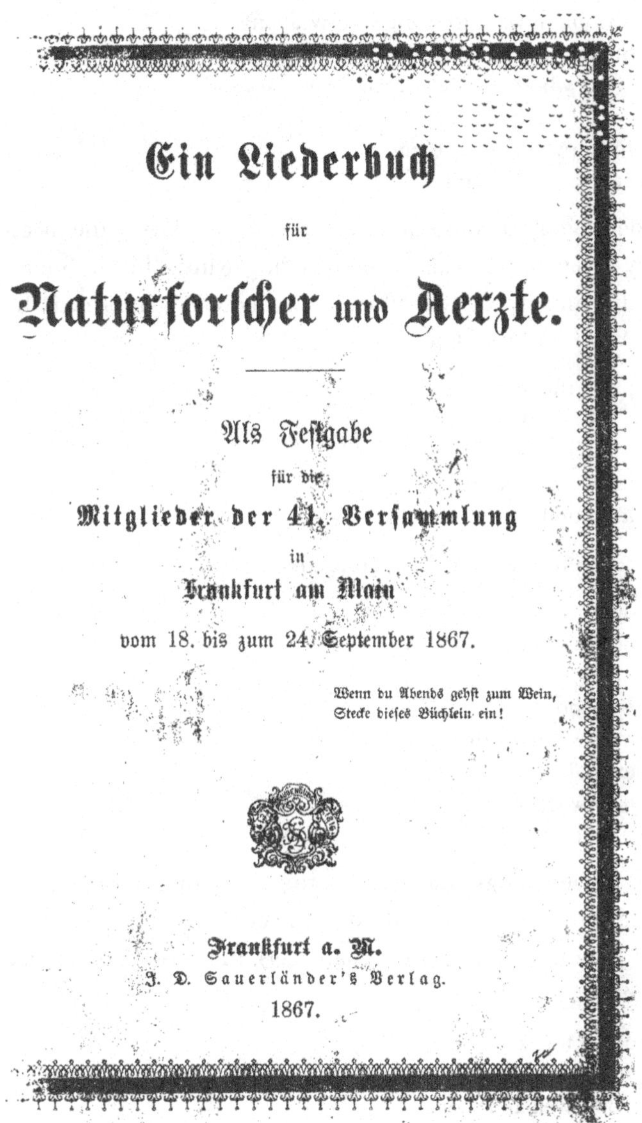

Ein Liederbuch

für

Naturforscher und Aerzte.

Als Festgabe

für die

Mitglieder der 41. Versammlung

in

Frankfurt am Main

vom 18. bis zum 24. September 1867.

Wenn du Abends gehst zum Wein,
Stecke dieses Büchlein ein!

Frankfurt a. M.

J. D. Sauerländer's Verlag.

1867.

Figure 3.3

A Songbook for Investigators of Nature and Physicians. Source: Hoffmann-Donner 1867, title page.

In forceful battle, you might want to swing your weapons
Wherever conflicts of opinion flare up.
For that befits everyone who honestly and freely grapples
With what he sees as his truth.
Yet at the meal, which lovingly links them,
The fighters gladly extend their hands.[49]

Other tunes underscored the importance of camaraderie during the meetings, particularly at evening meals while drinking wine. "In schlimmer Zeit," sung to the melody of the adored "Die Lorelei", informed the members of the Versammlung that

It has been proven that the one
Who is linked with his friend,
Happily carries the heaviest of burdens.
Now pass around the wine, quicker.
So that the song will sound clearer.[50]

"Ein Frühlings-Lied" emphasized that camaraderie evoked by singing

Befriended Comrades,
The heart has opened itself up.
The soul fundamentally
Makes itself known in song, . . .
Which reveals to you a similar striving
Forward to a serious goal.
Come, let us join hands,
Onward to a happy union.[51]

Many of the *Liedertafel* songs underscored the good humor that wine brought to the research of nature. Indeed, an overwhelming majority of the drinking songs spoke of wine (rather than beer), the beverage of choice among the higher social classes.

The wild animals of the forest,
The quiet ones from the stable:
Thanks to zoology, how well and soon
We shall understand all of you.
The magical fishes from the flood,
They truly are not missing.
Ichthyologists, hold on to them tightly.
And serve them on the dish!

But wait, the wine, you dear wine.
You give us a heavy head.
You aren't animal, you aren't mineral.

And you are no longer a plant.
We know what to do with you.
You are gladly slurped.
But despite all of that, who can help us here?
You do not fit into any classificatory system.

Wine—you are the fourth empire of nature.
Full of power and splendor,
This study is full of joy.
The investigators are ready.
The chilled collection lies in the cellar.
One certainly can meet them there.
Studying, until their heads become dizzy
Long live the study of wine![52]

It would, however, be a mistake to argue that all of the *Liedertafel* songs were so comic. Several of them were satiric, aimed for example at the speculative *Naturphilosophien*.[53] In a sense, these songs signified what counted as legitimate science by clearly delineating their theories and practices from the beliefs of more dubious characters. The tune "Die Klage des Gorilla," appropriately written on 1 April 1866 by the renowned Frankfurt psychiatrist and author of *Struwwelpeter*, H. Hoffmann-Donner, depicts a saddened and confused gorilla contemplating his existence as an imperfectly developed human.[54] Had he only been human, he could have been "a professor with many honors, and even a privy councillor," and might assist in the debate on German unity in the Frankfurt Parliament.[55] The gorilla becomes livid by the song's end and threatens to crush the skulls of Paul Belloni Du Chaillu (the French-American explorer who brought the first gorilla to America), Charles Darwin, and Carl Voigt, and "turn their brains into mush" for insinuating that gorillas were aborted humans.[56] The physicians, in particular, sang songs parodying certain forms of medical practice. Some physicians lifted their voices to criticize the increased use of instruments for medical diagnoses.[57] Others parodied medical fads of the mid-nineteenth century, such as the use of spa water as a panacea.[58]

There were also songs specializing in physiology, anatomy, anthropology, zoology, mineralogy, astronomy, botany, and chemistry. The tune "Repetitorium anatomicum" was described as a "necessary song" that had to be sung for the study of anatomy. It jovially described the effect of wine upon various organs of the human body.[59] "Vom Stoffwechsel" (On the Exchange of Matter) of 1866 offered a light-hearted approach to the chemical structure of proteins, wondering in the end how new cells were formed.[60]

Throughout the first two decades of the Versammlung, the German *Naturforscher* and physicians sang patriotic songs written during the Wars of Liberation or during skirmishes between the Prussian Rhineland and France. These songs included: "Das Lied der Deutschen" ("The Song of the Germans," also known as "Deutschland, Deutschland über alles!"), "Was ist das deutsche Vaterland?" ("What Is the German Fatherland?"), and "Wacht am Rhein" ("The Watch on the Rhine"). All three of these songs, of course, became powerful, sinister anthems of the Third Reich. During the late 1840s, '50s, and '60s, however, scientists sang the patriotic songs from the Wars of Liberation with less frequency and began to assert that science, not politics, should be the engine powering German unification. With the rise of professionalization and the political aspirations of the German territories, the goals (and songs) of the Versammlung changed. One thinks here, of course, of the leading physician and cellular pathologist of the mid- and late nineteenth century, Rudolf Virchow, who helped found and became leader of the Fortschrittspartei, which sought a peaceful means to unify the German territories and opposed Otto von Bismarck's militaristic orientation. Virchow claimed that science and other forms of German culture could unite the German Reich. Virchow's two keynote addresses to the Versammlung in 1866 and 1871 were entitled "Über die nationale Entwicklung und Bedeutung der Naturwissenschaften" (On the National Development and Meaning of the Natural Sciences) and "Über die Aufgaben der Naturwissenschaften in dem neuen nationalen Leben Deutschland" (On the Tasks of the Natural Sciences in the New, National Life of Germany). Science, according to Virchow, could help forge a national identity to which all Germans could relate, by alleviating the many forms of fragmentation that plagued the new Reich brought on Bismarck's imperialism and the Roman Catholic Church.

An example of a Versammlung tune arguing for science and the intellect to unify the fragmented German territories was "Den Gästen" (To the Guests).

The German federation is very varied.
One wants it restricted, the other extensive.
One wants it with corners, the other totally round.
One wants it long and narrow, that one wants it spacious.

Even language itself no longer seems
To be the symbol of unity.
In fratricidal battle, language became a weapon,
Which cut us in half.

If language fails, and borders fail
To establish unity
There is a bond that braids us.
It is the free German spirit [*Geist*].
It is and was at all times
The power of the German people.
It freed us from the path
Of deception and the darkest of nights.[61]

Along the same lines, "Ein fröhliches Verfassungswerk" (A Merry Constitutional Work), composed by the physicians members of the German Reichsversammlung in July of 1848, parodied the politicians' attempts to create a constitution:

Here no rightists, no leftists.
The order of the day: just drink, drink!
And drink to the end.
An oration well you can create.
Water, however, must not flow.
It is an unnecessary amendment.
A vice-regent you want
And a vice-regent you will find.
Almost as good as Johann!
Metternich has been forced out,
But Johannesberg remains
That is where the true man lives.

We are almost ready with the constitution.
Whatever is a nuisance just put it under the table.
Whatever is convivial, move it here.
Allow the wine to flow liberally,
We are all fresh, happy, and free.

Our flag—black/red/gold
That should say it all.
It's not difficult to figure out.
Gold and red wine
We slurp well into
The black of night.[62]

Oken incorporated *Liedertafel* evenings in the Versammlung because he believed that by singing (and drinking) together, investigators of nature would become better acquainted and friendlier. This camaraderie would result, Oken asserted, in scholars discussing their opinions with civility, rather than resentment, and would promote collaborations among German *Naturforscher*. Indeed, at the Hamburg Versammlung meeting in 1830, the initially bewildered Scotsman James Johnston began to comprehend the purpose of music by the meeting's conclusion: "Men learn to know, to esteem, and to judge each other more justly. Jealousies are removed, friendships are formed—and thus personal rivalry, harsh language and controversial sparring are diminished in philosophical writings."[63] Oken personally informed him that "you cannot speak so harshly or condemn in so unqualified a manner the theoretical speculations or experimental results of a man with whom you have held agreeable personal intercourse, as we are too prone to do of those whom we have never seen or conversed with."[64]

Johnston seemed to be impressed by this form of sociability and formation of identity of German *Naturforscher*. He noted how, after the opening day's feast, the orchestra played and was joined by "a score or two" of German savants, who had "lent their voices for the occasion."[65] He continued, "Nor shall I readily forget the energy and intensity of feeling with which the songs, whether Bacchanalian or Patriotic, were sung in this assembly of *savans*. One of the latter class of songs pleased me such much, that I amused myself, during dreary parts of the naval expedition, with turning it into hobbling rhymes."[66] The song was "Was ist das deutsche Vaterland?," a renowned choral anthem, the words of which had been penned by the nationalist poet, Ernst Arndt. Johnston proceeded to rewrite the lyrics parodying some of the aspects of the meeting.

Music also played a critical role in the Versammlung annual meeting in Stuttgart in 1834. Two mixed-sex choirs and an all-male choir entertained the guests with Wuerttemberg songs.[67] The *Schwäbischer Merkur* reported on 23 September 1834: "Our *Sängvereine* also felt obliged to prepare a welcome to our foreign guests. The choirs were set up in the auditorium, which offered room for many spectators. . . . Around 6 in the afternoon an all-male choir started the concert with a tune composed to a poem by Gustav Schwab. The all-male choir sang at times by themselves, and at other times were accompanied by an all-female choir. And soon several well-practiced singers performed quartets."[68] After the performance *Hofkammerrath* Waitz from Altenburg spoke of the purposes of both the Versammlung and the *Liederkränze*, so important to the Wuerttembergers' cultural and moral life, as "establishing the true morality among our *Volk*."[69] These *Liederkränze*, present at most Swabish festivals, were also responsible for the performance of older church pieces, which had fallen into neglect over the years. On 23 September 1834 the Stiftskirche (The Church Foundation) and the Verein für Kirchengesang (Association for Church Hymns) performed a large array of choral songs.[70]

The Versammlung Meeting in Berlin, 1828

In our times the taste especially for music is generally well distributed. Music is in these days, perhaps more so than ever before, loved, recommended, fostered, and supported and has become an object of general interest.
—Gottfried Weber, *Caecilia* 4 (1826)

What differentiates Berlin from all other cities in Germany is the cheerful and harmless sociability, in which friends are taken in with alacrity.
—Lorenz Oken, *Isis* 22 (1829)

By far the most impressive example of the role of music within the Versammlung was the Berlin Meeting of 1828. The Berlin gathering attracted the attention of both Charles Babbage and David Brewster. Brewster wrote to Lord Brougham on 14 March 1829:

Mr Babbage has written to me about the establishment of a great scientific association or society embracing all Europe. The idea has sprung out of the Congress of Philosophers at Berlin . . . and has been warmly embraced by many leading continental philosophers. The power of such a body to promote science, and give respectability to the various classes of men who sustain the intellectual character of the age is obvious. Foreign sovereigns would give it every encouragement. There never has been a more ardent patron of science than the King of Prussia is, and I am assured that the infection is spreading through the higher classes in every part of Germany. I wish you would give the subject a thought. Mr Babbage and I would take the oar if you would touch the helm.[71]

As is well known, the German Versammlung provided the impetus for the formation of the British Association for the Advancement of Science (BAAS).[72] The British were also interested in certain forms of sociability present at the German meetings, particularly the seating plans for the various dinners: "On the ticket delivered to the visitors at Hamburg the name, designation and numero of the recipient is written on one side, and on the other is a small plan of the floor and the seating of the place of meeting. The benches are all marked with engraved numbers at the extremities, and the particular place appointed for the bearer of the ticket is marked in red ink with his distinguishing numero, so that every one can find his seat without difficulty."[73] According to Johnston, the British preferred witty dinner speeches over the roar and merriment evoked by the Teutonic tunes: "you and I may have missed those after-dinner speeches, which, in our country, do sometimes nobly redeem the clatter of knives and forks, and give an intellectual character even to a city feast."[74]

Although the BAAS differed in critical ways from the German Versammlung, the organization and extravagance of the friendly evening meals were rather comparable. For example, the evening meals of the BAAS meeting in Newcastle-upon-Tyne in 1838 took place in a dining hall, which could accommodate 800 guests. The striped calico sides were ornamented with "coronals or chaplets of laurel and flowers, along its sides, of foreign and British names alternately, as Newton, Leibniz, Davy, Lavoisier, Ray, Linnaeus/Buffon, etc. Right or wrong the idea is friendly and is adopted. I had thought of mottos but 'qui vitam excoluere per artes' etc. is too common, and now I rather object to such things at all."[75]

Johnston, who echoed Brewster and Babbage's sentiments, reckoned that "this meeting in Berlin was by far the most splendid that has yet taken place . . . its being held in the capital of a powerful kingdom, where the government had shown a disposition to honour, and pay attention to scientific men."[76] The number of participants had soared from twenty-one members and approximately fifty *Liebhaber der Naturwissenschaften* (amateurs of science) during the first meeting in Leipzig to 466 members and over 280 more guests, dignitaries, and members of the Prussian Royalty during the Berlin meeting.[77] With such imposing numbers, several *Naturforscher,* including Oken, were beginning to fear that the reason for which these meetings were created—for participants to forge acquaintances with each other—was being threatened.[78]

That particular Berlin meeting was co-organized by Alexander von Humboldt, renowned naturalist and explorer, and Hinrich Lichtenstein, professor of zoology at the University of Berlin and founder of the Berlin Zoological Museum. Humboldt certainly had opposing views to Oken's as to what constituted proper *Naturforschung*, as evidenced by his winter lectures of 1828 delivered in the Singakademie after his return to Berlin from Paris.[79] Despite Oken's objections, Humboldt created seven disciplinary sections at the Berlin meeting with the hope that each subject would be discussed in greater detail than had been the case in previous meetings: geognosy-mineralogy, astronomy-geography, chemistry-physics, botany, zoology, anatomy-physiology, and practical medicine.[80] Oken protested by arguing that Humboldt had the wrong idea about the purpose of the Versammlung. He wrote to Friedrich Buchholz in 1829, "Our association cannot have the purpose of directly extending the natural sciences. Also, one does not have the view to indoctrinate, rather solely to discuss and to befriend."[81] Humboldt, however, underscored that the savants should not lose sight of the overarching unity of nature. Although Humboldt's methodology differed markedly from Oken's speculative *Naturphilosophie*, he, like Oken, emphasized the importance of music to the fledgling Versammlung.

On 17 September 1828, the evening before the first day of sessions, the visitors entered into the Schauspielhaus.[82] The names of renowned and deceased German investigators of nature and physicians glistened in the foreground projected on the sky-blue walls between the columns, "as a monument to the fame of the Fatherland"[83] (figure 3.4). Berlin's renowned neoclassical architect Karl Friedrich Schinkel proposed the design, which had been carried out by the theater inspector, C. Gropius.[84] Humboldt and Lichtenstein wished to "invent a tradition"—to borrow the phrase of Eric Hobsbawm[85]—that could serve as an ancestral lineage for the new, fledgling Versammlung. The most influential of these figures formed the middle column; they included Nicolaus Copernicus, Johannes Kepler, Gottfried Wilhelm Leibniz, Leonhard Euler, Immanuel Kant, and Abraham Gottlob Werner, with the more ancient scholars at the top. Savants from various fields composed the columns: the optician Joseph von Fraunhofer, the mathematician Georg Simon Klügel, Agricola, the inventor of the air pump Otto von Guericke, the chemist Georg Ernst Stahl, the "Romatic physicist" Johann Wilhelm Ritter, Bernoulli (most likely Daniel), the natural philosopher Christian Friedrich Wolf, and various physicians. Interestingly, the columns were not divided into subjects; rather, the point was to illustrate that the various scholars were intertwined, a united front investigating a unified nature. The two end-columns were flanked by quotes from the

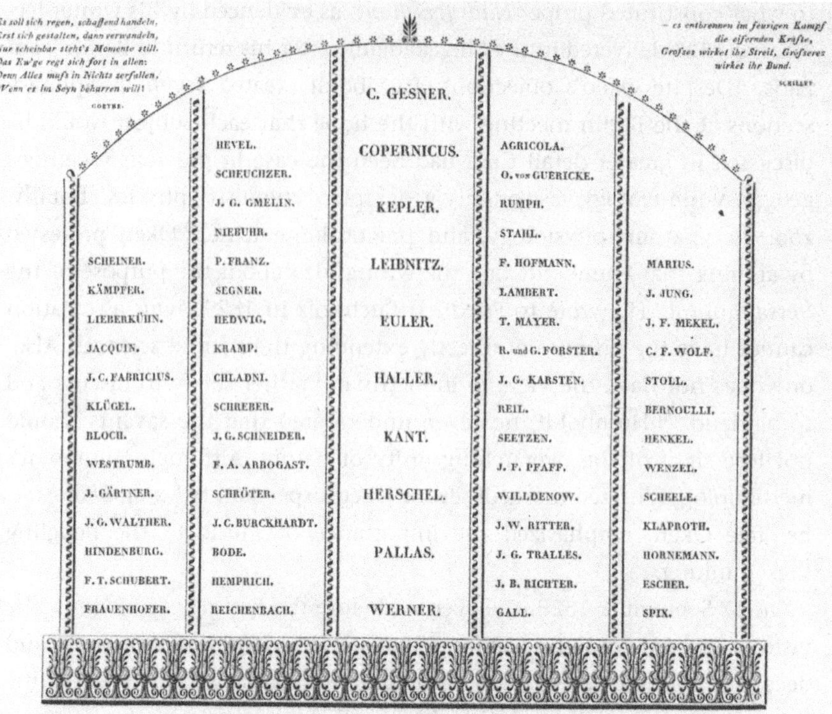

Figure 3.4
Names of renowned German investigators of nature and physicians projected on the walls of the Schauspielhaus at the Meeting of the Association of German Investigators of Nature and Physicians in Berlin, 1828. Source: Humboldt and Lichtenstein 1829, p. 19.

German territories' leading authors fascinated with the forces of nature, Goethe[86] and Schiller.

Zelter conducted several hundred male and female voices singing Georg Friedrich Handel's "Alexander's Feast" in the hall, where the scientific lectures and discussions took place later in the week.[87] Handel had composed this oratorio in 1736, setting to music the words of John Dryden's famous poem. Dryden's "Alexander's Feast" deals with the power of music to affect the full range of human emotions, from pacifying to igniting the passions of war. The piece commemorates the royal feast in honor of Alexander's conquest of Persia. A song comes from Jove on high, and "the listening crowd admire the lofty sound."[88] Bacchus, who is praised by the songs of "the sweet musician," comes to the feast, while the trumpets are sounded, and the drums are beaten.[89] Such inspiring music encourages the king to continue his battles, "Sooth'd with sound, the king grew vain, Fought all his battles o'er again, And thrice he routed all his foes, And thrice slew the slain.[90] Afterward, however, "[h]e chose a mournful Muse, Soft pity to infuse."[91] The first part of the oratorio ends with a choral refine, "The many rend the skies with loud applause: So Love was crown'd, but Music won the cause."[92] The oratorio's second and final part concludes with an appendix, which is not found in Dryden's original poem. Indeed, it was written by Newburgh Hamilton, and is present in both the 1736 and 1738 editions of the work. The message was rather relevant to the annual gatherings of the Versammlung:

Your voices tune, and raise them high,
Till th' eccho, from the vaulted sky,
The blest Cecilia's name.

Music to Heav'n and her we owe,
The greatest blessing that's below:
Sound loudly then her fame!
Let's imitate her notes above!
And may this ev'ning ever prove
Sacred to Harmony and Love.[93]

After the performance, which apparently did not please Zelter, who described it as "full of holes"[94] since two soloists were missing, Humboldt ascended to the podium to deliver the introductory address (figure 3.5). In the presence of the King Friedrich Wilhelm III of Prussia and Prince Albrecht, Humboldt exclaimed, "But what can be more gratifying to the soul—this picture of the common Fatherland—than this gathering, which we have received for the first time within these walls?"[95] He was pleased that participants that September morning came from the Neckerland, where Kepler and Schiller were born, from the Baltic region, from the Rhineland, "where the treasures of an exotic nature were gathered and researched."[96] Indeed, he welcomed everyone from "everywhere, where the German language is spoken, and where its sensual edifice works upon the spirit and the soul of the people in unison: from the Alpine region to the Vistula, where Copernicus brought astronomy to new levels—everywhere in the larger area of the German nation where Germans unite to track the secret effect of nature's forces."[97] Germany, according to Humboldt,

Figure 3.5
Portrait of Alexander von Humboldt (1769–1859). Photo: Deutsches Museum, Munich.

was revealing its spiritual and intellectual unity. Humboldt's patriotic yearnings were clear: "If, however, I, in view of this gathering, must hold back the expression of my personal feelings, I am at least permitted to name the patriarchs of German fame."[98] He named Goethe, Olbers, Sömmering, and Blumenbach.[99] The audience recalled how Humboldt stressed that "the main purpose of the society is the personal approaches of those who work on the same field of the sciences, the exchange of ideas—they can be about facts, opinions, or doubts—and the establishment of friendly relationships, which shed light on the sciences, and which grant life its gay charms, and customs, their patience and gentleness."[100] Humboldt also explained how the founders of the society sought "to unite all branches of physical knowledge" (the descriptive, metrological, and experimental sciences).[101] They were all committed to keep the true belief in the unity of nature.

Humboldt called upon a friend of the family, the nineteen-year-old Felix Mendelssohn-Bartholdy (figure 3.6), to compose a piece for this festive occasion, celebrating both the Versammlung and Berlin.[102] Humboldt was an active participant in the Mendelssohn family's musical soirees.[103] Mendelssohn did much to draw attention to past German composers. As an orchestral and festival director and founder of the Leipzig conservatory, he conducted the works of deceased German masters. Rather famously, in 1829, only months after composing this cantata for Humboldt, Mendelssohn conducted J. S. Bach's St. Matthew Passion, contributing to the Bach Revival.[104] And, his commitment to a national identity was evidenced by his own compositions, including the Reformation Symphony, the *Lobgesang* (song of praise), and the St. Paul and Elijah oratorios.[105] He wrote to his mentor Zelter, calling for an "end to our exaggerated modesty, which makes us accept everything that comes from others as good, and which keeps us from appreciating our own until it has first been recognized by others."[106]

Mendelssohn composed a secularized cantata in six movements in the key of D major (figure 3.7; music available for download at http://mitpress.mit.edu/ at this book's page). It was composed for an all-male choir, wind instruments, cellos, basses, a piano, and timpani. Some of the members of the Singakademie who sang the cantata were also *Naturforscher* and physicians belonging to the Versammlung—including, for example, Lichtenstein. Lichtenstein served on the governing body of the Singakademie from

Figure 3.6
Portrait of Felix Mendelssohn-Bartholdy (1809–1847), by James Warren Childe, 1829. Source: Bilderarchiv Preußischer Kulturbesitz-Berlin/Art Resources, NY.

1815 to 1823 and again from 1826 until his death in 1857.[107] And he was well aware of the musical scene in Berlin during the period, as evidenced by his lengthy correspondence with his godfather, the renowned German pianist and composer, Carl Maria von Weber.[108] Felix's sister, Fanny, critically remarked on the event, writing to her friend the diplomat Carl Klingemann on 12 September 1828: "Humboldt, the cosmopolitan, most ingenious, charming, scholarly Hofmann of his time, is hosting a festival, the likes of which this city has never seen. The venue is the Concert Hall, the number guests seven hundred, including the King. . . . Felix has been requested to compose a cantata for the reception (you see, Felix is coming

Figure 3.7

First page of Felix Mendelssohn-Bartholdy's *Festkantata* composed for Alexander von Humboldt at the Meeting of the Association of German Investigators of Nature and Physicians in Berlin, 1828. Source: Musikabteilung mit Mendelssohn-Archiv, Staatsbibliothek Preußischer Kulturbesitz-Berlin: mus. ms. autogr. Mendelssohn-Bartholdy 48, folio 1.

into fashion)."[109] Her sarcasm becomes most evident while critiquing the all-male nature of both the Versammlung and its choir:

As the *Naturforscher*-paradise has no women and is Mohammedan-like, the choir comprises only the best male voices from the area. And because Humboldt, who is no great musician, has limited his composers to a small number of musicians, the orchestra assumes a curious shape. There are basses and cellos, trumpets, horns and clarinets. A rehearsal took place yesterday, and it was said to have made a good impression. What aggravates me is that we were not there. You cannot imagine how on this occasion what a comical mixture of sleepy villagers [*Krähwinkelei*] and metropolis dwellers [*Großstädterei*] comes to light. The entire arrangement, the reception of foreign guests, the integration of all of these impressive names for one purpose (even if it is only social) is undeniably grandiose, but who knows . . . , just how much Beyermann has supplied for the aforementioned amount, and what the reception of his guests will cost Humboldt.[110]

Felix clearly was up to the task of composing for such a unique collection of masculine talent.[111]

The words to this *Festkanata* were penned by Berlin's leading music critic of the period, Ludwig Rellstab, who had just returned to Berlin from Spandau.

Willkommen! rufen wir Euch froh entgegen
Der Gruss der Freundschaft ists, der Euch erklingt;
Doch waltet über diesem Fest ein Segen,
Der uns mit grösster Weihe Kraft durchdringt
Mit Stolz und Rührung muss es uns bewegen
Das Heil, das des Beherrschers Huld uns bringt.
So möge ihm des Dankes Gruss ertönen,
Dem Schirmer alles Grossen, Guten, Schönen.

Aus alter grauer Nacht des Chaos
Entwirrte mühsam sich der Elemente Kraft;
Fest stellte sich die Erde, starr und trotzig
Bot sie dem Sturm der Höhen stolze Gipfel
Und warf des Ufers schroffe Felsenbrust
Dem Meer entgegen. Feuer, Luft und Wogen
Bekämpfen sie voll Wuth. Es bricht der Sturm
Die mächt'gen Blöcke aus dem festen Lager
Und donnernd stürzen sie ins Thal hinab.
Die Woge schäumt voll Ingrimm an den Damm
Der Berge, wühlt sich tiefe Klüfte aus.
Und furchtbar dringt des Feuers wilde Kraft
Zerstörend ein bis zu der Tiefe Schooss!

Laut tobt des wilden Kampfes Wuth
Die Zwietracht bringt Zerstörung
Es drohen Flammen, Sturm und Fluth
Mit grimmiger Verheerung.
Was Gott erschuf mit weiser Macht
Sinkt wieder in die alte Nacht.

Halt ein! tönt einer Wunderstimme Klang
Und plötzlich ist der Elemente Zorn
Gefesselt, Sturm and Wogen ruhen,
Zur stillen Glut senkt sich das Flammenmeer.
Da bricht des Lichtes wunderbare Klarheit
Aus Äthers Räumen wunderbar hervor;
Hell, offenkundig allen wird die Wahrheit
Versöhnt ist jetzt der Elemente Chor.
Gemeinsam wirkt der Kräfte eifrig Streben,
Denn Eintracht nur kann wahres Heil ergeben.

Jetzt wirken und schaffen
Verschwisterte Kräfte
Und bilden und bauen
Die herrliche Welt.
Es pranget die Erde,
Es schimmert das Feuer,
Und liebliche Lüfte
Bewegen die Fluth.
Hoch wölbt sich der Äther
Und blinkende Sterne
Ziehn goldener Kreise
Sanft strahlende Bahn.
So wirken und schaffen
Verschwisterte Kräfte
Und bilden und bauen
Die herrliche Welt.

Und wie der grosse Bau der Welt sich ordnet,
So bildet sichs auch in des Menschen Brust.
Es wohnt die wilde Kraft der Elemente
In seiner Seele, die verderblich wirkt
Wenn nicht ein grosses, leuchtend hohes Ziel
Zur Einheit schlichtet starrer Kräfte Zwist.
Dann mag der Trieb nach allen Seiten schwellen
Zu einem Stamm gehören alle Zweige
Und der Erkenntniss segensreicher Baum

Wird prangend in der vollen Blüthe stehn
Und segnend wird der Himmel ihn beschirmen.

Ja segne Herr, was wir bereiten,
Was die vereinte Kraft erstrebt
Dass in dem flüchtgen Strom der Zeiten
Das Werk uns gleich den Felsen steht.
Und wie sichs hebt und thürmt nach oben
In Würde, Pracht und Herrlichkeit,
So wird es nur Dich selber loben
Denn Deiner Grösse ists geweiht.[112]

Oken was clearly moved by the performance, noting that a "selection of the best Berlin musicians under the direction of Zelter and Felix Mendelssohn Bartholdy united to give those present reason to discuss the most meaningful, merriest, and the most fitting passages to this festival, and what poetry and music have brought forth in the German lands."[113] The following tunes were sung later that evening: Goethe's "Entschluß" and "Beherzigung"; Friedrich Ferdinand Flemming's "Ode XXII aus dem Horaz," in Latin; Schiller's "Der Besuch"; Schmidt's "Das deutsche Lied"; Theodor Körner's "Bundeslied"; Friedrich Rükert's "Die goldene Zeit"; and "Domine salvum fac regem." The evening's festivities lasted three hours.[114]

On 23 September, between 700 and 800 guests, including Versammlung participants, were pleasantly surprised by a lunchtime choral concert made up of two *Liedertafeln* under the direction of Zelter and his soon-to-be successor as director of the Singakademie, Carl Friedrich Rungenhagen. The performance took place in the dining hall of the Exercierhaus on Carlsplatz. Twenty lavishly decorated tables seated 480 guests, which included the investigators of nature and physicians, some of whom were accompanied by their wives and children. The two *Liedertafeln,* comprising a total of seventy-two men, were set up between the tables.[115] Zelter wrote to Goethe that "[o]ur choirs were in good form. . . . We then earned our keep one afternoon with a *Liedertafel* for the foreign and local guests. There were between 700 and 800 participants. Both Berlin *Liedertafeln* came together, and since the location in the new Exercierhaus is wide and high enough, the effect surpassed my expectations. Seventy male voices are so powerful that one could clearly hear the words outside, and the foreign guests assured us that they had never heard such a thing in their lives."[116]

Lichtenstein and Humboldt argued that the foreigners should be given a "dignified impression" of the high level of music present in Berlin.[117]

Fifteen songs were sung by both *Liedertafeln*, including Goethe's "Liederstoff"; Köhler's "Die Sänger an ihre Mahlgenossen"; Flemming's "Dulce cum sodalibus"; Opitz's "Der König"; Ardnt's "Das deutsche Vaterland"; Pfund's "Vorwärts!"; a sixteenth-century drinking song entitled "Nun laßt uns fröhlich seyn!"; Goethe's "Soldatentrost"; Spiker's "Schottisches Lied," a table song; Suetonius' "Cantus martialis romanus," "St. Paulus"; and a concluding song, "Jetzt schwingen wir den Hut." Oken was overjoyed to report that the music increased "the merriment and enthusiasm of all participants. Health rang out from all sides, from the singers, the locals and the foreign guests."[118] Indeed, the mood must have been most cordial, with visiting investigators of nature and physicians retiring after these concerts to the homes of the Berlin *Naturforscher* to discuss that day's sessions and activities. The personal relationships that Oken so desperately sought to create were beginning to be cultivated. Johannes Müller wrote of his getting acquainted with J. E. Purkyne, E. H. Weber, and A. Retzius, all of whom remained in close contact for decades.[119] Purkyne, who was married to the daughter of the Berlin professor of anatomy and physiology C. A. Rudolphi, spoke of how his father-in-law hosted a friendly gathering at his home after an evening meal. His guests included Berzelius, A. A. R. Retius, E. Mitscherlich, F. Wöhler, E. A. Fries, S. Nilsson, P. F. Wahlberg, F. Tiedemann, E. Horn, J. Elliot, A. W. Otto, G. H. Behn, C. G. Ehrenberg, K. W. U. Wagner, F. A. L. Thienemann.[120] Perhaps most important contact forged at this meeting occurred between Wilhelm Eduard Weber and Carl Friedrich Gauss.

In November 1830 Göttingen's professor of physics, Tobias Mayer, died. The university's professor of mathematics and astronomy, Gauss, detailed his opinion of the requisite qualifications of the successful replacement.[121] First and foremost, Gauss argued that the scholar should be "a physicist with a fundamental education in mathematics [*gründlich mathematisch gebildeter Physiker*]."[122] His letter discusses five possible candidates: Bohnberger of Tübingen, Brandes of Leipzig, Gerling of Marburg, Seeber of Freiburg, and Weber of Halle.[123] He discusses the merits of each savant, but what he says about Weber is particularly relevant to this chapter. Gauss comments that he had personally met Weber at the Versammlung meeting in Berlin in 1828 and "had been with him many times thereafter."[124] Gauss was impressed by Weber's lecture on the physics of reed pipes (detailed in chapter 5), calling it "a well-ordered and well-delivered lecture."[125] He added that he thought the young physicist was clearly talented. And perhaps just as important, Gauss added that Weber had given him the impression of "possessing a *liebenswürdiger* [charming] character, which exists more in the world of the scholar than in the external world."[126] The Versammlung meeting in Berlin also provided Weber with referees. In his letter to the *Staatsminister* expressing an interest in the Göttingen position, Weber lists four scholars who could adjudicate his work, all of whom he had met at that meeting: Gauss, Humboldt, Berzelius, and Oersted.[127] And Prof. Blume wrote a letter of support for Weber's application, arguing that he had heard only positive reports on Weber's experimental skill, having made a name for himself at the Berlin meeting.[128]

In conclusion, a major part of the Berlin meeting of the Association for German Investigators of Nature and Physicians was to form an identity. They wished to establish this cultural and national identity in the absence of a political one, and music played a critical role in the establishment of that identity. The desire to achieve cultural and national unity was reflected in the Versammlung's love of choral music. Among cultured persons of the early nineteenth century, the cultivation of one's own voice and the singing together within a close circle of friends had become rather fashionable. German folk songs demonstrated that the group function was regarded as more meaningful than the self-awareness of the isolated individual. These songs enabled *Naturforscher* and physicians to share emotions and values. They described events from the point of view of a group that worked in harmony toward common goals: Germans who sang together stayed together. Music contributed to the construction of a national identity and persona, as well as to the sociability of the group. German societies, such as the German Association of Investigators of Nature and Physicians and the *Liedertafel*, as the late German historians George Mosse and Thomas Nipperdey have argued, were self-representations of a democratically controlled nation, objectifying the ideals for which the nation was supposed to stand.[129] Male camaraderie, so aptly depicted by Mosse and Ute Frevert[130] in particular, was a hallmark of the many choral societies and the Versammlung. This form of sociability within the Versammlung was to last well into the twentieth century.

4 The Organic versus the Mechanical

The previous two chapters have mapped two very different aspects of music and their relationships to investigators of nature prevalent in the late eighteenth and early nineteenth century: musical-instrument manufacture and choral music. The former was very much a mechanical practice based on the laws of vibrations, waves, aerodynamics, adiabatics (as we shall see in the next chapter), and—to a much greater extent—on the tinkering of craftsmen in their workshops. Communal folk singing, on the other hand, celebrated the organic, unifying spirit of the *Volk*. In 1830 J. G. Kaestner gestured to the contrast between the artisanal and the spiritual in music.[1] When *Handwerk* referred to the technical aspects of musical-instrument manufacture, it admittedly had little in common with the spiritual facets of music.[2] However, when *Handwerk* was used to mean the constant, meticulous practicing of the art, it became "the interaction between the human body and spirit" that produced the true, "free genius."[3] The genius constantly needed to ply his craft; otherwise, he would lose his power.

Clearly, on one level, these two realms, the technical and the spiritual, can coexist in harmony. On a deeper level, however, tensions arise, particularly when a number of mechanicians, physicists, and even musicians turn to mechanical principles in order to improve the art of music. The ensuing chapters detail how the mechanical and organic interacted, sometimes peacefully, sometimes antagonistically. Some scholars, such as E. T. A. Hoffmann and G. L. P. Sievers, viewed music as quintessentially exemplifying organic unity, a product mastered by the organic genius, independent of any mechanical, physical laws. At the same time there were movements within music, physics, and instrument making to standardize aesthetic characteristics, including the use of chronometers and later metronomes to demarcate precise intervals of time, and calls for a

European standard concert pitch, which would be both musically appropriate and acoustically determined. And this period also witnessed an ongoing fascination with musical automata, vast improvements on Jacques de Vaucanson's flute player of 1730, which rivaled their organic counterparts in both precision and expression, or an instrument's ability to alter its volume.

The Organic

German savants of the late eighteenth and early nineteenth centuries were devoutly antimechanist and antireductionist, celebrating what they felt to be the *Geist* (spirit) and *Seele* (soul) of humankind. No German better represented his generation's opposition to a mechanistic depiction of nature than the writer, Privy Councilor, and *Naturforscher* Johann Wolfgang von Goethe (figure 4.1).[4] Goethe's work on morphology was fundamentally committed to vitalism. Although he readily admitted that the principles of physics and chemistry could assist in the elucidation of the properties of living organisms, he vehemently opposed any attempt to reduce biological phenomena to physico-chemical principles. In his essay "Bildung und Verwandlung" (Formation and Transformation), Goethe argued that "analytical efforts, if continued indefinitely, have their disadvantages. To be sure, the living thing is separated into its elements, but one cannot put these elements together again and give them life."[5] In his "Vorarbeiten zu einer Physiologie der Pflanzen" (Preliminary Notes for a Physiology of Plants), written in the mid-1790s, Goethe declared: "From the physicist in the strict sense of the term, the science of organic life has been able to take only the general relationship of forces, their position and disposition in the given geographical location. The application of mechanical principles to organic creatures has only made us more aware of their perfection; one might almost say that the more perfect living creatures become, the less can mechanical principles be applied to them."[6]

Goethe was, of course, attacking the mechanical philosophy of French rationalism, which had dominated European thought for over a century and a half. Organisms, contrary to the French depiction, were the antithesis of complex machines. In his autobiography *Dichtung und Wahrheit* (*Poetry and Truth*) Goethe recalled his reaction to the mechanistic nature presented throughout Baron d'Holbach's *Système de la nature*, published in

Figure 4.1
Portait of Johann Wolfgang von Goethe (1749–1832). Photo: Deutsches Museum, Munich.

1770: "We found ourselves deceived in the expectation with which we had opened the book. A system of nature was announced, and therefore we truly hoped to learn something of nature, our idol. . . . But how hollow and empty we felt in this melancholy, atheistic half-night [*atheistische Halbnacht*], in which the earth vanished with all its images, the heavens with all its stars."[7] This mechanical ideology of the French had driven the young Goethe to Shakespeare, where he discovered the notion of genius and the creative spirit of both humankind and nature.[8]

Goethe was hardly alone in condemning the French *Weltanschauung*. His *Sturm und Drang* cohorts, and later the early Romantics, echoed Goethe's disdain and indignation. All of them fundamentally opposed the notion that mechanical laws could solely explain all of nature's phenomena. Nature was to the late-eighteenth- and early-nineteenth-century German *Naturforscher* a living entity.

German Romantics were fascinated by the tension between the mechanical and organic. Heinrich von Kleist's eerie allegory "Über das Marionettentheater" (On the Puppet Theater) contrasts the "soul of a dancer" with the "mechanical forces" governing the motions of a marionette.[9] The operatic dancer, who frequents marionette performances, surprises the narrator by insisting that studying the mechanics of the dancing puppets can actually improve his own routines. According to the dancer, the mechanical devices possess two advantages over humans: first, "they never put on airs; then affectations appear, as you know, when the soul (vis motrix) finds itself in some point other than the center of gravity. Since the mechanician has no other point at his disposal, by means of the crank or string, all the other limbs are, as they should be, dead, pure pendula and merely following the law of gravity; a superb trait, for which one looks—in large part in vain—in our dancers."[10] Second, somewhat paradoxically, they are "anti-gravity. They know nothing of material's inertia, what dancers strive against more than all other properties [of nature]."[11]

One of the leading music critics of the early nineteenth century was also one of the German territories' foremost Romantic authors, E. T. A. Hoffmann. He was a regular contributor to the *Allgemeine musikalische Zeitung*, offering intelligent reviews of the works of J. S. Bach, Ludwig van Beethoven, Carl Maria von Weber, and Robert Schumann.[12] And a number of his stories had musical themes, particularly musical automata. "Der Sandmann" (The Sand Man) details a tormented college student who unwittingly falls in love with a piano-playing automaton named Olimpia, the "daughter" and creation of the student's professor of physics Spalanzani, "a skillful mechanician and fabricator of automata" with the assistance of Coppelius, "an honest mechanician and optician."[13] This story was immortalized by Jacques Offenbach in act 1 of his opera, *The Tales of Hoffmann*.

Another work of Hoffmann's, "Die Automata" (The Automata), is particularly significant. First appearing in the *Zeitung für die elegante Welt* (The Newspaper for the Elegant World) in 1814, the story was divided into two parts, one a ghost story and the other the recollections of Ferdinand, the main character, of an automaton called the Talking Turk, "a masterpiece of mechanical skill," which possessed clairvoyant abilities.[14] Ferdinand is haunted by his interaction with the Talking Turk, as the automaton foretells a horrific prophecy involving his secret beloved. Ludwig, Ferdinand's college companion, expresses his disdain of such mechanical contraptions: "what I detest most of all is the mechanical imitation of human motions."[15] Professor X, the creator of the Talking Turk, offers the two men a concert with musical automata, which he had also constructed: a man playing a flute, a woman seated at a piano-like instrument, and an orchestrion.[16] The room's walls, bedecked with mechanical musical clocks, adds to the eerily contrived ambience. After a brief performance of his automata, Ferdinand and Ludwig hastily depart. Ludwig repeats, this time much more eloquently, his loathing for the mechanical mimicking of the organic and spiritual:

> All that machine music (in which I include the Professor's own playing) makes every bone in my body ache. I am sure I do not know when I shall get over it! The fact of any human being's doing anything in association with those lifeless figures which counterfeit the appearance and movements of humanity has always, to me, something fearful, unnatural, I may say terrible, about it. I suppose it would be possible, by means of certain mechanical arrangements inside them, to construct automata which would dance, and then to set them to dance with human beings, and twist and turn about in all sorts of figures; so that we should have a living man putting his arms about a lifeless partner of wood, and whirling round and round with her, or rather it. Could you look at such a sight, for an instant, without horror? At all events, all mechanical music seems monstrous and abominable to me; and a good stocking-loom is, in my opinion, worth all the most perfect and ingenious musical clocks in the universe put together.[17]

Specifically, Ludwig questions whether mechanical operations such as breathing and fingering techniques are all that is necessary to summon forth beautiful music. For him, the human spirit, which could not be mechanically reconstructed, is the primary trait that guides the organs of the body:

For is it the breath, merely of the performer on a wind-instrument, or the skillful, supple fingers of the performer on a stringed instrument which evoke those tones which lay upon us a spell of such power, and awaken that inexpressible feeling, akin to nothing else on earth—the sense of a distant spirit world, and of our own higher life in it? Is it not, rather, the mind, the soul, the heart, which merely employ those bodily organs to give forth into our external life what we feel in our inner depths so that it can be communicated to others, and awaken kindred chords in them, opening, in harmonious echoes, that marvelous kingdom, from which those tones come darting, like beams of light? To set to work to make music by means of valves, springs, levers, cylinders, or whatever other apparatus you choose to employ, is a senseless attempt to make the means to an end accomplish what can result only when those means are animated and, in their minutest movements, controlled by the mind, the soul, and the heart. The gravest reproach you can make to a musician is that he plays without expression; because, by so doing, he is marring the whole essence of the matter. Yet the coldest and most unfeeling executant will always be far in advance of the most perfect machines. For it is impossible that any impulse whatever from the inner man shall not, even for a moment, animate his rendering; whereas, in the case of a machine, no such impulse can ever do so. The attempts of mechanicians to imitate, with more or less approximation to accuracy, the human organs in the production of musical sounds, or to substitute mechanical appliances for those organs, I consider tantamount to a declaration of war against the spiritual element in music; but the greater the forces they array against it, the more victorious it is. For this very reason, the more perfect that this sort of machinery is, the more I disapprove of it; and I infinitely prefer the commonest barrel-organ, in which the mechanism attempts nothing but to be mechanical, to Vaucanson's flute player, or the harmonica girl.[18]

A key point Ludwig makes is the expression of a human performer. Mechanical musical instruments, cold, lifeless, lacking any feeling or expression, could never match the emotions evoked by instrumentalists and vocalists. Or could they? As will be discussed in this chapter, the goal of mechanicians was precisely to accomplish what Ludwig thought impossible—to construct mechanical musical instruments that would be just as evocative as human performers. The ideal of Romantic music could indeed be achieved by machines.

Ferdinand, also critical of the automaton's lifeless music, wondered why skillful mechanicians did not instead apply their talents to the construction of musical instruments for living performers. Ludwig concurred, arguing that musical instruments are key to unlocking "the marvelous acoustical secrets which lie hidden all around us. . . ."[19] Instruments that evoked mysterious-sounding music from glass and metal cylinders, glass threads and slips—much like Chladni's euphone and clavicylinder—were at the forefront of divulging "the mysterious tones of nature."[20]

The philosopher G. W. F. Hegel shared Ludwig's concern that mechanical music was destroying music's spiritual element. "The executant artist . . . must submit himself entirely to the character of the work. . . . [But] he must not, as happens often enough, sink to being merely mechanical, which only barrel-organ players are allowed to be. If . . . art is still to be in question, the executant has a duty to give life and soul to the work."[21] Things apparently had not changed all that much by the 1840s. After emigrating to Paris and becoming caught up in the city's Liszt-mania, the German poet Heinrich Heine lamented that "technical perfection, the precision of an automaton, the identification of the musician with stringed wood, the transformation of a man into an instrument of sound, is what is now praised and exalted as the highest art."[22] Live performers were mimicking automata, rather than vice versa.

While febrile imaginations feared the creation of musician-replacing automata that would flawlessly execute the composition ad infinitum, in the cacophonous world of the Industrial Revolution's new factories machines were in fact busy replacing workers. The economic ideology of mass production of the Industrial Revolution, epitomized by the chemist and philosopher of manufacture Andrew Ure, celebrated the replacement of skilled labor with machines run by unskilled workers, thereby increasing a factory's efficiency. With this new economy came a new aesthetic. Rather than marveling at the exquisitely beautiful, unique product of the master craftsman or the virtuoso, an aesthetic arose based on the perfect replication of an artifact many times over, an aesthetic of precision and high fidelity to a perfect type. In addition, the Scottish natural philosopher David Brewster argued for another, more disturbing, aspect of a machine's precision in his 1819 work announcing his new invention, the kaleidoscope: "There are few machines, indeed, which rise higher above the operations of human skill. It will create, in a single hour, what a thousand artists could not invent in the course of a year; and while it works with such unexampled rapidity, it works also with a corresponding beauty and precision."[23] The kaleidoscope used repeated mechanical processes but did not produce the same image twice; rather it improved upon the artist's creativity. It was a machine that could mimic the artist's creative genius, potentially replacing it.

Automata and mechanical musical instruments enjoyed the advantage of not being subject to the performer's whims. This period saw the conservatoires' emphasis on personal interpretation and ornamentation of the performer. Automata by contrast lacked the ability to take liberties with the composer's work. Also, automata could generate more tones than a single, organic performer. J. N. Mälzel's panharmonicon, for example, could generate the tones of a small orchestra. Certain automata could carry out with ease passages that surpassed anatomical limitations—such as extreme stretches of the fingers on a keyboard, rapid double trills, or speedy cadenzas. Hence, despite the vastly improved mechanical skilled technique of the performers of the early nineteenth century, the composer was no longer restricted to the shortcomings and foibles of human virtuosi.

The Mechanical

Who were the builders of these mechanical devices? Far removed from the world of Goethe, Kleist, Hegel, and Heine, they were skilled artisans, mechanicians, who were generally trained as clockmakers, the most skilled craftsmen of the period. Many of the practices and techniques in clock making could be transferred with minimal alteration to musical instruments, as evidenced by the plethora of musical clocks employing flute pipes and harps. A number of textbooks of the period on clock making include sections on musical automata.[24] The average Berlin apprentice in clock making, for example, learned his trade in four years, if he paid his master, or five to six years in order to remunerate with his labor.[25] Often these mechanicians traveled throughout Europe selling their wares. After successfully building a reputation, they could lead more sedentary and lucrative lives as court or university mechanicians, the latter of whom often repaired experimental apparatuses used in physics and chemistry labs. Some, such as Christian Hoffmann of Leipzig and J. August D. Oertling of Berlin, discussed in detail in the following chapter, enjoyed membership in regional scientific societies.

The mechanicians discussed in this chapter influenced music in three ways. First, they constructed automata, such as Friedrich Kaufmann's trumpeter, which mechanically mimicked human performers. Second, they produced many musical instruments, which, powered by humans, mechanically reproduced music, such as musical clocks and barrel organs. Third, mechanicians constructed numerous keyboard reed instruments, extremely popular during the first half of the nineteenth century. And crucially, organ builders began to incorporate relevant techniques of improved reed-pipe construction from these mechanicians.

Gleaning information on mechanicians (specifically those relevant to musical-instrument manufacture) is a rather difficult task, for their traces are much harder to find than those left behind by savants. They rarely published works, as their contributions to the scientific and musical enterprises were their instruments. There are, however, several sources that begin to shed light on the lives of these barely visible, yet critically important, actors. We know, for example, a bit about the mechanician and court instrument maker from Cassel, Johann Heinrich Völler.[26] Born in the village of Angersbach near Darmstadt, he was the fourth son of a farmer, whose uncle taught him mathematics. As a result of his private tutoring, he surpassed all of his classmates. At the age of fifteen, it became clear that he was particularly gifted in algebra, penmanship, and music, specifically violin playing.[27] He expressed his interest in studying to become a joiner, much to the dismay of his father, who expected him to help out on the farm. After convincing his father that his future lay in joinery, he started his apprenticeship, where he met an organ builder named Scholtmann, who wanted to hire the young lad as an apprentice. Scholtmann offered him an apprenticeship for 100 florin, a price he simply could not afford. After much pleading, the boy's grandfather paid a portion of the fee, and convinced the father to chip in the rest. So, in 1786 Johann entered into the apprenticeship as an organ builder.[28] Johann quickly learned the requisite skills for repairing and building organs. He finished his apprenticeship with Scholtmann two years later, and traveled to Cassel in 1789 to continue working on organs with a new master, Zitzmann, instrument maker and organ builder.[29] Johann served as Zitzmann's apprentice for ten years, honing the mechanical skills relevant to musical-instrument manufacture and repair.

After his apprenticeship was completed, he embarked on a prolonged journey throughout Europe selling his instruments and repairing organs.[30] He returned to Cassel and became the court instrument maker.[31] In addition to repairing musical instruments and pocket watches, he invented a number of musical automata, including the apollonion, a piano-like instrument with flute pipes and bellows. He also invented household mechanical devices not directly related to music: a door that would chime when opened, a dollhouse filled with miniature, wooden Hessian soldiers in complete uniform, and a so-called perpetual-motion machine composed of a spiral-shaped candle with a spoon attached to its base. If one placed a small ball on the candle, it would descend down the spiral track and touch the spoon, which would send it back to the top.[32] And the link with music was not unique to Johann. He paid for the musical education of his oldest brother Johannes, who eventually became the music director in Stettin. And his brother Hermann studied music in Worms.[33]

A critical part of mechanical musical instruments and automata was the reed pipe, the proper construction of which represented a good test of the mechanician's skill.[34] During the final years of Catherine the Great's reign, the mechanician Christian Gottlieb Kratzenstein of St. Petersburg had tinkered with the idea of attaching free-vibrating reeds to pipes in order to produce sounds similar to the human voice for his speaking automaton.[35] As a result, he is considered to be the inventor of the free-vibrating reed pipe.[36] In 1791, after twenty years of labor, Wolfgang von Kempelen, the inventor of the infamous Turkish chess player, published a treatise detailing a mechanical apparatus using ivory as his reed, which could reproduce a human voice.[37] Eight years later Johann Jakob Schnell's *Äolsklavie* was reported in Vienna to sound very similar to a human voice.[38] During the 1780s, the Danish organ builder Kirsnik (also spelled Kirschnigk) of Copenhagen applied Kratzenstein's idea to the pipes of his newly invented instrument, which he called the organochordium, an organlike instrument with a keyboard for the right hand and a bellows that was pumped by the left. The pipes of "Kirsnik's harmonica," as it would come to be known, produced pleasant tones without the resulting shrill or rattling that often accompanied earlier reed instruments.[39]

Kirsnik's instrument drew the attention of the music theoretician and composer Abbot Vogler (Georg Joseph Vogler), who envisaged an organ with free-vibrating reeds, whereby the reed was mounted on the side of a trumpet-shaped wooden resonator.[40] In 1788 Vogler had traveled to St. Petersburg to witness Kirsnik's harmonica. A year later he designed his own instrument, the orchestrion (not to be confused with mid-nineteenth-century musical automata of the same name), a chamber organ much larger than Kirsnik's device, comprising four keyboards of sixty-three keys each and a pedal board of thirty-nine notes. The contraption was actually built by the Stockholm organ builder Rackwitz, who had been apprenticed to Kirsnik from 1763 to 1770.[41] Vogler's orchestrion debuted in Amsterdam, and then traveled to London in 1790. It could increase and decrease in volume by means of swell shutters.[42] In that same year Vogler wrote to Rackwitz inquiring whether the Swede could assist him in building an organ in Rotterdam with beating-reed pipes. In 1791 Rackwitz and Vogler traveled to Frankfurt am Main, where Rackwitz built more free-vibrating reed pipes and attached them to an organ in a Carmelite cloister, the first German organ with reed pipes.[43] Five years later, Vogler built his renowned orchestrion, which debuted in Stockholm and used free-vibrating reed pipes.[44] Using Vogler's orchestrion as a model, the Prague instrument maker Leopold Sauer built a large piano-like device with strings and reed pipes, which was subsequently purchased by Count Leopold von Klinsky.[45] In 1804 Sauer completed a similar instrument, and a Viennese organ builder by the name of Ignaz Kober built a similar device for the Schottenkirche. Around 1807 Vogler added free-vibrating reeds to the organ in Neuruppin as well as to Berlin's Catholic Church.[46] Vogler presented his acoustical research involving organ construction and free-vibrating reeds to a gathering of the Gesellschaft der naturforschenden Freunde (Society of Friends Investigating Nature) of Berlin on 15 December 1800 and commenced his address by arguing that "acoustics has yet to be applied to organ building."[47] He stressed the importance of the delivery of the wind to the pipes, rather than the construction of the pipes themselves, as this was generally already well known to organ builders.[48]

The mechanician Strohmann of Frankenhausen in Thuringia was independently working on free-vibrating reeds of a similar design to Vogler's from around 1808 onward. He, too, received a contract to build an instrument that was to reproduce the sounds of a small orchestra.[49] The most difficult parts of this machine to build, he conceded, were the ones responsible for producing tones of the reed instruments, particularly the clarinet, oboe, and bassoon. The problem, in short, was to construct a reed pipe that could tolerate "the softest piano as well as the strongest forte and [could] express crescendo and decrescendo without the pitch becoming higher or lower."[50] He built free-vibrating reed pipes much along the same lines as Vogler did, but whereas Vogler manufactured them only for the lower bass tones, Strohmann built them for a much larger range of pitches.[51] Not being an organ builder, Strohmann was convinced that such pipes could be easily adapted to an organ, so that the instrument could be rendered expressive: "Here is where one comes to the characteristic that one can increase or decrease the tone via a strong or weak flow of the wind. And because the mechanism of this is simple, in the future one must be able to perform on such organs with the most beautiful expression and delicacy without any unusual difficulty."[52] Apparently, he underestimated how difficult a task it was to construct such pipes for large church organs, where the amount of air needed for crescendos was far greater than was required for the small musical instrument he designed. The mechanician Uthe was working on the same technical problem, proffering different dimensions for the free-vibrating reeds than Strohmann.[53] Little did either Strohmann or Uthe know that the French organ builder Gabriel Joseph Grenié (discussed below) was toying with precisely that idea.

Perhaps the most renowned mechanician relevant to the production of mechanical musical instruments and automata was Johann Nepomuk Mälzel, born in Regensburg on 15 August 1772, the son of a mechanician and organ builder. His father saw fit to have his son learn to play the piano. And he became a gifted player, offering private lessons from age sixteen to twenty.[54] He labored in his father's workshop, learning the craft of a mechanician and musical-instrument maker and demonstrating an impressive range of skills relevant to the construction of automata, which mimicked the tones of clarinets, trumpets, violas, and cellos by means of a revolving cylinder and bellows. Drawing upon Vogler's and Rackwitz's inventions, he used free-reed pipes in his aforementioned panharmonicon, an automaton that could produce the tones of various instruments of an orchestra and which debuted in Paris in 1807. The consummate entrepreneur (he is credited with inventing—or pirating—the metronome; see chapter 7 of this work), Mälzel desperately sought out a composition for his new machine, the panharmonicon. He turned to Ludwig van Beethoven, who promised to compose a piece for his mechanical musical instrument commemorating Wellington's victory over Napoleon in 1813. Mälzel showed Beethoven an outline of an appropriate composition for his panharmonicon, and Beethoven composed a piece, entitled "Wellington's Victory or the Battle of Vittoria," in October of that year.[55] The piece contained parts for piccolos, flutes, oboes, clarinets, bassoons, horns, bass and kettledrums, timpani, and a triangle. Mälzel attempted in vain to transfer the composition to his panharmonicon's cylinder. After numerous failures he suggested to Beethoven that the composer rewrite his work for a large orchestra accompanied by cannons. Beethoven agreed, and his piece debuted in front of a packed audience in the main auditorium of the University of Vienna. The proceeds of the concert went to the victims of the Battle of Hanau. Mälzel organized the concert, which included Antonio Salieri and Joseph Weigl conducting orchestral works. Giacomo Meyerbeer and Ignaz Moscheles were also present. Mälzel had his trumpet automaton perform two marches by Jan Ladislav Dusik (also spelled Dussek).[56] Mälzel's younger brother, Leonhard, was also an accomplished musician who built musical automata. He constructed a harmonicon, which he called the Orpheus-harmonicon, and which received praise from Beethoven and Salieri. And he built an orchestrion, called the metal harmonicon, for Count Thun of Prague.[57]

Johann Gottfried Kaufmann provides another, rather typical, example of a clockmaker and mechanician of this period. Born in Siegmar near Chemnitz on 12 April 1752, he began his apprenticeship as a stocking weaver, but after three years left for Dresden to pursue a more mechanical field. He had found a *Meister* who taught him clock making. A year and a half later, the *Meister* died, leaving Johann with the business, his widow and their children. In 1779 he had married his former employer's daughter, a common practice among *Handwerker*. She bore him a child, Friedrich, on 5 February 1785.[58] After manufacturing numerous clocks, he had begun to tinker with the construction of musical clocks, despite having never received a musical training. He then turned his attention to organ building. He had been the first in Saxony to combine a flute clock with a harp clock.[59]

During the first decade of the nineteenth century, the elderly Johann and his apprenticed son Friedrich turned their attention to the production of musical automata. Clockmakers, it turns out, often built musical automata with pulsating or free vibrating reeds.[60] The young Friedrich, whose skills surpassed his father's, witnessed in 1804 a performance of Mälzel's automatic trumpeter. Inspired by Mälzel's invention, they began to work on an automaton of their own, the bellonion (also spelled belloneon), completed in 1805 (figure 4.2).[61] It was made up of twenty-four trumpets and two bass drums. The sound was generated by pulsating reeds,

Figure 4.2
Johann Gottfried and Friedrich Kaufmann's Bellonion, 1805. Photo: Deutsches Museum, Munich.

which were set into motion by a revolving barrel and bellows, and whose tones were channeled through real trumpets. Hence the tonal color of the contraption, which was controlled by clockwork, closely resembled numerous trumpeters playing together.[62] The instrument could also alter the volume of its tones. The bellonion simulated an army's band so well that when the one in Berlin's Charlottenburg Castle was set off accidentally, Napoleon, who was quartered in the Castle with a number of his troops, allegedly jumped out of his bed fearing a surprise attack by the Prussian army.[63]

Mälzel's trumpeter served as the prototype for Friedrich Kaufmann's *Meisterwerk,* also known as the trumpeter, an automaton dressed in a Spanish soldier's garb (see figures 4.3, 4.4, and 4.5). A visible hand crank controlled two spiral springs, covered by the decorative shirt, powering two bellows and the spiral, wooden cylinder containing two sets of pegs. Four levers and two toothed segments transferred the motion of one of the sets of pegs to two rotating brass trumpets containing six pulsating reeds. The turning of the cylinder resulted in the movement of both sets of pegs, which determined the rhythm of the piece performed. The cylinders also controlled the movement of air from the bellows through the reeds. In 1812 Carl Maria von Weber traveled to the Kaufmanns' workshop and was pleasantly surprised by the automaton, claiming that it possessed a "beautiful tone" and surpassed Mälzel's original instrument.[64] Weber was most impressed with the trumpeter's ability both to produce double tones with the same volume and clarity and with its execution of double trills.[65]

In 1816 father and son took their automata to London and Paris demonstrating the fruits of their combined acumen to scores of savants and businessmen. Two years later, Johann died while touring Frankfurt, leaving the Dresden workshop in the hands of his very able son. One year later Friedrich completed the chordaulodion, an automaton that produced most of the instruments of an orchestra by combining piano keys and reed pipes powered by a bellows. It debuted on 7 May 1819 in Berlin, where it stunned onlookers with its performance of the overture from Mozart's *Titus* and a duet from Paer's *Sargino*.[66] In 1811 Johann and Friedrich had suggested a new method for improving the expression of reed pipes. Previously, they had altered the volume of the pipes by opening and closing the box containing them by means of nag's-head and Venetian-blind swells, traditionally used in harpsichords and organs.[67] Wind chests were fitted with a

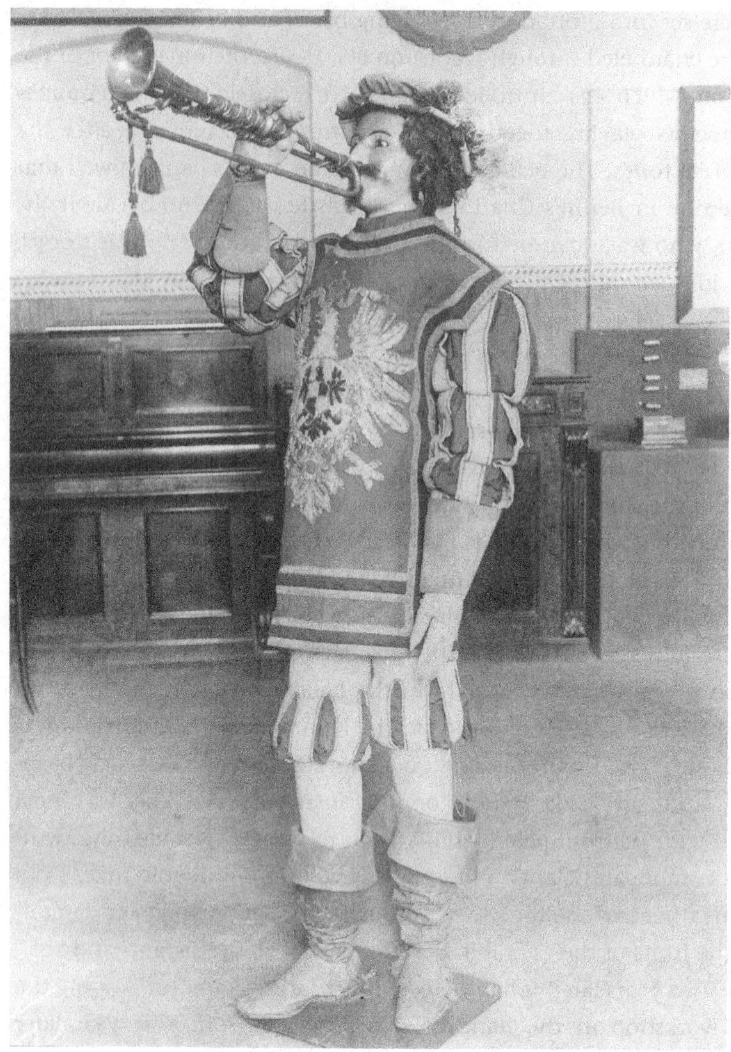

Figure 4.3
Friedrich Kaufmann's trumpeter-automaton, 1810. Photo: Deutsches Museum, Munich.

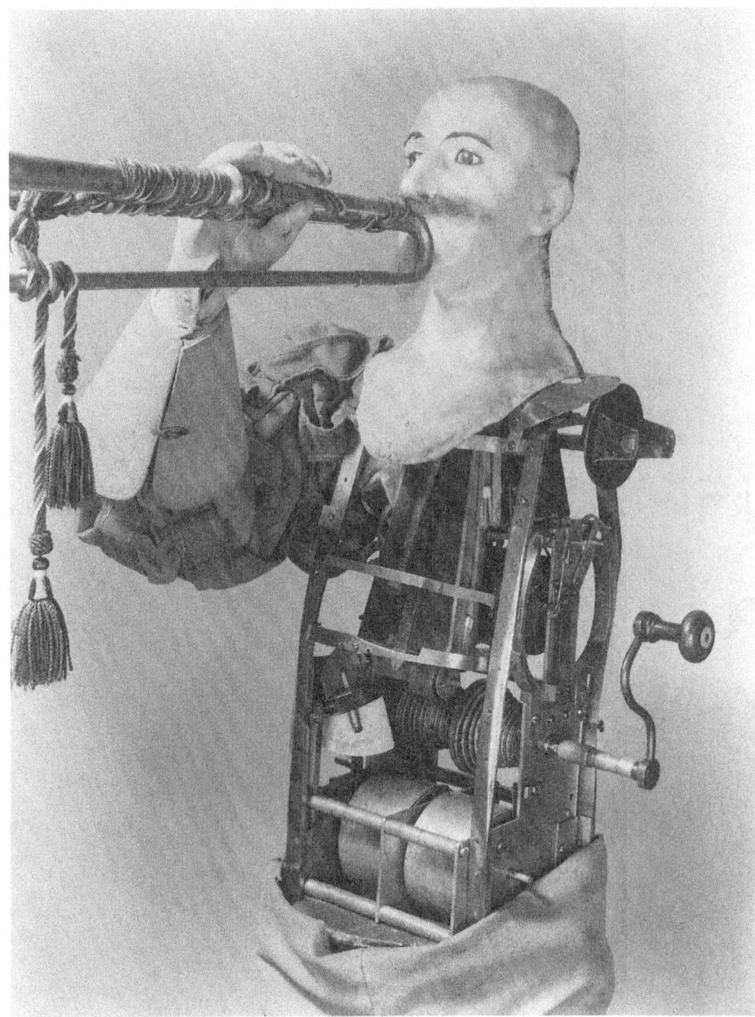

Figure 4.4
Friedrich Kaufmann's trumpeter-automaton, 1810. Front view. Photo: Deutsches Museum, Munich.

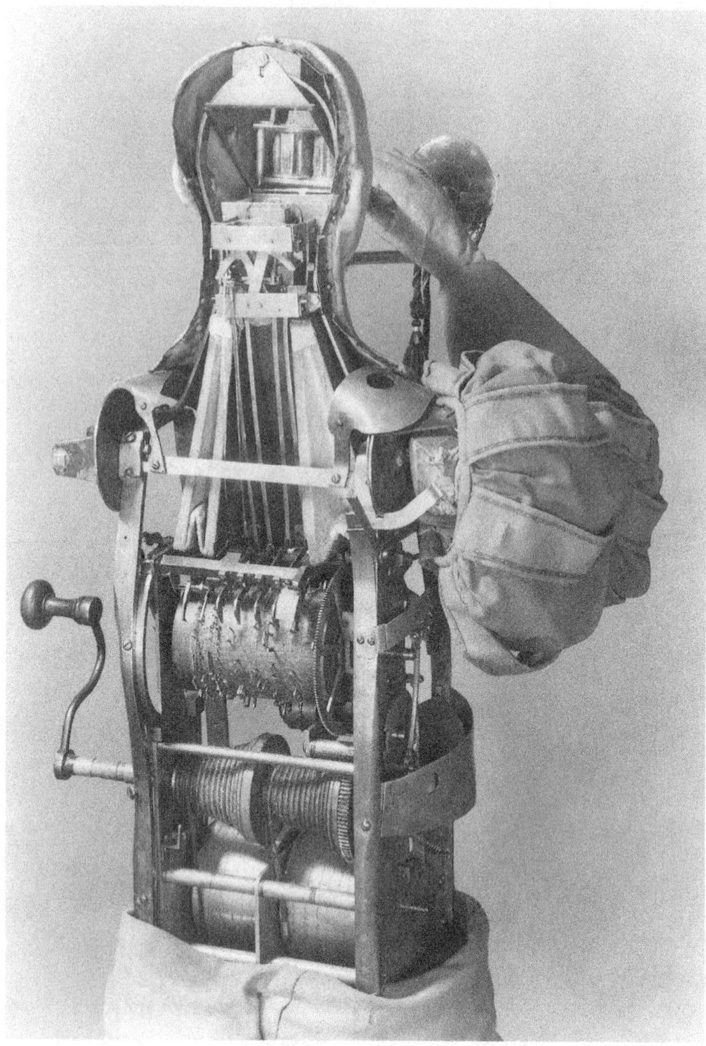

Figure 4.5
Friedrich Kaufmann's trumpeter-automaton, 1810. Back view. Photo: Deutsches Museum, Munich.

sliding shutter, which would be raised or lowered by a pedal. Pressing the pedal down raised the shutter, thereby decreasing the dampening effect, while releasing the pressure closed the shutter over the pipes' openings, decreasing the organ's volume throughout the performance. This form of swell was called the nag's-head swell.[68] The other technique of increasing the instrument's volume, the Venetian-blind swell, was accomplished by suspending narrow pieces of wood, approximately three inches in width, on center pins around which they could swivel, opening and closing them, much like Venetian blinds. Small levers connected to a pedal controlled the degree of dampening. When the wooden strips were aligned abreast, the volume decreased dramatically. Turning them 90° increased the volume to the instrument's full strength.

While building the chordaulodion, they stumbled upon a new procedure to enhance expression. Their initial problem had been the perceptible increase in the reed-pipes' pitches[69] as the volume of the instrument increased. To correct this sharpening of pitch, Friedrich employed a type of Venetian-blind swell such that the pipes decreased in pitch when the blinds were closed (i.e., with a pianissimo volume) by the same amount they increased in pitch when the blinds were opened (i.e., with a fortissimo volume).[70] To achieve this in 1811, Friedrich had employed two bellows: one large and one small, called an assisting bellows and which was one-twentieth the size of the larger. The larger bellows supplied the force for the fortissimo. To decrease the tone's volume, the vent was closed off, prohibiting wind from entering the wind canal, and the small bellows alone supplied the air to the pipes by siphoning air from the larger bellows by means of another vent.[71] Four years later, Friedrich invented a new method of reed-pipe compensation for the chordaulodion, which did not involve a Venetian-blind swell. He drilled a small hole at the end of each reed pipe, which could be opened and closed, rendering the pipe's compensation more effective.[72] In 1819 Friedrich displayed his compensated chordaulodion in Paris, demonstrating the bellows mechanism to Gabriel Joseph Grenié, but keeping the method of compensation a secret.[73] In 1838 a vastly enlarged and improved chordaulodion, the symphonium, made its debut. It was another automaton, comprising piccolos, flutes, clarinets, tympani, drums, and a piano.[74]

The chordaulodion, symphonium, bellonion, and the trumpeter formed the backbone of the Kaufmanns' Acoustical Cabinet, one of Dresden's leading attractions of the second half of the nineteenth century. All of these instruments possessed a critical mechanical and musical quality: they could increase and decrease in volume without affecting the tones' pitches, such that "more spirit [*Geist*] and life could be brought to the music than any other previous machine could."[75] The expression of the Kaufmanns' various instruments had been noted by Carl Maria von Weber back in 1812.[76] Commenting specifically on the chordaulodion, Theodor Hell (the pseudonym of Karl Gottfried Theodor Winkler, prominent editor of the Dresden *Abendzeitung*, theater director, playwright, publisher, and translator) underscored that "[t]he strength of this instrument's performance lies in the mimicking of human expression and feeling. It is nearly incomprehensible how this dead machine can summon forth not only all of the nuances of piano and forte, crescendo and decresendo, but also every portamento, every accelerando and rallentando, every accentuation of each tone, which we expect only from the animate breaths of a person."[77]

In December 1841 Friedrich Kaufmann toured Prague with his mechanical instruments. The following report appeared in the *Allgemeine musikalische Zeitung*: "In the last few days we heard three concerts from Hr. Friedrich Kaufmann, this renowned acoustical genius. . . . The machines appear to be a living being [*lebendiges Wesen*], thinking and feeling. Wonderful and spiritual is the instruments' interplay, the performance full of feeling, the expressive display of the finest nuances of forte, piano, crescendo and decrescendo, etc."[78]

An editorial in Leipzig's *Illustrirte Zeitung*, too, expressed fascination with how "unmachine-like" these instruments sounded: "One totally forgets that one is hearing a machine here, as Herr Kaufmann has succeeded in eliminating the stiffness and uniformity of mechanical musical instruments, and in its place has provided for the ear of the public a living, warm [*lebenswarmige*], moving piece of music."[79] Clearly, it was an advantage to produce a machine that had the properties that were just the opposite of what was commonly thought to be "machine-like." The earlier claim of E. T. A. Hoffmann's character Ludwig that "the coldest and most unfeeling executant will always be far in advance of the most perfect machines" was beginning to ring hollow. Friedrich's last *Meisterwerk* was his orchestrion, completed in 1851, after five years of intense labor. Unlike nearly all previous orchestrions, Friedrich's device was a musical automaton.[80] It was composed of trumpets, horns, clarinets, oboes, flutes, bassoons, piccolos, bass drums, kettle and military drums, and a triangle. Each group of instruments had its own bellows, designed specifically for that instrument, enabling it to crescendo and decrescendo. In that same year Friedrich exhibited his wares at Buckingham Palace for Queen Victoria, Prince Albert, and other members of the royal family.[81] The *Illustrated London News* provided a detailed description of the automata, praising the superior machine-like quality of precision, which at times was necessary for the execution of certain pieces (See figure 4.6).

> There can be no mistake—all the instruments [the orchestrion, the bellonion, the chordaulodion, the symphonion, and the trumpet automaton] depicted in our illustration actually emit sound, and are by no means decorative. How the maker has so ingeniously contrived that the cylinders move with such mathematical exactitude, and that the supply of wind (of course, varying for each tube) should be so precisely regulated, is scarcely to be conceived even by those thoroughly initiated in matters of mechanics and acoustics. For instance, it is almost miraculous to hear the slight shade of this invisible instrumentation, to mark the just gradations of crescendo, diminuendo, and sforzando, besides the usual fortes and pianos. We have never heard anything so perfectly astounding as the finale of "Don Giovanni": shutting one's eyes, it seemed as if the famed vocal and orchestral forces of Costa were exclaiming at one time, with portentous effect, "Trema!" In the dance music, the three different times going on in the finale were observed with unerring precision, the mechanical agents doing what the living artists will rarely accomplish—keep together. Nothing could be finer than the "Coronation March" from Meyerbeer's "Prophete." Godfrey's Coldstream Band must look to their playing, for the "Orchestrion" is a formidable rival.[82]

The symphonion was praised as yet "another triumph of mechanical skill";[83] "the precision with which the chromatic scale, ascending and descending is attained, would dismay a Richardson, or a Remusat."[84] Likewise, the trumpet automaton was celebrated as an "extraordinary piece of mechanism" producing "double sounds of equal strength and purity. . . ."[85] The article concluded by congratulating Friedrich and his son, Friedrich Theodor: "Their difficulties must have been enormous; first, in the just investigation of sound; and, secondly, in its application by mechanical means. To construct such instruments without models, for they are quite original, the maker must be a musician, a mechanic, a mathematician, and a philosopher."[86]

As a result, in part, of the Kaufmanns' superior musical automata, one began to query whether such machines were indeed superior to performers playing a specific instrument:

Figure 4.6
"Kaufmann and Son's Grand Musical Performance at Buckingham Palace," 1851. Photo: Deutsches Museum, Munich.

We ask ourselves: Is the organist really in a position to play better than a machine could? Quite the contrary, in addition to the fact that a machine plays with much greater perfection and efficiency [*Sicherheit*], it contains its own registers and provides its own wind. And the machine can bring about more variation and shading in its performance than the artiste can. . . . Consider a harmonium—nearly every organist possesses one. We find here a register called "expression." If this register is not used, the tones of the harmonium resemble an organ's, but are stronger. If this register is used, one can simply achieve the desired volume by strong or weak pedaling allowing the tones to swell and diminish. So one can play on a harmonium with multiple voices and in a soulful manner, which normally only a full orchestra can achieve.[87]

An important shift had occurred here. Previously, musical automata had been considered inferior to living performers. The *Illustrirte Zeitung* of 1847 featured an article on mechanical music, asserting that early musical automata of the eighteenth century suited amateurs, but not those who truly were informed (*Kenner*) about music, "because these [instruments] are machines and cannot feel what they are playing."[88] More recent musical automata, the article confessed, not only possessed greater "precision, force, fullness and purity of tone," but also "tenderness."[89] They were just as expressive as musicians, but they surpassed their human counterparts in efficiency.

Reed Instruments in the Early Nineteenth Century

A number of mechanicians responsible for constructing musical automata also built new types of musical instruments, which offered yet another type of aesthetic in addition to musical automata. The German public could no longer be satisfied with the type of instruments played in early and mid-eighteenth-century courts and churches. Romantic expression seemed to be the key. Audiences did not want to be entertained; they now wished to be *moved*. During the late eighteenth and early nineteenth centuries a new aesthetic of musical expression arose, documented, analyzed, and perpetuated by the literary works of Johann Gottfried Herder, Wilhelm Heinrich Wackenroder, Ludwig Tieck, August Wilhelm and (especially) Friedrich Schlegel and E. T. A. Hoffmann. The numerous articles penned in the periodicals by the musicians Carl Maria von Weber, Hector Berlioz, and Robert Schumann elaborated upon that aesthetic well into the nineteenth century.[90] These works were quite distinct from the early theories of music affections, as had been formulated by Johann Georg Sulzer in his two-volume tome, *Allgemeine Theorie der schönen Künste* (General Theory of the Beautiful Arts) (1771 and 1774), and by Heinrich Christoph Koch in his three-volume work, *Versuch einer Anleitung zur Composition* (Attempt at Instruction on Composition) (1782, 1787, and 1793). Sulzer and Koch had dedicated themselves to analyzing the internal content of musical works and the techniques for summoning forth listeners' emotions with a view "to inspire nobler resolutions." As Koch had written: "The fine arts in general, and thus also music, possess a unique property which enables them through artistic means [*künstliche Veranlassungen*] to awaken feelings in us. They awaken pleasure through the enjoyment of a good represented through art and fear through an evil brought forth by it. Thus, if the fine arts make use of their special power to have the feelings they arouse inspire noble resolutions, to affect the education and ennoblement of the heart, then they serve their highest purpose and show themselves in their proper worth. If deprived of this noble function, if used to another end, then the fine arts are degraded, they are dishonoured."[91]

Koch was echoing Sulzer's earlier sentiment here:

> Just as philosophy and science have knowledge as their ultimate goal, so the fine arts have the goal of sentiment. Their immediate aim is to arouse sentiments in a psychological sense. Their final goal, however, is a moral sentiment by which man can achieve his ethical value.[92]

> We wish to warn the artist who would arouse sensibilities not to do this based on some general ideal. . . . Any sentiment that is to be truly effective must have a real and particular subject.[93]

The new aesthetic of expression of the early nineteenth century centered on the listener's response to the music, or as Max Paddison puts it, "the inner world of the listener shaped by the dynamic temporal unfolding of the music, as 'sounding inwardness.'"[94] It was a romantic aesthetic that stressed the organic listener over the mechanical structure of the musical work. This view was espoused particularly by Wackenroder and Hoffmann, and subsequently by the philosopher Arthur Schopenhauer.[95] The shift in aesthetics from the composition to the listener was characteristic of the increased emphasis on subjectivity so emblematic of the early Romantic movement.

Mechanicians and musical-instrument makers of the period strove to invent new instruments that would satisfy this new craving for sentimentalism. Musical instruments with free-vibrating reeds were just such instruments, including the aeolodicon, aeolsclavier, aeol-harmonica, choraleon, aeolmelodicon, physharmonica, aeoline, and melodicon, all precursors to the renowned harmonium invented by François Debain in Paris in the early 1840s.[96] As we have seen, none of these aforementioned keyboard instruments with reeds was an automaton; performers played them as they would an organ or piano. The first attempts at building free-vibrating reed pipes were by mechanicians and organ builders wishing to improve upon the types of sounds their organs, keyboard instruments with reeds, and mechanical automata could generate (see figure 4.7).

In 1817 the instrument maker Johann David Buschmann of Friedrichroda embarked upon a tour exhibiting his two new musical instruments with reed pipes, the uranion and the terpodion.[97] During his trip to Königshafen, he met up with Bernhard Eschenbach, a Bavarian accounts official, who around 1815 had designed a free-reed instrument, which he called an aeoline.[97] Eschenbach's cousin, the instrument maker Johann Casper Schlimmbach, manufactured the instrument and applied the reeds to an organ in Stadt-Lauringen near Schweinfurt.[99] The Schweinfurt organ builder Johann Michael Voit improved upon Eschenbach and Schlimmbach's aeoline, calling his invention the aeolodicon.[100] Aeolines (and aeolodicons, which are essentially the same instrument) were organ-like instruments, in which the tones were elicited by a bellows, activated by the performer's knee or foot, setting metal reeds in vibration.[101] They possessed a range of up to six octaves: the upper octaves producing flute and clarinet–like tones, the tones of the middle range were French horn–like, and the lower tones were reminiscent of a contra-bassoon.[102] By altering the pressure of one's knee on the swell pedal, the aeoline could alter its volume. The tones were rather ethereal, well suited for romantic, spiritual music. Aeolodicons were used in regional churches to accompany choirs (figure 4.8).[103]

One of the earliest records of free-vibrating reeds in Berlin dates back to 1814, when the clockmaker Christian Möllinger, renowned for his astronomical clocks with flute registers (*Flötenregister*)—one of which he had presented to King Wilhelm III in 1791—built a keyboard instrument with free-vibrating reeds.[104] Wilhelm Vollmer built a similar instrument, which he called a melodica, and wrote to the Prussian Royal Minister von Bülow on 6 April 1821 that his instrument produced tones similar to those of the

Figure 4.7
Harmonium by Hermann Alert, Berlin, circa 1860. Source: Bilderarchiv Preußischer Kulturbesitz-Berlin/Art Resources, N.Y.

Figure 4.8
Aeolodicon (Harmonium), by Georg Schmid, circa 1840. Photo: Deutsches Museum, Munich.

French horn, bassoon, and clarinet, and that he could alter the tone's volume without changing its pitch.[105] The melodica was in principle an aeoline or aeolodicon, also having a range of up to six octaves.[106] In 1821 Johann D. Buschmann and his two sons Eduard and Friedrich set up shop at Friedrichstraße 183 in Berlin, with the view to build terpodions and improved aeolodines (which were also called aeolines). Johann built a uranion, which was similar to an aeolodicon but could alter its volume.[107] Friedrich also specialized in the design of the aura, the precursor of today's mouth harmonica.[108] Throughout the 1820s Friedrich produced hand-aeolines and concertinas, with a bellows, double-pulsating and free reeds, and a keyboard.[109]

Aeolodicons were also similar to the more popular physharmonica. In 1818 Anton Haeckl (Häckl)[110] of Vienna built an early version of his physharmonica—a keyboard instrument with free reeds, which had a range of four octaves with the reeds beneath the keyboard and played with the right hand, while the piano was played with the left. The ability of its tones to change volume was a vast improvement upon the aeolodicon.[111] Later versions of Haeckl's physharmonica were solo instruments: freestanding, no longer attached to pianos. The Viennese instrument maker received a five-year exclusive privilege for manufacturing in 1821.[112] Unlike the aeolodicon, the physharmonica possessed brass plates with slit openings in which brass reeds were screwed.[113] The Viennese instruments possessed the characteristic of expressing an impressive range of volume controlled by a pedal attached to the bellows.[114] They were often used as substitutes for larger, more expensive church organs.[115] In 1832 Friedrich Buschmann set up a piano and physharmonica firm in Hamburg. Four years later, he constructed a physharmonica with a new air-suction system, rendering the instrument more expressive.

The early physharmonicas served as a prototype for Thuringian Johann Friedrich Sturm's version of the aeolodicon. Sturm, who moved to Berlin in order to carry out his work on musical instruments, decided to mount the free-reed pipes of an organ onto a piano-like instrument.[116] In October 1828, four years after he produced his first aeolodicon, Sturm wrote that for the past ten years he had thought about building and possessing an instrument that was neither a pianoforte nor an organ, but somewhere in between, "because constantly playing the tones with the same volume tires the ears. The feelings of the best players can be conveyed through the music via piano and forte."[117] His aeolodicon enjoyed a solid tone with both diminuendos and crescendos; thus, according to its inventor, it was particularly well suited for choral melodies and adagios.[118] Sturm's goal was to produce tones mimicking the existing reed instruments, namely the clarinet, oboe, and bassoon, the tones of which are elicited by wind passing through a reed. His instrument caught the attention of the composer Spontini.[119] By 1829 the interest in reed instruments was increasing rapidly in Berlin, as evidenced by Sturm's plans to create an aeolodicon company, which was to employ twelve to sixteen laborers.[120]

The ethereal tones produced by the aeolodicon and physharmonica clearly appealed to the Romantic German audience. A contemporary critic pointed out in 1829 that there were at least twenty aeolodicons being actively played between Gotha and Schweinfurt, and this despite (or perhaps, as a result of) King Friedrich Wilhelm III's well-known aversion to reed instruments in general.[121] The aeolodicon found a home in Biedermeier bourgeois culture, popular with a large spectrum of professionals ranging from choir directors, organists, and music directors, and even forest directors and painters.[122] As the choral director Schüler from the Prussian city of Suhl reported, "this instrument, which is increasingly receiving recognition, is a credit to its inventor and our age. It possesses so much that is attractive that those who get to know it can no longer separate themselves from it. When, for example, I have tolerated the burdens and heat of the day, I always recover with my aeoline."[123] Sturm's aeolodicons were so important to Berlin's music scene that they merited an official assessment of their efficacy as church instruments. On 7 May 1829 one of his aeolodicons was tested in Berlin's Spital Church on St. Gertraud Street. C. F. Zelter, who had initially praised Sturm's instrument, judged the Spital Church's aeolodicon during a church service. After having been named professor of the Royal Academy of Art, Zelter was responsible for the inspection of new and restored organs and organlike instruments.[124] On this particular occasion, Zelter complained that the aeolodicon was not powerful enough to accompany the choir. He added, "The sound of the aeolodicon is somewhat ethereal, however, over time it has a soporific effect. Choir music should be stimulating and lively in order to encourage participation among the members of the congregation."[125] Other competent musical adjudicators agreed. Bernhard Klein, however, who was the leader of Berlin's Youth *Liedertafel* and cofounder of Cologne's Institut für

Kirchenmusik (Institute for Church Music), underscored the role of the aeolodicon as a *Hausinstrument*: "Your Excellency might allow me to make the following remark. I recognize the perfection of the tranquil and excellent mild tones created by the beautiful crescendo of the aeolodicon and am pleased with it. If its introduction in churches does not seem purposeful, the instrument is highly recommended for domestic devotions and for the accompaniment of chorales in halls."[126] One notes that a critical point in judging these reed instruments was their ability to increase and decrease the volume of their tones. Expression was one of the most desired attributes of early nineteenth-century instruments.

Organs

Expression, however, was important not only to these newly invented instruments, but also to the older, more established ones as well, particularly the organ. Different registers on German organs received the same amount of air from the bellows well into the 1820s. Hence, labial pipes often received too much air and sounded shrill, while the reed pipes received too little to afford a healthy crescendo and diminuendo.[127] German organ builders slowly (and some very reluctantly, it seems) began to build more expressive organs, which were vast improvements on Vogler's attempts some three decades earlier. German and Austrian organ builders began to tinker with their organ construction with the explicit aim of increasing the instrument's expressive ability. The Viennese piano and organ builder Michael Rosenberger received on 28 April 1826 a five-year privilege for his "improvement on the reed works on organlike instruments, whereby these instruments remain protected from the influence of the external air, thereby keeping an ever-steady pitch."[128] Ten weeks later his privilege was increased to encompass his new invention, the Polyharmonium, in which he increased the airflow through reed pipes. The result was that his instrument produced full tones similar to an organ's.[129] Interestingly, whereas French use of reed pipes for expression was carried out initially on small organs, German musical-instrument makers used them first on smaller harmonium-like reed instruments and mechanical musical instruments, scaling up their technological application to organs during the late 1820s.[130] The organ builder Johann Friedrich Schulze was the first German to toy with the idea of increasing the volume of his organs by employing a steam engine in 1829 to replace the bellows.[131]

Prussia in particular was experiencing the rebirth of the importance of the organ to church services. Music in Prussia during the first decade of the nineteenth century was in a deplorable state. Choirmasters and organists, which had been coveted positions during the reign of Frederick the Great, were now filled with "unfit subjects," rendering church services worthless, "cold and dirty," "distorting and tearing apart the noble and high choral in the lowest and most absurd manner."[132] J. S. Bach's renowned biographer, Johann Nikolaus Forkel, expressed these sentiments in 1801, bemoaning the decline of music's importance to education and worship. "But so much comes out of the fact that it is mainly the lack of knowledge in musical things, that brings about this musical disaster, that has led so many men, and still leads them, to desire an ever greater reduction of music in churches and schools, and will finally take things so far that it will either completely rob the church of its most powerful means of devotion or at least bring it so low that no enlightened Christian can any longer hold it for a means of devotion."[133] J. S. Bach, J. J. Quantz, J. P. Kirnberger, and J. F. Fasch were now distant memories tearfully evoked by those indulging in remorseful nostalgia.

C. F. Zelter was able to reverse this trend and restore music education and training to its prized place among members of the *Bildungsbürgertum*. In 1809 he composed a detailed plan addressed to Wilhelm von Humboldt[134] for improving music education in Prussia, arguing that "one would have to say that among Germans music has had the greatest and quickest effect on the education of the nation. All philosophers are at least united on this point: that none of the *beaux-arts* has had such an unlimited power on human nature as music. We have known from experience that music is an effective art."[135] This proposal was a response to the official request of the Prussian minister and trustee of the Academy of Arts Karl August Fürst von Hardenberg's official request of 3 May 1803 for recommendations on how the Academy could "have a sure and measurable impact on the spirit of the age and the productivity of this epoch" and "awaken the artistic energies of the country."[136] The renowned Singakademie was the model for Prussia's musical future, "arising on German soil through German patriotism. It can hardly be matched by any other city in Europe in that the character of the society is built on the national character of German stability."[137] In his view, if the abysmal level of church and social singing could be raised, the morality of the German people would improve. He strongly intimated that a major cause of the decline in popular singing was the princely patronage of frivolous and even immoral court opera, an elitist musical form that could not unite all social strata of early-nineteenth-century Prussia. Recall from chapter 3 that he excluded opera from his view of serious music.[138] Zelter's reform aimed at improving music education in Prussian schools and music training in Prussian churches, which included first and foremost the education and adjudication of organists and choirmasters, the supervision of city musicians, and the reorganization of choirs from which a seminary for schoolmasters would emerge.[139] Wilhelm von Humboldt, the brother of Alexander, concurred, criticizing the neglect of music institutes affiliated with church services. During the ephemeral reorganization of the Holy Roman Empire from 1803 to 1806, the number of sovereign entities under its control was cut from several hundred to approximately forty.[140] This reshuffling, combined with the financial crisis that accompanied the Wars of Liberation, greatly reduced the number of available cantorships throughout the German territories. This reduction, in turn, greatly exacerbated the overall decline in the importance of music within the Protestant territories, a trend that had started back in the mid-eighteenth century.[141]

As Applegate astutely observes, during the first years of the nineteenth century, musicians were attempting to secure their future in this time of tremendous social and political turmoil. They wisely reached out to the neohumanist movement of *Bildung,* as so eloquently enunciated by W. von Humboldt, with the hope that members of the *Bildungsbürgertum* would learn to appreciate music as a critical cultural art form.[142] According to W. von Humboldt, music raised the standards of both public church services in particular and national education in general. And he echoed Zelter's desire to improve music instruction in Prussian schools.[143] King Friedrich Wilhelm III approved of Zelter's plan, thanks in large part to von Humboldt's endorsement. The King declared on 17 May 1809: "I require explicitly, however, that the plan be regulated by the dignified clergymen, that this music is particularly directed toward song and organ, and that—for this reason—singing in schools and the examination of choirmasters and organists be taken care of."[144]

In 1813 an anonymous report in the *Allgemeine musikalische Zeitung* railed against the state of music in churches, particularly those in small villages, the lack of skill of both choir members and organists, and the low quality of organs.[145] A year later another anonymous article, this time from

Saxony, appearing in the *Allgemeine musikalische Zeitung* echoed the shock and dismay over the quality (or lack thereof) of church organists, while a third article published in 1819 placed the blame squarely on the decline of church attendance by youth as well as the decrease in the number of choirs in the Protestant territories.[146] In 1821 Prussian organ inspector and Royal Music Director of the small Prussian city of Neuruppin C. Friedrich G. Wilke detailed the deplorable state of organs.[147] The cantor C. F. Hermann complained that organists were playing too loudly, forcing choral and congregational members to scream,[148] while August Ferdinand Häser explained that the most important duty of organists in Protestant churches included the accompaniment of the congregation's choir. They needed to be well acquainted with the various hymns in order to convey the proper feelings and passions to the congregation.[149] Catholic churches, on the other hand, had not enjoyed the tradition of congregational singing for as long as the Lutheran churches had. Indeed, the Catholics took up congregational singing just during this period, whereas Lutherans had been singing since the mid-eighteenth century.[150] Within this context of music in Lutheran services, Friedrich Schleiermacher penned his Berlin *Gesangbuch* (Hymnal) of 1829, which stressed the importance of choral singing in support of sermons.[151] Music was particularly important for the various Catholic festivals, such as the Birth of Christ, the Feast of the Glorious Resurrection, and the Feast of the Holy Light. The organ had a less important role in Catholic churches, as it was used only for the introduction and conclusion of passages or for suitable interludes.[152] Because of the organ's strength, it was not suitable for the celebration of the more important festivals, as it often drowned out the choir.[153] As a result, instrumental music using string and wind instruments was the preferred accompaniment of choral music appropriate to masses, oratorios, psalms, and offertories.[154] Fröhlich argued that even symphonic works, quartets, and overtures could be effectively used as choral accompaniment.[155] The organ was often accused of overpowering the choral response to the priest, hence rendering it inaudible.[156]

In 1830 the Bremen music instructor W. C. Müller lamented what he claimed was the death of church music throughout the German territories. The secularization of the Catholic states combined with political revolution compromised the previous level of music used in worship. And the Protestant territories, he argued, deemed it necessary to save money after the political troubles brought on by the French occupation. As a result, many Protestant churches stopped using music in Sunday worship altogether.[157] Similarly, in 1834 Pellisov complained that despite being "the only instrument that is authorized by this [Roman Catholic] Church to support hymns," "the use of the organ has been rather limited."[158] And he noted that "in the Papal Church in Rome, no organ can be found."[159]

As a direct consequence of Zelter's reforms, a plethora of manuals on organ building appeared throughout the German territories during the early nineteenth century.[160] In particular, church-organ builders sought to make the organ more expressive by constructing new organs, or rebuilding older ones.[161] In 1823 Wilke penned an article summarizing the ways known at the time in which organists could alter the volume of their instruments.[162] Three of the methods would only dampen the organ's tones. By releasing their effect, the organ could crescendo back to its normal volume. The first technique was the "echo swell." The pipes were surrounded by wooden walls, or the "echo box," resembling small desks with lids. The first version of the echo swell had been invented during the 1670s by English organ builders.[163] Originally, the organist would open up the lids of the desks manually before playing the piece, thereby increasing the volume to the organ's normal strength. Clearly, one could not alter the volume during the playing of the work. By 1712 Abraham Jordon of London had invented the "nag's-head swell" detailed above.[164] Unfortunately, in organs the shutter was often so heavy that the loud crashing sound of the wooden shutter would mar performances. In 1769 Burkat Shudi, a London harpsichord builder, had received a patent for improving the volume of his instruments by means of the aforementioned Venetian-blind swell.[165] Shudi's Venetian-blind swell was first applied to the organ by Samuel Green of London some twenty years after Shudi's invention.[166]

The third technique was the wind swell, which was based on two opposing actions. The first diverted wind away from the pipes for a diminuendo, and the second led wind to the pipes for a crescendo.[167] The main wind canal of the organ contained a moveable flap, which the organist could control with a lever. If the flap was lying in a position perpendicular to the canal's upper and lower walls, then the canal was closed off. If the flap was in a position parallel to the upper and lower walls, the air could flow freely. In the first case, the tone decreased considerably, since most of the air was blocked off. Indeed, the tone was so weak that it was often inaudible, hence revealing the drawback of this particular technique. In the second case, the organ played at its full strength.

The fourth technique of altering an organ's volume was the keyboard swell (*Crescendokoppel*). The keys of the main register of the organ were depressed in such a fashion that the keys of the second register, to which they were coupled, were depressed simultaneously and sounded, and when the first set was depressed even further, a third register of keys sounded as well, thereby increasing the volume. The instrument was constructed by the organ builder Moreau for the Church of St. John in Gouda, Holland, in 1736. Unfortunately, such a technique did not permit a gradual crescendo or decresendo, merely a sforzando or rinforzando.[168]

By far the most impressive technique for altering an organ's volume was the compression swell, a technique developed during the early nineteenth century. Compression bellows could send different amounts of air to different pipes, thereby controlling the volume. The initial problem with this particular technique was that some pipes could tolerate an increased flow of air better than others. Labial pipes and pipes with beating reeds, for example, could not tolerate such a swell, as the resulting tone was painfully shrill, or the rattling of the beating reed became disturbingly audible.

The French organ builder Gabriel Joseph Grenié is credited with being the first to render some of an organ's pipes expressive by means of compression bellows. He desired to build an *orgue expressif* (a small expressive organ), whereby the volume of the instrument would be altered—precisely what Strohmann had wanted organ builders to do. The major problem all organ builders of the period needed to overcome was to develop a system whereby each pipe would receive the proper amount of air in order to alter its volume. A large, sixteen-foot pipe, for example, clearly needs more air for a crescendo than a one-foot pipe. And, what turns out to be a greater technological challenge, the organ builder needed to ensure that the range of volume was the same for the highest notes (with the shortest pipes) and the lowest (with the longest). In 1810 Grenié accomplished these feats with his small expressive organ. Later in the year, he began constructing larger organs, whose volume could also be changed. He constructed foot pedals, which set in motion compressors that regulated the airflow from the bellows through the wind chest to the pipes. The labial pipes and pipes with beating reeds had their own wind chest connected to the normal bellows, rather than the compressor. The quality of the wind flow remained the same for these pipes. The free-reed pipes had their own wind chest and compressor bellows, which sent different amounts of wind to the pipes.[169] These free-reed pipes could change their volume, but only to a limited degree.[170] By altering the pressure of his feet on two pedals, the organist could alter the volume of the pitches.[171] By 1814 Grenié had finished his larger expressive organ, which had been commissioned by the Royal Conservatory of Music in Paris. The *orgue expressif* was too small to be effective in large Parisian churches, but it found its place on the stage accompanied by orchestras.[172] François-Louis Perne, composer and Inspector of the Royal Conservatory of Music in Paris, noted that an attentive public could appreciate "the soft, pleasant and melancholic harmonies" produced by Grenié's instrument.[173] The steady crescendo of the *orgue expressif,* combined with the orchestra, resulted in a "passionate" performance.[174] As a direct result of Grenié's technological advances, the organist could sforzando and rinforzando, gradually crescendo and diminuendo, and even increase the volume of the notes played by one hand while decreasing the volume of the notes played by the other.[175] Grenié's work impressed not only the Parisian musical public, but France's scientific elite as well. The French physicist Jean-Baptiste Biot described the mechanics of Grenié's *orgue expressif* in his *Précis élementaire de physique expérimentale.* Biot explained, among other things, that the tone's pitch depended mostly on the length of the reed, from the point at which it was attached to the pipe. The pitch was also determined (to a lesser extent) by the reed's elasticity, mass, and degree of concavity.[176]

But increasing an organ's expression posed other problems for organ builders. Although Grenié and, to a lesser extent, Friedrich Kaufmann were able to solve the problem of different pipes tolerating different amounts of airflow by constructing compression bellows for free-vibrating pipes and regular bellows for labial pipes and pipes with beating reeds, there remained a critical acoustical problem. Even though free-reed pipes could tolerate an increase in airflow without producing harsh tones, they still increased in pitch when receiving greater quantification of air. Increasing the air pressure through a pipe results in a perceptible increase in the pipe's pitch.[177] This acoustical challenge was tackled by the renowned German experimental physicist Wilhelm Weber, who, as we shall see, viewed the problem as one belonging to the domain of adiabatic phenomena.

5 Wilhelm Weber, Reed Pipes, and Adiabatic Phenomena

Most physicists and historians of science associate Wilhelm Eduard Weber with his unparalleled experimental precision in electrodynamics, particularly his renowned collaboration with Carl Friedrich Gauss in Göttingen (figure 5.1). What is generally not well known, however, is that Weber's experimental prowess and his introduction to the wave theory originated in his early fascination with acoustics. Having been raised in the same house where E. F. F. Chladni was domiciled proved critical to Weber's early career.

Returning to Wittenberg in early 1814 after one of his many journeys, Chladni was greeted by the news that his house had been burned to the ground. In September of 1813 Prussian soldiers had bombarded the house in an attempt to dislodge the occupying French troops during the Wars of Liberation. Low on cash, he could not afford to purchase another house, so he decided to take up lodgings in the "Golden Sphere," owned by the professor of physics and natural history of the University of Wittenberg, Christian August Langguth. The house must have been enormous, as it contained other lodgers at the time of Chladni's arrival: the family of the professor of theology, Michael Weber, whose house and property had been confiscated during the war. As the Webers had twelve children, one more occupant in the "Golden Sphere" seemed hardly noticeable. One of the dozen offspring included the precocious nine-year-old, Wilhelm Eduard (1804–1891). The Weber family decided to move out of the house later in 1814, first to Schmiedeberg and then to Halle in 1817, where the university had just joined with the University of Wittenberg.[1] Chladni was a frequent and welcomed guest in the Weber household. In Halle Wilhelm attended the Pädagogium der Frankeschen Stiftung, studying ancient

Figure 5.1
Portrait of Wilhelm Eduard Weber (1804–1891). Photo: Deutsches Museum, Munich.

languages and experimenting in the natural sciences with his older brother, Ernst Heinrich (1795–1878), who had already begun his career as professor of anatomy at the University of Leipzig.[2] In 1822 Wilhelm enrolled at the joint University of Halle-Wittenberg as a mathematics student. Three years later, he and Ernst Heinrich completed an impressive tome, *Wellenlehre* (Doctrine of Waves), which was dedicated to their aged friend and mentor, Chladni, to whom they referred as "the founder of experimental acoustics and inventor of a new class of musical instruments."[3]

Weber's acoustical work was fundamentally concerned with improving the design of musical instruments and the lives of those who built and played them. And, his notoriously liberal political stance as a member of the Göttingen Seven of 1837 was also evident in his acoustical research.

Like many of his generation, Weber saw science in general, and physics in particular, as assisting German manufacture, which in turn would fuel German unification. He envisaged skilled artisans working in concert with physicists to improve manufacture. Artisans, shedding their craft secrecy, were to share their valued experience with physicists, while physicists were to proffer scientific laws and equations as guidelines for musical-instrument manufacture. Similar to his work on deriving equations for organ builders discussed below, his *Elektrodynamische Maassbestimmungen* (Determinations of Electrodynamic Units) sought in part "to give definite and precise instructions for future use," with the hope that mechanicians would no longer base their work on trial and error, wasting valuable time and money repeating the same procedure time and time again.[4] There was never a fear that physics and mathematics might replace the labor of artisans; rather a partnership was sought that would simultaneously increase Germany's national economic ambitions as well as avoid an extremely industrialized and socially fragmented state, such as was the case with Great Britain. The reform-minded milieu generated by the Prussian government, searching for ways for science and technology to bolster their Kingdom, was conducive to such research. These officials were particularly keen to assist projects relevant to music, as Prussia was undergoing the most thorough and comprehensive music reform in its history.

Weber's earliest work on the acoustics of organ pipes dates to 1825 with the publication of *Wellenlehre*.[5] Interested in the influence of the pressure and density of a sound wave on a pipe's pitch, the Weber brothers investigated the properties of a standing vibration[6] in a tube with one end closed, since standing vibrations were responsible for the sounding of organ pipes, wind instruments, and the human voice box.[7] Wilhelm proffered a physical analysis of the production of a longitudinal wave via its superposition as a result of reflection at different phases.[8] Of particular interest to him was the standing wave of air in reed pipes.[9] After performing an extensive series of experiments on reed pipes, the Webers were now in a position to test the claim of the music critic and amateur musician, Gottfried Weber, who had argued in 1824 that the dimensions of the reed, and not of the air column, determine the pitch of the reed pipe.[10] The Weber brothers conducted a series of investigations altering the dimensions of both the reed and air column and determined that the pitch of reed pipes with short air columns depends only on the length, thickness, and elasticity of the reed.[11] But as the air column increases past a particular length, it too plays a role in determining the pitch of the reed pipe, contrary to what Gottfried Weber had claimed.[12] The Weber brothers continued their acoustical research throughout the 1820s and early 1830s. The physician Ernst Heinrich concentrated his efforts on the physiology of the eardrum, while Wilhelm Eduard focused his experimental prowess on the acoustics of organ reed pipes.

The Physics of Reed Pipes

On 19 September 1828, at the meeting of the Versammlung deutscher Naturforscher und Aerzte in Berlin (discussed in chapter 3), Wilhelm Weber offered the fruits of his research on reed pipes in a lecture, which was subsequently published in J. C. Poggendorff's *Annalen der Physik und Chemie*,[13] to the physics and chemistry section on the compensation of organ pipes. This work was based on his doctoral dissertation and *Habilitationsschrift* (a second major work required of those wishing to teach at German universities) at the University of Halle-Wittenberg, where he studied under Johann Salomo Christoph Schweigger, editor of the *Journal für Chemie und Physik* (also known as the *Jahrbuch für Chemie und Physik*).[14] The largest and most perfect of all musical instruments, the organ, Weber argued, suffered from one major disadvantage: its tones could not gradually swell and diminish. Thus, the range of expression organists could achieve was rather limited. Despite various attempts by artisans and savants to improve pipe design, this problem still remained. The organ pipe, in essence, is a longitudinally vibrating column of air and the volume of its tone increases and decreases by intensifying and diminishing the flow of air respectively. But, by swelling or lessening the flow, the organist would actually slightly change the pitch of the note as well.[15] Even the improved reed pipes, Weber pointed out, manufactured by Christian Gottlieb Kratzenstein for his speech machine, by Abbot Georg Joseph Vogler for his orchestrion, by the Dresden mechanician Friedrich Kaufmann, and by Gabriel Joseph Grenié for his *orgue expressif* still suffered from this defect.[16] After having performed a lengthy series of experiments, Weber determined the laws by which reed pipes sound, and claimed to "be in a position to produce organ pipes, which, regardless of how strong or weak the flow of air, the tone produced will always have the same pitch."[17]

Weber reminded his colleagues on that mid-September afternoon in the Prussian capital that the initial tone produced by a tuning fork is slightly lower in pitch than the tone generated near the end of its resonance. Transverse vibrations oscillate in a plane that is perpendicular to the direction of the propagation of waves that occur within their substance. Strings and tuning forks are good examples. It is the property of all real (and not ideal) transversally vibrating bodies that their frequency of oscillation—and so the pitch of the longitudinal vibration that is thereby generated in the surrounding air (and whose direction of propagation is the plane of oscillation of the tuning fork)—is a bit lower with a strong vibration and a bit higher with a weak vibration. That is, the pitch of the sound generated in air by a tuning fork immediately after it is struck is lower than the pitch when the volume dies down. Weber conceded that this phenomenon was well known to the musicians of the period.[18] By contrast, longitudinal vibrations oscillate in the same direction that they propagate; air columns are good examples. And real-world longitudinally vibrating bodies possess the opposite property to transverse vibrations: unlike the sound generated by the vibrating string or tuning fork, increasing the amplitude raises the pitch of a resonating column of air. A horn player, Weber informs us, blowing harder into his instrument will increase the note's pitch. He cleverly used the opposite properties of these two kinds of vibrations to cancel out the effects of amplitude on pitch.[19] He proceeded to set up the point of the lecture: "If it were possible to connect a resonating plate of metal [or a reed], which vibrates transversally, and a resonating air column, which vibrates longitudinally, in such a way that they both produce simultaneously vibrations of the same velocity, it would also be possible to produce a musical instrument from these materials, which would not change its pitch, regardless of how weakly or strongly it is agitated. I have, as a matter of fact, been able to construct such an instrument."[20]

Weber had the mechanician J. August D. Oertling of Berlin weld a transversally vibrating metal reed onto a tube containing a vibrating column of air. As shown in figure 5.2, *AC* is the air column bearing the longitudinal vibrations that constitute sound, while *ab* is the metal reed, which oscillates transversally. Both column and reed are set into vibration by blowing through crack *a*. Without touching the reed *ab*, Weber blew into end *A*, causing the reed to vibrate, and this in turn opens and closes the square hole of the space to which *b* is fastened at the end. The exterior air can as

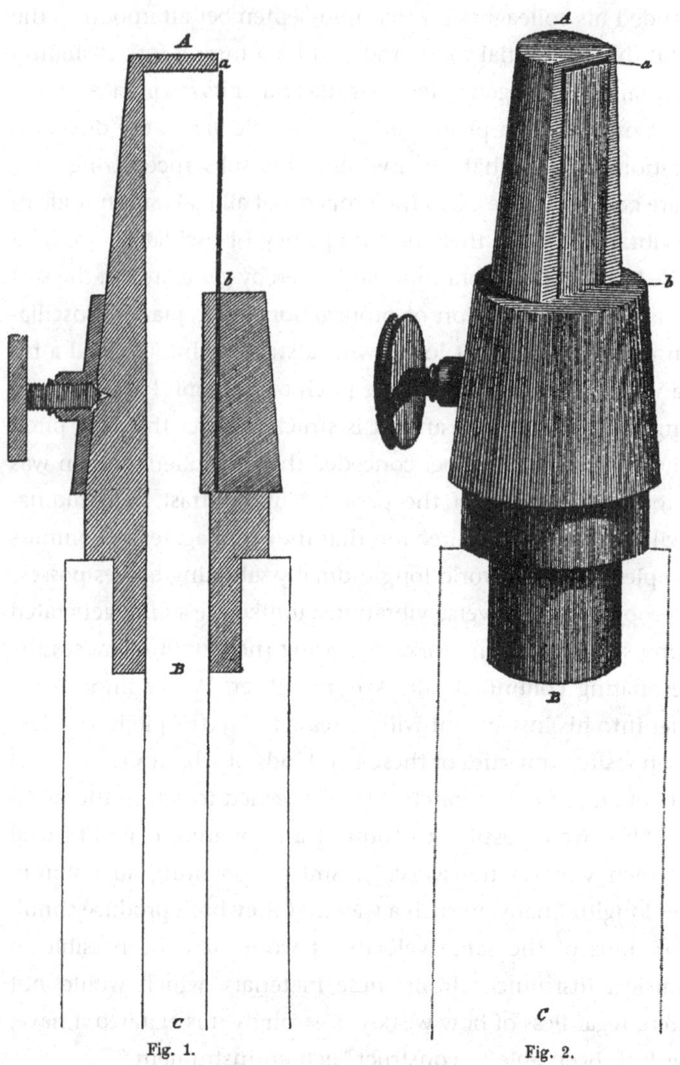

Figure 5.2
Oertling's reed pipe made for Weber in 1828. Source: Wilhelm Weber, *Werke* (1892–1894), vol. 1 (1892), p. 253.

a result penetrate only periodically, as pulses. The rapidity with which the pulses succeed one another determines the pitch of the sound.[21] In short, if the column and reed are adjusted such that the speeds of the reed and of an air particle are always in synch, then each is compensated exactly by the other. At any moment the amplitude of an air particle in the resonating column would always be in the same ratio to the amplitude of the vibrating reed, and since the corresponding frequency of the generated air wave increases with the amplitude in the one and decreases in the other, they reverse one another's effects.

Under certain conditions the vibrations of the air column *AC* that occur in the absence of the reed will give way to the air vibrations that are generated by the oscillations of the metal reed *ab*. In this case the pitch of the reed pipe *decreases* with an increase in airflow.[22] Under different conditions, however, the air vibrations that are generated by the metal reed give way to the vibrations of the air column proper, and the pitch of the reed pipe then *increases* with an increase in airflow. Weber consequently reasoned that there must be a third case when one blows increasingly harder into the pipe, in which the vibrations of the reed lower the pitch by precisely the same amount that the vibrations of the air column raise it.[23] After determining, by careful experimentation, the point at which the reed pipe is compensated for different sized reeds made from diverse metals, Weber decided to determine the laws that governed the relationship of these vibrations, "so that it [the construction of the pipes] can be carried out in practice with ease."[24] He argued that in order to fulfill completely the criterion of compensated organ pipes, the artisan needed to construct a pipe for each tone of the scale by employing those physical laws that Weber himself provided.[25]

In 1829 Weber published a second article on reed pipes, entitled "Ueber die Construction und den Gebrauch der Zungenpfeifen" (On the Construction and Use of Reed Pipes).[26] Whereas his previous article detailed the method of constructing compensated organ pipes, the first part of his second essay concerned itself with the myriad of uses for reed pipes. One crucial application was their use in providing a constant pitch for tuning musical instruments. As will be discussed in chapter 7, he recommended the reed pipe and his tonometer for the scientific determination and standardization of concert pitch throughout Europe. Weber also recommended the use of reed pipes to determine accurately the velocity of sound in air.

Drawing upon both the earlier work of Daniel Bernoulli of the mid-eighteenth century, who had measured the speed of sound in air with covered labial (also known as flue) pipes—that is, with organ pipes whose tones are produced by air passing across the edge of a fissure or lip—and the vastly improved measurements that Pierre Louis Dulong made in 1828 of the speed of sound through different types of gases in order to determine their specific heats, also using labial pipes, Weber (ignoring the influence of the pipe's mouth on the pitch) calculated the speed of sound through air using his reed pipes.[27] It turns out that the pitch of a reed pipe with a long air column is very close to the pitch of a covered labial pipe of the same length.[28] Using Oertling's reed pipes, Weber's average value differed from the accepted value of the speed of sound (as determined by the French Academy) by less than 1.8 percent.[29] As discussed below, Weber provided a correction for the reed pipe to obtain even more accurate results, particularly for reed pipes with short air columns, where the influence of the reed on the pitch is greatest.

Finally, Weber suggested here that the reed pipe be used as a scientific instrument to examine the properties of two vibrating bodies (such as a reed and an air column) that are attached to each other.[30] Bodies that are capable of generating and being stimulated by sounds resonate together when placed near to each other. This resonance is relevant to the design of musical instruments, as the timbre of nearly all such instruments depends on their resonating parts. The timbre of a violin, for example, changes when the resonating fingerboard is modified, and the muting of a violin is achieved not by touching the strings, but by clamping down the bridge. For Weber the reed pipe could be used as an analytical tool to investigate sympathetically vibrating bodies, with a view to improving musical-instrument manufacture.[31]

Such precision measurement required special reed pipes to meet Weber's rigorous criteria. Since he was a physicist and not a mechanician, he could not build such an instrument himself. Indeed by the nineteenth century, scientific instruments were rarely built by physicists. He therefore turned to two skilled artisans, the mechanician Christian Hoffmann (also spelled Hofman) in Leipzig and the Berlin mechanician J. August D. Oertling in Berlin,[32] both of whom specialized in (among other things) the production of scientific and mathematical instruments. It is noteworthy that Weber turned to mechanicians, rather than organ builders, even though skilled organ builders lived in both Leipzig and Berlin. The reeds for Weber's pipes needed to be made with a precision required of acousticians, not organists.[33] The expenses Weber incurred from having these two mechanicians work on his reed pipes were deferred in part by Prussia's Minister of Culture Karl Freiherr vom Stein zum Altenstein, who was interested in supporting applied science, particularly when it was relevant to musical-instrument manufacture.[34]

Hoffmann was the son of Johann Christian Hoffmann, *Universitätsmechanikus und Optikus* (University Mechanician and Optician) and honorary member of the Naturforschende Gesellschaft (Society of Natural Researchers) of Leipzig. Christian took over his father's warehouse of mathematical, optical, and physical instruments on the Klöstergäßchen in 1824.[35] In 1826 Christian became an ordinary member of the Leipzig Naturforschende Gesellschaft.[36] As is the case with most skilled mechanicians, we know very little about Christian Hoffmann, as so little information is left behind in the archives. We know that in 1844 he applied for, and received, a privilege for a table scale, whereby no one was permitted to build a scale of similar construction for the ensuing five years.[37] Some eleven years later, Christian turned his attention away from physical, optical, and mathematical instruments and toward publishing. In 1855 he applied to the city council for a concession to keep his journeymen. These journeymen included a carpenter, a metal pourer, a blacksmith, a goldsmith, and a locksmith.[38] A list of instruments that his mechanical institute produced in 1855 included printing presses, steel and copper pressure presses, bookbinding machines, ploughs, cylinders for paper glazing, and paper and stamp presses.

In addition to Hoffmann, Oertling was the other mechanician to whom Weber turned to build his instruments. He enjoyed a much more successful career manufacturing scientific precision instruments than Hoffmann. Originally residing in Dorotheastraße of Berlin in the 1820s, he moved to Jerusalemstraße 1, where he lived throughout the 1830s, Krausenstraße 53 in 1840, Markengrafstraße 31 in 1843, finally settling in Oranienburgerstraße 57, living there during the 1840s and '50s. He had three sons, all of whom were opticians and mechanics: August, C[hristian?]., and Friedrich.[39] C[hristian?]. later on became a piano manufacturer and owned an instrument shop at Potsdamerstraße 95. The father was renowned for his manufacture of precision optical instruments, many of which were used at the Potsdam Observatory during the mid-nineteenth century, such as high-precision thermometers and a reflecting circle (*Spiegelkreis*). His most influential work dealt with the production and testing of plano-parallel objectives for scientific instruments.[40]

In his third essay on reed pipes, "Versuch mit Zungenpfeifen" (Trial with Reed Pipes),[41] Weber returned to the theme of his first paper, namely the compensation of organ pipes. He illuminated how organ builders, following his cue, could go about constructing compensated pipes for differing pitches. Weber argued that he based his work on the "sure method of calculations so that it could easily lead to practical execution."[42] First, he determined the laws governing the compensation for two sets of vibrations produced by sounding bodies. Second, he uncovered the laws governing the reed pipes' dimensions (*Mensuren*) and their corresponding pitches.[43] In this case Weber analyzed how the pitch of each reed pipe depended on the dimensions of its components, namely the reed *and* (contrary to what Gottfried Weber had claimed) the air column. Weber discovered that the frequency of the reed itself depended on its elasticity as well as on the changing pressure of the vibrating air in the pipe.[44]

Weber's fourth essay on reed pipes appeared later in 1829, in the *Annalen der Physik und Chemie*.[45] It offered more detailed analyses and derivations than his previous essay concerning the laws governing reed pipes, the data of which he had presented at the 1828 meeting of the Versammlung deutscher Naturforscher und Aerzte in Berlin. In the first section, Weber detailed the theory of an ideal and real reed pipe, with the view of determining the law governing its pitch.[46] He set up the problem of deriving an equation for the pitch as one resembling two swinging pendulums: one being on top of the other and attached in such a way that the top pendulum is fastened to a stationary object and carries the other pendulum on its moving plumb bob.[47] The reed, separated completely from the air column, swings in a pendular fashion with a different period than when it is attached to (and therefore influenced by) the air column. The laws of oscillation of elastic bodies with one fixed end and one free end had been determined by Euler in 1779. The reed corresponds to the upper pendulum. The air column corresponds to the lower pendulum; it works as such through waves that strike one end of the reed and are then reflected back to the column.[48] D. Bernoulli and Euler had each deduced the laws and conditions according to which the plumb bobs of a double pendulum vibrate synchronously, which Weber drew upon for his study. His derived law for a real reed pipe was

$$n^2 = n'^2 + \frac{2g\mu kpn'}{\pi\rho c}\tan\frac{\pi ln'}{c} \qquad \text{(eq. 1)}$$

where

n = the number of vibrations made by an isolated reed in one second;

n' = the number of vibrations the reed pipe generates in one second;

p = the weight of a mercury prism whose unit area is its base and whose height is the barometric height of the average pressure of air in the air column;

k = the ratio of the increase in pressure to the increase in density in a sound wave (according to Gay-Lussac and Welter's calculations, $dp/d\rho$ which Weber takes to be $c_p/c_v = 1.375$);

l = the length of the reed pipe;

μ = the ratio of the surface area of the vibrating part of the reed to the cross section of the vibrating air column;

g = gravitational constant;

c = the velocity of sound in air; and

ρ = the weight of the reed, based on the size of its unit area. This is determined by dividing the unit area of the entire reed by its entire weight (see figure 5.3).[49]

In the second part of the essay, Weber put his law to the test of experiment. In order to determine the precise pitch of his reed pipes, Weber had a monochord constructed. To ensure that the monochord produced accurate values, Weber tuned it to a tuning fork of known pitch.[50] He then employed Taylor's formula[51] to determine the number of vibrations per second, using the known weight of the string, its length, its elasticity, and any additional weight added to determine the vibrations of the monochord's string. He would bow the string of the monochord, eliciting a pitch. If the pitch of the pipe was higher than the string's, he would add weights to the string until he could detect no beats between the two pitches. If the pitch of the pipe was lower, he would subtract weights from the string until he could hear no beats between the two pitches. When no beats were detected, the waves of both pitches were perfectly in phase; they

230

Die Theorie der Zungenpfeifen hat uns zu folgender Gleichung geführt, wenn, wie in allen Zungenpfeifen unserer Orgeln, die Aufsenseite der Platte mit einem Behälter von verdichteter Luft communicirt:

$$nn = n'n' + \frac{2g\mu kpn'}{\pi\varrho c} tang \frac{\pi ln'}{c}$$

Um unsere Versuche mit dieser Theorie zu vergleichen, müssen wir alle in dieser Gleichung vorkommenden Gröfsen, bis auf *eine*, aus unsern Versuchen bestimmen, und die letzte Gröfse endlich aus ihnen mittelst der Gleichung berechnen. Wir wollen zu dem Vergleichungspunkte der Theorie und Erfahrung die Gröfse n wählen, oder die Zahl der Schwingungen der isolirten Platte in einer Secunde, welche für alle Versuche *einer* Zungenpfeife, die sämmtlich mit derselben Platte und verschieden langen Luftsäulen gemacht wurden, gleich war.

Bei Bestimmung aller in der Gleichung vorkommenden Gröfsen, aufser n, sehen wir, dafs folgende für alle Versuche einerlei Werth behielten, nämlich:

$$g,\ \mu,\ k,\ p,\ \pi,\ c;$$

und dafs folgende

$$\varrho,\ l,\ n'$$

bei jedem Versuche einen eigenthümlichen Werth hatten, den man für l und n' in der Tabelle verzeichnet findet. Wir wollen jene erstern Werthe zu bestimmen suchen.

Wenn wir alle Längenmaafse in Pariser Linien, alle Gewichte in Grammen ausdrücken, den Barometerstand 28 Zoll hoch nehmen, und das specifische Gewicht des Quecksilbers zu dem des Wassers 13,593 oder das Quecksilber, Biot's und Arago's Versuchen gemäfs, bei 28 Zoll Barometerstand und bei der Temperatur des schmelzenden Eises 10494,8 Mal schwerer als atmosphärische Luft setzen, wenn wir endlich das Verhältnifs der Spannkraft des Wasserdampfs in der Luft zur Spannkraft des Gemenges aus Luft und Dampf, wie es in der Luftsäule unserer Zungenpfeife stattfand, Dalton's Versuchen ge-

Figure 5.3

Weber's equation for the compensation of reed pipes. Source: W. Weber 1829c, pp. 230–231.

231

mäfs, in unserm Falle 0,0322 annehmen; so erhalten wir aufser den bekannten Werthen von

$g=2174$ Linien nach Borda's Versuchen,

$k=$ 1,375 nach Gay-Lussac's und Welter's Versuchen,

$\pi=$ 3,14159 . . .

folgende Werthe von

$$\mu=\frac{2{,}956\,.\,14{,}06}{\frac{1}{4}\pi(4{,}141)^2}$$

$$p=12\,.\,28\,.\,13{,}593\,.\,0{,}2256^3$$

$$c=\sqrt{\left[2g\,.\,kp\,.\,12\,.\,28\,.\,10494{,}8\,\frac{1+0{,}00375\,.\,28}{1-0{,}375\,.\,0{,}0322}\right]}$$

nach Laplace's Theorie. Der Werth für ϱ endlich ist für die fünf Platten der Reihe nach

$$\frac{2{,}670}{2{,}956\,.\,26{,}54},\ \frac{2{,}295}{2{,}956\,.\,26{,}54},\ \frac{1{,}783}{2{,}956\,.\,26{,}54},\ \frac{2{,}458}{2{,}956\,.\,26{,}54},\ \frac{2{,}3045}{2{,}956\,.\,26{,}54}$$

Ich habe im Laufe dieser Abhandlungen über Zungenpfeifen mehrmals erwähnt, dafs, wenn man Zungenpfeifen mit gleichen Platten und Luftsäulen von verschiedener Länge betrachtet, die Erfahrung lehre, dafs bei kurzen Luftsäulen der Ton der Zungenpfeife dem Tone der isolirten Platte sehr nahe komme, bei langen Luftsäulen der Ton der Zungenpfeife demjenigen sehr nahe komme, welchen die Luftsäule, wenn sie an dem einen Ende offen, am andern verschlossen wäre, geben würde, und nur in geringem Grade vom Tone der isolirten Platte abhänge. Nach dieser Erfahrung ist, wenn unsere Theorie der Zungenpfeife damit harmoniren soll, zu erwarten, dafs die aus den beobachteten Schwingungen der Zungenpfeife *berechneten* Schwingungen der isolirten Platte der *wahren* Zahl der Schwingungen der isolirten Platte bei *kurzen* Luftsäulen fast eben so nahe komme, als die *beobachteten* Schwingungen der Zungenpfeife der *wahren* Zahl ihrer Schwingungen; dafs aber bei *langen* Luftsäu-

Figure 5.3 (continued)

possessed the same pitch. Weber tested five reed pipes and compared their pitches in vibrations per second according to both the theoretically predicted value and the experimentally determined result.

Reed pipe	Observation	Prediction
No. 1	793.5	786.6
No. 2	565.2	563.1
No. 3	406.1	444.3
No. 4	1142.0	1188.7
No. 5	779.0	787.2[52]

By using the accepted value for the speed of sound at a given temperature determined by the French Academy, Weber concluded that the experiments confirmed his theory of reed pipes.

Later in the essay, Weber described the use of his reed pipes and newly determined law to measure the velocity of sound in air and in other gases,[53] the air pressure present in the sound wave, and the specific heats of elastic fluids.[54] Weber's law (eq. 1) is, however, unwieldy as it stands because the speed of sound both multiplies and appears within a tangent expression. To make the relationship tractable, Weber relied on his experimental knowledge, for he knew that the pitch of a reed pipe (and therefore the speed of sound in it) approximates that of a covered labial pipe of the same length as long as the air column becomes longer.[55]

Since the speed in the covered labial pipe is just twice the product of its length by the sound's frequency, Weber could approximate by assuming that, in the reed pipe formula, c is near to, but not quite, $2ln'$. As $(\pi ln')/c$ (call it y) thereby approaches $\pi/2$, it becomes nearly equal to its tangent, and the small difference $\pi/2 - y$ becomes $1/(\pi/2 - y)$, in which case the tangent can be approximated as:

$$\tan y = \frac{1}{\frac{\pi}{2} - y}, \text{ where } y = \frac{\pi ln'}{c}.$$

Therefore, substituting in eq. (1), the speed of sound through air in a reed pipe is

$$n^2 = n'^2 + \frac{2\left[\frac{2g\mu p}{\pi\rho}\right]kn'}{\pi(c - 2ln')},$$

Solving for c then provides the proper correction term:

$$c = 2ln' + \frac{2\left\lfloor \frac{2g\mu p}{\pi\rho} \right\rfloor kn'}{\pi(nn - n'n')}.^{56}$$

Weber tested the velocity of sound through air using three reed pipes produced by Oertling. After substituting the variables l and n' from three of Oertling's reed pipes, as well as for the correction using 1.375 for k, he compared his results with the values obtained from Laplace's theory, which yielded 1066.3 Parisian feet. His closest result differed only by 0.06 percent from Laplace's formula. His least accurate result was a mere 0.83 percent different.[57]

Next, Weber tackled another aspect of Laplace's law for the propagation of sound.[58] Laplace's theory is based on two critical relations. The first is Mariotte's (or Boyle's) law, which asserts that at constant temperature, the pressure of a given quantity of gas is proportional to its density. This law, Weber argued, had been supported by a plethora of scientific experiments. The second relationship used by Laplace was not so solid experimentally. According to Weber, Laplace, using experiments done by his colleagues, had deduced that the ratio of the increase in pressure to the increase in density of a sound wave (i.e., $dp/d\rho$) is both constant across temperature and greater than the ratio of the pressure to density at equilibrium. Weber briefly mentioned the attempts by François Delaroche and Jacques Étienne Bérard and by Nicolas Clément and Charles Bernard Desormes of 1812 and Joseph Louis Gay-Lussac and Jean Joseph Welter of 1822 (described below) to measure this value, none of which he found totally acceptable.[59] The problem with a precise, direct measurement was that in a sound wave, air is compressed and dilated very rapidly. One simply did not have the means to experiment directly on the sound wave itself.

Weber realized that by rearranging this equation and solving for k, the ratio of the increase in density to the increase in pressure of a sound wave, one had a direct experimental means to examine Laplace's claim. He argued that "a test and confirmation of Laplace's law . . . can be taken from the reed pipes," and that they could "measure the specific heat of elastic fluids," improving upon Dulong's work.[60]

$$k = \frac{c}{\frac{2g\mu p}{\pi\rho} n' \tan y}(n^2 - n'^2);\ y = \frac{\pi l n'}{c}$$

Weber's k is a constant of proportionality, much like Biot's and Dulong's notation discussed below, as well as Joseph Louis Lagrange's, who had investigated the possibilities of setting the pressure of air in a sound wave proportional to some power of its density.[61] Since Weber was solving for k here in order to "test the merit" (*bewähren*) of Laplace's theory with respect both to the speed of a sound wave and the measurement of the specific heats of elastic fluids, he could not rely on a measurement of the velocity of sound based on that theory.[62] He drew upon his earlier calculations on the speed of sound through a reed pipe with a long air column, since (recall) its pitch is very close to the pitch of a covered labial pipe of the same length. Hence, he could base his calculations on D. Bernoulli's method for covered labial pipes and apply it to his own observations with reed pipes.[63]

In order to put Weber's experimental work in its proper scientific and technical context, a brief summary of the history of the relationship between the speed of sound and the specific heat of air is necessary. Laplace had suggested to Jean-Baptiste Biot that Newton's experimentally incorrect expression for the speed of sound through air was based on his neglect of the changes in temperature due to longitudinal compression and rarefaction of the sound wave.[64] Newton had demonstrated in book II, proposition 1 of his *Principia* of 1687 that the velocity of sound in air was $\sqrt{\frac{P}{\rho}}$, where P was the *vis elastica* (or pressure) and ρ was its density. In 1802 Biot had published a paper based on the assumption that sound waves are adiabatic, and therefore that heating occurs in compression and cooling in rarefaction.[65] In this work Biot suggested the following relationship for the density of air in a sound wave, $\rho' = \rho\,(1 + s)$, where ρ is the density of undisturbed air and s is the fractional change in density. Assuming isothermal conditions, the pressure of the air in the sound wave $P' = P(1 + s)$, where P is the pressure of the undisturbed air. However, Biot argued that the temperature does not remain constant, believing that the effect of fractional compression s would heat the air and therefore increase the pressure by a factor $(1 + ks)$. His expression for the pressure of a gas experiencing adiabatic changes in volume was, in full, $P' = gHn\rho\,(1 + s)(1 + ks)$, where H is the reading on a mercury barometer, g is the gravitational constant, and n is the ratio of the density of mercury to air. Neglecting the minimal value s^2, Biot obtained the velocity of sound in air, $v = \sqrt{gHn(1+k)}$.[66] It is crucial to point out that an independent experimental confirmation of k was not available in 1802. However, by drawing upon the results on the velocity of sound experimentally determined for the Académie des Sciences in 1738 by a group under the direction of Cassini de Thury, Biot could solve for k, since $\sqrt{gHn(1+k)} = 1107\,\text{ft/s}$. Hence, k became = 0.2869 (which yields a value of 1.2869 for c_p/c_v, or γ in modern notation).[67]

Having not been convinced by Biot's contribution, in 1808 Siméon-Denis Poisson published a lengthy essay offering his own attempt to bring the theoretical expression for the velocity of sound into accord with its experimental value.[68] To determine the assumed rise in temperature to give k its experimentally observed value, he reckoned that condensation in a sound wave is thermally equivalent to heating at constant pressure to the requisite temperature, followed by an isothermal compression to the initial density.[69] By comparing his (uncorrected) theoretical value for the speed of sound at 6°C, the temperature at which de Thury's group had measured the velocity back in 1738, with their actual measurement, he determined that the experimental value of k must be 0.425. With that value, he was able to conclude that a dilation or compression of 1/116 would result in an increase in temperature of 1°C. Still, no one could proffer direct experimental evidence for the law governing adiabatic heating.[70]

Direct experimental confirmation for k needed to come from work on specific heats, or the amount of heat needed to raise a particular quantity of gas a certain temperature at constant pressure or volume. On 7 January 1811 the First Class of the Institute announced its prize competition in physics: "Determine the specific heats of gases, in particular those of oxygen, hydrogen, azote, and some compound gases, comparing them with the specific heat of water. Determine, at least approximately, the change in specific heat that is produced when the gases expand."[71] The determination of specific heats was relevant to the study of chemical combination, the production of efficient steam engines, and the calculations both of the speed of sound and absolute zero.[72]

Although a number of essays were submitted, two are of particular importance to this chapter. The first paper, which was runner-up in the competition, was submitted by the First Professor of Applied Chemistry at the Conservatoire des Arts et Métiers in Paris—Nicolas Clément—and Charles Bernard Desormes, who had been an assistant to Guyton de Morveau and was to be elected a corresponding member of the chemistry section of the Académie des Sciences in 1819. Both were interested in industrial chemistry. The second relevant essay, which won the prize, was written by a friend of Biot's, François Delaroche, and the Montpellier chemist, Jacques Étienne Bérard.[73]

In their first set of experiments Delaroche and Bérard deduced the relative values of the specific heats of common gases at constant pressure from the temperatures at which these gases kept a calorimeter above ambient temperature.[74] Their experiments were based on a constant-flow method. The gas being examined flowed through a spiral tube about one meter in length inside a copper calorimeter filled with cold water. A number of liters of gas passed through the calorimeter at standard temperature and pressure, decreasing in temperature during its passage. At equilibrium, the gas heated the water enough to preserve it at a temperature greater than the surrounding air.[75] They obtained the following specific heats at constant pressure for the common gases.

	By volume	By weight
Air	1.0000	1.0000
Hydrogen	0.9033	12.3401
Carbonic acid	1.2583	0.8280
Oxygen	0.9765	0.8848
Nitrogen	1.0000	1.0318[76]

In a second set of experiments, Delaroche and Bérard proposed to ascertain the dependence of the specific heat of a gas on its density. Hence, they modified their first experiment by allowing air to pass through their apparatus at two different pressures. The initial temperature was the same in both cases. They obtained a volume-specific heat of 1.2396 : 1, and the ratio of pressures was 1.3583 : 1.[77]

Whereas Delaroche and Bérard's experiments were a straightforward determination of specific heats, Clément and Desormes used an indirect method, as they were primarily interested in the heat capacity of a vacuum as well as determining the value of absolute zero. The gas under investigation in a 28-liter spherical vessel was heated in a water bath. It expanded due to the heat, and traveled via a glass tube into another, smaller vessel containing oil. The heat caused oil to rise up into a second glass tube against the pressure in a third vessel, which was lower than atmospheric. They reckoned that the time it took for the oil to climb between two marks on the second tube was proportional to the specific heat of the gas.[78] They then took a second reading, namely the time the oil required to descend between those two marks while cooling. Their volume-specific heats for the common gases were similar to those of Delaroche and Bérard.[79]

In 1816 Laplace had returned to his interest in adiabatic phenomena.[80] In December he read a paper to the Académie des Sciences, which correctly claimed that Newton's and Lagrange's work had assumed that the elastic motions of air particles occur isothermally, substituting the pressure of air for its elasticity, which is the reciprocal of compressibility.[81] He continued by arguing that one could calculate the effect of small temperature changes occurring in the passage of a sound wave by multiplying Newton's expression for the velocity of sound, namely $\sqrt{\frac{P}{\rho}}$, by $\sqrt{\frac{c_p}{c_v}}$.[82] The best available data for Laplace's calculation were those that had been generated in Delaroche and Bérard's second set of experiments, which (recall) began with two samples of gas at the same initial temperature, having the same volume but different pressures (and so different densities). They measured the heats ΔQ_2 and ΔQ_1 liberated during cooling at constant pressure to a final common temperature. Laplace reasoned that the heat contained in a quantity of air under constant pressure is proportional to its volume, from which Laplace could calculate that $\Delta = c_p/c_v = \dfrac{\frac{P_2}{P_1}-1}{\frac{\Delta Q_2}{\Delta Q_1}-1}$.[83] Drawing upon Delaroche and Bérard's data, Laplace took "the quantities of heat which two equal volumes of atmospheric air liberate with a decrease in temperature of approximately 80°. The one is compressed by the weight of the atmosphere [P_1], while the other is compressed by that weight plus 0.36 [P_2]. They [Delaroche and Bérard] found that the heat liberated under the greater pressure is 1.24 times [ΔQ_2] the heat released under the lesser pressure [ΔQ_1]."[84] Hence, $\dfrac{\left(\frac{1.36}{1}\right)-1}{\frac{1.24}{1}-1} = 1.5$ for c_p/c_v or γ.[85] Laplace concluded his short essay requesting that

> physicists . . . determine by direct measurement the ratio of the specific heat of air at constant pressure to that of constant volume. . . .

In lieu of the difficulty of direct measurement of specific heats, the velocity of sound is the more precise method of obtaining the ratio.[86]

In 1822 Gay-Lussac and Welter, whose research on the topic was most likely encouraged by Laplace himself, had offered a value for γ that seemed to support Laplace's speculations.[87] They had shown experimentally that the temperature of air that, having escaped from a container under a given pressure and then entering a partially evacuated receiver, does not change, reasoning that the increase in temperature produced by exiting the small hole is equal to the cooling caused by expansion. From this they calculated a value of 1.3748 and noted that γ was nearly constant over an impressive range of both temperatures and pressures. Laplace used their value in his 1822 calculation for c_p/c_v. "This important ratio may be deduced with great exactitude from the interesting experiments which MM. Gay-Lussac and Welter are carrying out at this moment on the compression of air. Four of these experiments, done at an atmospheric pressure of 757 mm and which these learned physical scientists [*physiciens*] have kindly communicated to me, have given this ratio as 1.3748. The extreme results only differ from this mean result by one part of 136."[88] As Gay-Lussac and Welter used air in flasks containing anhydrous calcium chloride to eliminate any water vapor, Laplace needed to correct for γ while calculating the speed of sound through air. His deduced calculation with the correction was 337.14 m/s for 16°C, close to the experimental value of 340.9 m/s measured at 16°C.[89] Gay-Lussac and Welter's value for γ was to be used two years later by Sadi Carnot, in his seminal work of 1824, *Réflexions sur la puissance motrice du feu* (Reflections on the Motive Power of Fire).[90]

In May of 1828 Dulong read a paper to the Académie des Sciences describing his experiments determining, albeit indirectly, c_p/c_v.[91] Given that Laplace's calculation of the speed of sound was Newton's expression multiplied by the correction factor, Dulong thought one should calculate c_p/c_v by accurately measuring the speed of sound through various gases, rather than experimentally determining the specific heats themselves. He measured the pitch (and therefore the speed of sound) of a tone sounded in a 60-cm organ labial pipe filled with various common gases. He then combined his measurements with the following formula to find c_p/c_v for a given gas: $\frac{n}{n'} = \sqrt{\frac{(1+\varepsilon t)}{(1+\varepsilon t')}}\sqrt{\frac{k}{k'}}\sqrt{\rho}$. Here t is the temperature of air, t' the temperature of the gas, n the frequency of the pitch produced by the pipe filled with air, n' the frequency of the pitch produced by the pipe filled with the gas, ε Gay-Lussac's coefficient of expansion (0.00375), and ρ the density of the gas. Setting the air's density to 1.000, k is then c_p/c_v for air, which had already been determined from observations made on the velocity of sound to be 1.421.[92] Hence k', or c_p/c_v of the gas, is the only unknown, and therefore can be solved for. Dulong obtained an average value of 1.415 for c_p/c_v for oxygen, 1.407 for hydrogen, 1.367 for carbonic acid, 1.428 for carbon dioxide, and 1.343 for nitrous oxide.[93] He concluded that c_p/c_v was the same for all elementary gases.[94] And he noted that if air was heated at constant pressure from 0 to 1°C, it would expand 1/267 of its initial volume.[95] From these experiments, Dulong could draw the influential conclusion that at constant pressure and temperature, equal volumes of any elastic fluid (simple or compound) will liberate or absorb the same amount of heat when suddenly expanded or compressed by the some fraction of its initial volume.[96] Weber was in a position to corroborate Dulong's results for specific heats without assuming (as Dulong himself had) the correctness of Laplace's theory.

> The more Laplace's theory is used in other physical investigations, and the more [research] is built upon it, the more desirable it is that it be justified in every respect. As long as Laplace's theory is only used in acoustics, and only applied to the speed of sound in air, Gay-Lussac and Welter's experiments suffice as a confirmation of Laplace's law. Dulong concluded in accordance with this [Laplace's] theory the same from his observations on *specific heats*. The simplicity of Dulong's discovered laws, which stand or fall [*bestehen und vergehen*, or literally, "exist and disappear"] with the correctness of Laplace's theory, is a marvelous trial [*Bewährung*] of the propagation of sound. Had Dulong measured the speed of sound in various gases with our reed pipes, rather than with other [labial] organ pipes, one would have been in a position to draw two conclusions from these experiments: namely, the speed of sound, which one can determine as least as exactly as with the best [labial] organ pipes, and the ratio of the *increase in pressure* to the *increase in density* of a sound wave, which one can obtain neither from other organ pipes, nor in any other manner, and which calculated on its own does not rely on any hypotheses from acoustical experiments to draw conclusions about the *specific heats* of elastic fluids.[97]

Weber calculated the ratio of the increase in pressure and density of the sound wave directly, and obtained a value of 1.372.[98]

Weber next turned his attention to the theory of other wind instruments with reeds, particularly the clarinet, oboe, and bassoon. According to

Weber, the investigation of reed instruments needed to be divided into two parts: the vibrations of cylindrical and conical air columns with side openings and the influence of the mouthpiece on the pitch. Weber speculated that the first area of concentration would be analogous to his previous studies on the reed pipe, but the influence of the mouthpiece still needed to be researched thoroughly. He blamed musical-instrument makers for constructing the instruments' reeds in such an imprecise way that no exact investigation of the mouthpiece's influence on the instrument could be undertaken, clearly underestimating the tacit nature of the craft of reed making, as oboists will tell you.[99] As soon as the reeds could be attached to these instruments in the same way as they were in his reed pipes, the physicist could develop a theory, which in turn could enable musical-instrument makers "to build a large number of musical instruments according to the most secured theoretical principles as is the case with optical instruments."[100]

In 1830 Weber focused his attention on the acoustical properties of other reed instruments, particularly the clarinet and its generation of harmonic tones.[101] After a series of experiments comparing the pitches generated by a clarinet with various tube lengths with the pitches generated by open and covered labial organ pipes, Weber contradicted Gottfried Weber's earlier claim that clarinets must exhibit the same acoustical properties as a covered labial pipe, since—Gottfried argued—both the clarinet and covered labial pipe produce the same harmonic tones, unlike an open labial pipe. Wilhelm demonstrated, *contra* Gottfried, that the clarinet and in essence all reed pipes actually represent a third type of pipe, half open and half covered.

Weber's final contribution to acoustics appeared in *Annalen der Physik und Chemie* in 1833.[102] He had already started his collaboration with Carl Friedrich Gauss in Göttingen some two years earlier, devoting nearly all of his time and energy to the electric telegraph and geomagnetic research. Indeed, Weber had first met Gauss after his lecture "On the Compensation of Reed Pipes" at the Berlin meeting of the Versammlung deutscher Naturforscher und Aerzte.[103] After the meeting, Weber sent a copy of his *Habilitationsschrift* to Gauss, who replied in a letter of 28 April 1830.

I was always of the opinion that acoustics belongs to those parts of mathematical physics, where the most scintillating advances are to be made. Acoustics actually only deals with space and time relations, and one should be able to make it subservient to mathematics, but how little, how extremely little that is the case, we all know! All of our experiments up to this point have been limited to the length and the speed of the propagation of the sound wave. From the relationship of these two parameters one can determine the pitch. But the differences between and the properties of the tones, what we call the timbre for the singular instances, and what becomes most wonderfully apparent in clearly articulated tones—in so far as mathematics can offer a clear insight—this has totally remained terra incognita. It appears that this is only dependent on the form of the sound waves. What an infinite field still lies before us! I am totally convinced that the human spirit will provide here the same clarity as it has given to the optical sciences. One only needs to learn what the specific differences for the differing tones are, not in the resonating body, nor in the ear, but in the elastic medium between the two.[104]

Three years after Gauss's letter, Weber began his essay by reviewing the role played by acoustics in other disciplines. As one can see, Gauss's predictions did not seem to come to fruition. "The greatest interest and the most meaningful influence that acoustics has had in recent times on theoretical-physical research is more indirect than direct. Acoustical investigations have served as an excellent standard in investigating other areas of physics . . . but no new path has been paved in acoustics, no new element has been added. The attention given this science has been aimed at the many existing, but still largely unknown, modifications of the phenomena of sound."[105]

The goal of this paper as well was to apply acoustical principles to the improvement of musical-instrument design. His first suggestion was applying the theory of nodal points of rods to string instruments. The theory of vibrating elastic rods had been known since the works of D. Bernoulli and Euler and the empirical studies by both Chladni and Felix Savart which confirmed (and sometimes corrected) that theory. Weber argued, however, that what had eluded the attention of the physicist was the utility such a theory could offer to the construction of musical instruments.[106] Weber was interested in determining the position of nodal points on freely vibrating elastic rods, for which Chladni discovered the "empirical rules," which "are quite vague and imprecise."[107] The position of the nodal points, Weber added, "should not be sought by empirical rules, but by laws calculated *a priori* with a precision, which one could not achieve by merely [experimental] searching."[108] One major advantage of determining nodal points in this fashion was that such a technique could be applied to instrument manufacturing much more easily than Chladni's method, which necessitated a "practical and skilled hand" and did not lend itself to manufacturing processes (*fabrikmäßig betrieben werden*).[109] Although a highly skilled experimental physicist, Weber clearly realized that manufacturing was not conducive to the precision and accuracy, which physicists demand. Nevertheless, general laws could serve as guidelines for musical-instrument makers.

Weber's second suggestion for the application of acoustical principles to musical-instrument design dealt with string instruments. By changing the shape of the bridge across which the string is suspended, one alters the string's tension and therefore its pitch. Furthermore, this modification, once thoroughly investigated, could be used to derive practical rules from the theory for the manufacture of string instruments. Weber's final suggestion for improved musical-instrument design dealt with the tones generated by strings. Weber argued that a string does not produce merely one tone, but at least two, the fundamental tone and another, and possibly even more.[110] And these tones can be unpleasant to the ear. These peculiar tones can be predicted theoretically only if one does not presume a string to be a thread, but an elastic, tightened rod. Euler had already provided a differential equation for the vibrations of elastic, tightened rods, and Poisson had refined the calculations.

Weber concluded this paper with a discussion of the application of acoustical principles to the production of wind instruments. He argued that the development of the theory of wind instruments had reached a standstill, not because the results of such theory had not been rigorously determined, but rather because the rigor of the execution never led to profitable results. Neither Lagrange's nor Poisson's calculations ever influenced the building of organ pipes, and the application of the theory of wind instruments had not led to any new phenomena since the work of D. Bernoulli.[111] It was, in Weber's eyes, time to determine the laws governing the fundamental forces of nature by rigorous experimentation relevant to musical acoustics. "Physics," he continued, "offers assistance in the many cases where [mathematical] theory abandons us, particularly when it comes to the manufacture of musical instruments."[112]

The Musicians' Response to Weber

Weber's work on the compensation of reed pipes was greeted with enthusiasm by German musicians. In 1826 Chladni summarized the Weber brothers' *Wellenlehre,* with its relevance and application to music, in

Caecilia.[113] A year later he did the same for the *Allgemeine musikalische Zeitung.*[114] Two years later, Gottfried Weber published in his *Caecilia* a shortened version of Wilhelm Weber's essay on the compensation of reed pipes.[115] The organist and organ inspector from Neuruppin, Friedrich Wilke, called Weber's work on the compensation of organ pipes "very important, worthy of praise and thanks."[116] But the inspector challenged Weber to compensate pipes with pulsating reeds (such as those used in the trumpet and trombone pipes of organs), which were more important than free-vibrating pipes.[117] *Das Universal-Lexicon der Tonkunst* (The Universal Encyclopedia of Music) proclaimed in 1837 that "Weber's discovery of compensated organ pipes, must, when its application is made easier in the future, bring about a worthwhile perfection of the organ."[118] The piano and organ builder Carl Kützing included a brief discussion of Weber's work on the role of the reed and air column in generating the pitch of an organ reed pipe in his *Beiträge zur praktischen Akustik als Nachtrag zur Fortepiano- und Orgelbaukunst* (Contributions to Practical Acoustics as a Supplement to the Art of Piano and Organ Building).[119] And Germany's leading nineteenth-century organ builder (and one of its best organists), Johann Gottlob Töpfer, drew upon Weber's work on reed pipes.

Töpfer, professor of music of the Grand Duchy's Seminary for Teachers (*Schullehrerseminar*) and organist of the city church of Weimar was the son of a village musician, who ensured that his son received training in piano, organ, and violin playing. After numerous tragedies at home, a family friend, Frau Räthin Jegemann, invited the young boy to complete his education with her in Weimar, where he continued to study piano. He studied at one of Weimar's *Gymnasia*, and later attended the seminary for teachers, furthering his organ studies. After demonstrating a superior ability in organ playing and composition, his friends and teachers encouraged him to leave the seminary and pursue his passion for organ playing. He began to study music theory and often offered private lessons to children of wealthy parents. Particularly fascinated by the construction of the organ in Weimar's city church, he spent many hours in the workshops of organ builders, traveling throughout Saxony, Bavaria, Austria, and Bohemia and gleaning information on organ construction and renovation.[120]

Töpfer first alerted fellow organ builders to the acoustical work of Chladni and the Weber brothers in his *Die Orgelbau-Kunst, nach einer neuen Theorie dargestellt und auf mathematische und physicalische Grundsätze gestützt* (The Art of Organ Building, Represented by a New Theory and Supported by Mathematical and Physical Principles).[121] Of particular importance was the brothers' work on the acoustics of organ pipes, detailed in their *Wellenlehre* and summarized by them in an essay in the *Allgemeine musikalische Zeitung* in 1826.[122] A year later in his *Erster Nachtrag zu Orgelbau-Kunst* (First Supplement to the Art of Organ Building), Töpfer summed up Wilhelm Weber's paper "On the Theory of Reed Pipes": "The causes whereby the pitch of a free-vibrating reed, in conjunction with pipes of varying lengths and widths changes, have been so carefully researched by Prof. Wilhelm Weber, and the laws according to which the changes themselves follow are stated so precisely and are so solidly grounded that I refer anyone who wants to learn about this point to these treatises, of these [particularly] the first-rate 'Theory of Reed Pipes,' which one finds in Poggendorf's [*sic*] *Annalen für Physik* [*sic*], vol. 17."[123] Töpfer added "In his 'Theory of Reed Pipes,' Herr Prof. Weber has developed equations, the help of which the given sizes of the free-vibrating reed pipes can be determined for certain cases."[124] As Oertling's reed pipes were composed of materials different from those used in organs, Töpfer discussed the application of Weber's work with the relevant materials.[125] He provided his readers with numerous charts of the surface areas, widths, lengths, and thicknesses of reeds for the various semitones of four octaves (see figures 5.4 and 5.5).[126] Töpfer later discussed Weber's equation and gave examples of solving it with given dimensions of the pipe and its reed (see figures 5.6 and 5.7).[127] In his abridged treatise on the organ, *Die Orgel* (The Organ) of 1843, Töpfer summarized Weber's explanation of how a tone arises in a reed pipe.[128] Weber's research resonated with Töpfer's, as the organ builder, disappointed that a number of key aspects of organ building were still predicated on the principle of "trial and error," strove to base as much of the art on mathematical and physical principles as possible in order to increase efficiency.[129]

Weber's equation was meant to serve as a guideline for organ builders. It certainly did not replace the tinkering necessary for the artisans to determine the required measurements. Today's organ builders note that each organ is unique, being built to fit its particular environment. The builder needs to take into consideration such key elements as the dimensions of the church or hall where the instrument will be housed, the range of temperatures the organ will experience throughout the year, and of course the materials used to construct it.[130] And, Weber's "practical law"—as the late-nineteenth- and early-twentieth-century organ architect Audsley labeled it—was applicable to a specific range of reed pipes, not to all reed pipes on the organ.[131] The discussion and use of Weber's equation were nevertheless sufficiently relevant to organ building that it merited inclusion in Töpfer's seminal *Lehrbuch der Orgelbaukunst* (Textbook on the Art of Organ Building) of 1855.[132] Töpfer argued that to construct a large reed pipe with great precision, so that the tube would have the same effect in all the octaves of the fundamental tone, one needed to implement Weber's equation listed above (see figure 5.8). As Töpfer's texts were the ones most often consulted by organ builders during the mid-nineteenth century,[133] Weber's equation became a part of the organ-building process. Indeed, his acoustical research, though itself without lasting practical influence, signaled a collaboration among German physicists, musicians, and musical-instrument makers that was to last until at least the end of the nineteenth century.[134]

Just as Weber's interest in acoustics waned, so too did the Germans' interest in expressive organs. During the 1830s and '40s, the English and French began to build organs with compressed bellows with higher wind pressure and more efficient swell boxes, not surprisingly perhaps, as their interest in expression predated the Germans'. And, during the 1860s Aristide Cavaillé-Coll of Paris, who had invented the precursor to the harmonium, the *poïkilorgue*, back in 1830, perfected the electric powered, fully pneumatic tracker-organ.[135] Such an invention unleashed unprecedented volumes and enabled unparalleled expression.[136] This concert organ went a long way in rendering most of the seventy-seven stops expressive, whereas the swell organs of the 1830s had only twenty expressive stops.[137]

On the other hand, the expression of German organs built in the 1830s and '40s seemed to suffice for German audiences. In 1877, after listening to an organ performance in the concert room of the Public Halls in Glasgow, the renowned German pianist Hans von Bülow wrote: "I never met with an Organ so good in Germany, the instruments there not having the same amount of expression and flexibility—most delicate and exquisite nuances—that hearing the *diminuendi* and *crescendi* was to me a new sensation. If I would longer listen to an Organ like this . . . I would, were I not grown too old, jeopardise my pianistical [*sic*] career, and begin to study the Organ, where certainly I would be able to display much more

— 66 —

Mensurtabelle für die Zungen der freischwingenden und aufschlagenden Zungenstimmen.

Tonhöhe und Dicke der Zungen.							Breite der Zungen.	Länge der Zungen.
I	II	III	IV	V	VI	VII		
c^4 0''',0495							0''',458	2''',602
h^3 0,0510							0,479	2,718
b^3 0,0525	c^4 0''',0589						0,500	2,838
a^3 0,0540	h^3 0,0606						0,522	2,964
gis^3 0,0556	b^3 0,0624	c^4 0''',0700					0,545	3,095
g^3 0,0572	a^3 0,0642	h^3 0,0721					0,569	3,232
fis^3 0,0589	gis^3 0,0661	b^3 0,0742	c^4 0''',0833				0,595	3,375
f^3 0,0606	g^3 0,0681	a^3 0,0764	h^3 0,0858				0,621	3,524
e^3 0,0624	fis^3 0,0700	gis^3 0,0786	b^3 0,0883	c^4 0''',0991			0,648	3,680
dis^3 0,0642	f^3 0,0721	g^3 0,0809	a^3 0,0909	h^3 0,1020			0,677	3,843
d^3 0,0661	e^3 0,0742	fis^3 0,0833	gis^3 0,0935	b^3 0,1050	c^4 0''',1178		0,707	4,014
cis^3 0,0684	dis^3 0,0764	f^3 0,0858	g^3 0,0963	a^3 0,1081	h^3 0,1213		0,739	4,191
c^3 0,0700	d^3 0,0786	e^3 0,0883	fis^3 0,0991	gis^3 0,1112	b^3 0,1248	c^4 0''',1401	0,771	4,377
h^2 0,0721	cis^3 0,0809	dis^3 0,0909	f^3 0,1020	g^3 0,1145	a^3 0,1285	h^3 0,1443	0,805	4,571
b^2 0,0742	c^3 0,0833	d^3 0,0935	e^3 0,1050	fis^3 0,1178	gis^3 0,1323	b^3 0,1485	0,841	4,773
a^2 0,0764	h^2 0,0858	cis^3 0,0963	dis^3 0,1081	f^3 0,1213	g^3 0,1362	a^3 0,1528	0,878	4,984
gis^2 0,0786	b^2 0,0883	c^3 0,0991	d^3 0,1112	e^3 0,1248	fis^3 0,1401	gis^3 0,1573	0,917	5,205
g^2 0,0809	a^2 0,0909	h^2 1,1020	cis^3 0,1145	dis^3 0,1285	f^3 0,1443	g^3 0,1619	0,958	5,435
fis^2 0,0833	gis^2 0,0935	b^2 0,1050	c^3 0,1178	d^3 0,1323	e^3 0,1485	fis^3 0,1666	1,001	5,676
f^2 0,0858	g^2 0,0963	a^2 0,1081	h^2 0,1213	cis^3 0,1362	dis^3 0,1528	f^3 0,1715	1,045	5,927
e^2 0,0883	fis^2 0,0991	gis^2 0,1112	b^2 0,1248	c^3 0,1401	d^3 0,1573	e^3 0,1766	1,091	6,190
dis^2 0,0909	f^2 0,1020	g^2 0,1145	a^2 0,1285	h^2 0,1443	cis^3 0,1619	dis^3 0,1817	1,139	6,464
d^2 0,0935	e^2 0,1050	fis^2 0,1178	gis^2 0,1323	b^2 0,1485	c^3 0,1666	d^3 0,1871	1,189	6,750
cis^2 0,0963	dis^2 0,1081	f^2 0,1213	g^2 0,1362	a^2 0,1528	h^2 0,1715	cis^3 0,1926	1,242	7,049
c^2 0,0991	d^2 0,1112	e^2 0,1248	fis^2 0,1401	gis^2 0,1573	b^2 0,1766	c^3 0,1982	1,297	7,361
h^1 0,1020	cis^2 0,1145	dis^2 0,1285	f^2 0,1443	g^2 0,1619	a^2 0,1817	h^2 0,2040	1,358	7,687
b^1 0,1050	c^2 0,1178	d^2 0,1323	e^2 0,1485	fis^2 0,1666	gis^2 0,1871	b^2 0,2100	1,414	8,027

Figures 5.4–5.5

Charts of various dimensions of organ reed pipes and their corresponding pitches. Source: Töpfer 1834, pp. 66–67.

Tonhöhe und Dicke der Zungen.							Breite der Zungen.	Länge der Zungen.
I	II	III	IV	V	VI	VII		
a^1 0''',1081	h^1 0''',1213	cis^2 0''',1362	dis^2 0''',1528	f^2 0''',1715	g^2 0''',1926	a^2 0''',2161	1''',477	8''',382
gis^1 0,1112	b^1 0,1248	c^2 0,1401	d^2 0,1573	e^2 0,1766	fis^2 0,1982	gis^2 0,2225	1,542	8''',754
g^1 0,1145	a^1 0,1285	h^1 0,1443	cis^2 0,1619	dis^2 0,1817	f^2 0,2040	g^2 0,2290	1,611	9''',141
fis^1 0,1178	gis^1 0,1323	b^1 0,1485	c^2 0,1666	d^2 0,1871	e^2 0,2100	fis^2 0,2357	1,682	9''',546
f^1 0,1213	g^1 0,1362	a^1 0,1528	h^1 0,1715	cis^2 0,1926	dis^2 0,2161	f^2 0,2426	1,756	9''',968
e^1 0,1248	fis^1 0,1401	gis^1 0,1573	b^1 0,1766	c^2 0,1982	d^2 0,2225	e^2 0,2497	1,834	10''',410
dis^1 0,1285	f^1 0,1443	g^1 0,1619	a^1 0,1817	h^1 0,2040	cis^2 0,2290	dis^2 0,2570	1,915	10''',871
d^1 0,1323	e^1 0,1485	fis^1 0,1666	gis^1 0,1871	b^1 0,2100	c^2 0,2357	d^2 0,2645	2,007	11''',352
cis^1 0,1362	dis^1 0,1528	f^1 0,1715	g^1 0,1926	a^1 0,2161	h^1 0,2426	cis^2 0,2723	2,096	11''',855
c^1 0,1401	d^1 0,1573	e^1 0,1766	fis^1 0,1982	gis^1 0,2225	b^1 0,2497	c^2 0,2803	2,188	1''0''',380
h^0 0,1443	cis^1 0,1619	dis^1 0,1817	f^1 0,2040	g^1 0,2290	a^1 0,2570	h^1 0,2885	2,285	1''0''',928
b^0 0,1485	c^1 0,1666	d^1 0,1871	e^1 0,2100	fis^1 0,2357	gis^1 0,2645	b^1 0,2970	2,386	1''1''',500
a^0 0,1528	h^0 0,1715	cis^1 0,1926	dis^1 0,2161	f^1 0,2426	g^1 0,2723	a^1 0,3057	2,492	1''2''',098
gis^0 0,1573	b^0 0,1766	c^1 0,1982	d^1 0,2225	e^1 0,2497	fis^1 0,2803	gis^1 0,3146	2,602	1''2''',722
g^0 0,1619	a^0 0,1817	h^0 0,2040	cis^1 0,2290	dis^1 0,2570	f^1 0,2885	g^1 0,3238	2,718	1''3''',374
fis^0 0,1666	gis^0 0,1871	b^0 0,2100	c^1 0,2357	d^1 0,2645	e^1 0,2970	fis^1 0,3333	2,838	1''4''',054
f^0 0,1715	g^0 0,1926	a^0 0,2161	h^0 0,2426	cis^1 0,2723	dis^1 0,3057	f^1 0,3431	2,964	1''4''',765
e^0 0,1766	fis^0 0,1982	gis^0 0,2225	b^0 0,2497	c^1 0,2803	d^1 0,3146	e^1 0,3531	3,095	1''5''',507
dis^0 0,1817	f^0 0,2040	g^0 0,2290	a^0 0,2570	h^0 0,2885	cis^1 0,3238	dis^1 0,3635	3,232	1''6''',282
d^0 0,1871	e^0 0,1200	fis^0 0,2357	gis^0 0,2645	b^0 0,2970	c^1 0,3333	d^1 0,3741	3,375	1''7''',092
cis^0 0,1926	dis^0 0,2161	f^0 0,2426	g^0 0,2723	a^0 0,3057	h^0 0,3431	cis^1 0,3851	3,524	1''7''',937
c^0 0,1982	d^0 0,2225	e^0 0,2497	fis^0 0,2803	gis^0 0,3146	b^0 0,3531	c^1 0,3964	3,680	1''8''',820
H_o 0,2040	cis^0 0,2290	dis^0 0,2570	f^0 0,2885	g^0 0,3238	a^0 0,3635	h^0 0,4080	3,843	1''9''',742
!B_o 0,2100	c^0 0,2357	d^0 0,2645	e^0 0,2970	fis^0 0,3333	gis^0 0,3741	b^0 0,4200	4,014	1''10''',704
A_o 0,2161	H_o 0,2426	cis^0 0,2723	dis^0 0,3057	f^0 0,3431	g^0 0,3851	a^0 0,4323	4,191	1''11''',709
Gs^0 9,2225	B_o 0,2497	c^0 0,2803	d^0 0,3146	e^0 0,3531	fis^0 0,3964	gis^0 0,4449	4,377	2''0''',759
G^0 0,2290	A_o 0,2570	H^0 0,2885	cis^0 0,3238	dis^0 0,3635	f^0 0,4080	g^0 0,4580	4,571	2''1''',355

Figures 5.4–5.5 (continued)

— 80 —

Längen der freischwingenden Zungenpfeifen, für die im §. XXXVII. gegebenen 7 Mensuren berechnet.

	I			II			III			IV			V			VI			VII		
Tonhöhe.	Fuß.	Zoll.	Linien.	Fuß.	Zoll.	Linien.	Fuß.	Zoll.	Linien	Fuß.	Zoll.	Linien.	Fuß.	Zoll.	Linien.	Fuß.	Zoll.	Linien.	Fuß.	Zoll	Linien.
c^3		3	8		3	8		3	8		3	9		3	9		3	9		3	9
h^2		3	10		3	10		3	11		3	11		3	11		4	0		4	0
b^2		4	1		4	1		4	2		4	2		4	2		4	3		4	3
a^2		4	4		4	4		4	5		4	5		4	5		4	6		4	6
gis^2		4	7		4	7		4	8		4	8		4	9		4	9		4	9
g^2		4	10		4	11		4	11		4	11		5	0		5	0		5	0
fis^2		5	2		5	2		5	3		5	3		5	3		5	3		5	4
f^2		5	5		5	6		5	6		5	6		5	7		5	7		5	7
e^2		5	9		5	9		5	10		5	10		5	11		5	11		5	11
dis^2		6	1		6	1		6	2		6	2		6	3		6	3		6	3
d^2		6	5		6	6		6	7		6	7		6	7		6	8		6	8
cis^2		6	10		6	10		6	11		6	11		7	0		7	0		7	0
c^2		7	2		7	3		7	4		7	4		7	5		7	5		7	6
h^1		7	7		7	8		7	9		7	9		7	10		7	10		7	11
b^1		8	1		8	2		8	3		8	3		8	3		8	4		8	4
a^1		8	6		8	7		8	8		8	8		8	9		8	10		8	10
gis^1		9	0		9	1		9	2		9	2		9	3		9	4		9	5
g^1		9	6		9	8		9	8		9	9		9	10		9	11		10	0
fis^1		10	1		10	2		10	3		10	4		10	5		10	6		10	6
f^1		10	8		10	9		10	10		10	11		11	0		11	1		11	2
e^1		11	3		11	5		11	6		11	7		11	8		11	9		11	10
dis^1		11	11	1	0	1	1	0	2	1	0	3	1	0	4	1	0	5	1	0	6
d^1	1	0	7	1	0	9	1	0	10	1	0	11	1	1	1	1	1	2	1	1	3
cis^1	1	1	4	1	1	6	1	1	7	1	1	8	1	1	10	1	1	11	1	2	0
c^1	1	2	1	1	2	3	1	2	5	1	2	7	1	2	8	1	2	9	1	2	10
h^0	1	2	10	1	3	0	1	3	2	1	3	4	1	3	6	1	3	7	1	3	8
b^0	1	3	8	1	3	11	1	4	1	1	4	3	1	4	5	1	4	6	1	4	7
a^0	1	4	7	1	4	10	1	5	0	1	5	2	1	5	4	1	5	6	1	5	7
gis^0	1	5	6	1	5	10	1	6	1	1	6	3	1	6	4	1	6	6	1	6	7
g^0	1	6	6	1	6	10	1	7	1	1	7	3	1	7	5	1	7	7	1	7	8
fis^0	1	7	7	1	7	11	1	8	4	1	8	6	1	8	7	1	8	9	1	8	10
f^0	1	8	8	1	9	0	1	9	4	1	9	7	1	9	9	1	9	11	1	10	1
e^0	1	9	10	1	10	2	1	10	6	1	10	9	1	11	0	1	11	2	1	11	4
dis^0	1	11	0	1	11	5	1	11	9	2	0	1	2	0	4	2	0	7	2	0	9
d^0	2	0	4	2	0	10	2	1	3	2	1	7	2	1	9	2	2	0	2	2	2
cis^0	2	1	9	2	2	2	2	2	6	2	2	11	2	3	3	2	3	6	2	3	8
c^0	2	3	2	2	3	8	2	4	1	2	4	6	2	4	10	2	5	1	2	5	4

Figures 5.4–5.5 (continued)

— 81 —

	I			II			III			IV			V			VI			VII		
Ton-höhe.	Fuß.	Zoll.	Linien.	Fuß.	Zoll.	Linien.	Fuß.	Zoll.	Linien	Fuß.	Zoll.	Linien.	Fuß.	Zoll.	Linien.	Fuß.	Zoll.	Linien	Fuß.	Zoll.	Linien.
H_0	2	4	8	2	5	3	2	5	9	2	6	2	2	6	6	2	6	9	2	7	0
B_0	2	6	3	2	6	10	2	7	5	2	7	10	2	8	3	2	8	7	2	8	10
A_0	2	7	11	2	8	7	2	9	2	2	9	8	2	10	1	2	10	5	2	10	9
Gis_0	2	9	8	2	10	5	2	11	0	2	11	7	3	0	0	3	0	5	3	0	9
G_0	2	11	6	3	0	4	3	1	0	3	1	7	3	2	1	3	2	6	3	2	10
Fis_0	3	1	6	3	2	4	3	3	1	3	3	9	3	4	3	3	4	9	3	5	1
F_0	3	3	7	3	4	6	3	5	4	3	6	0	3	6	7	3	7	1	3	7	6
E_0	3	5	9	3	6	9	3	7	8	3	8	5	3	9	0	3	9	7	3	10	0
Dis_0	3	8	0	3	9	2	3	10	1	3	10	11	3	11	7	4	0	2	4	0	8
D_0	3	10	5	3	11	8	4	0	8	4	1	7	4	2	4	4	2	11	4	3	6
Cis_0	4	1	0	4	2	4	4	3	5	4	4	5	4	5	2	4	5	11	4	6	6
C_0	4	3	8	4	5	1	4	6	4	4	7	5	4	8	3	4	9	0	4	9	8
H_1	4	6	5	4	8	0	4	9	4	4	10	6	4	11	5	5	0	3	5	1	0
B_1	4	9	4	4	11	0	5	0	5	5	1	8	5	2	9	5	3	8	5	4	5
A_1	5	0	4	5	2	3	5	3	10	5	5	1	5	6	4	5	7	4	5	8	2
Gis_1	5	3	7	5	5	7	5	7	4	5	8	9	5	10	6	5	11	2	6	0	0
G_1	5	6	11	5	9	1	5	11	0	6	0	6	6	2	0	6	3	2	6	4	2
Fis_1	5	10	6	6	0	10	6	2	11	6	4	7	6	6	2	6	7	6	6	8	7
F_1	6	2	3	6	4	10	6	7	1	6	8	9	6	10	7	7	0	0	7	1	2
E_1	6	6	2	6	9	0	6	11	4	7	1	3	7	3	4	7	4	10	7	6	0
Dis_1	6	10	4	7	1	4	7	4	0	7	6	0	7	8	3	7	9	10	7	11	2
D_1	7	2	8	7	6	0	7	8	11	7	10	11	8	1	4	8	3	2	8	4	8
Cis_1	7	7	3	7	10	10	8	1	11	8	4	3	8	6	10	8	8	10	8	10	5
C_1	8	0	2	8	4	0	8	7	4	8	9	9	9	0	8	9	2	9	9	4	6
H_2	8	5	0	8	9	2	9	0	10	9	3	6	9	6	8	9	8	11	9	10	11
B_2	8	10	0	9	2	7	9	6	7	9	9	7	10	1	0	10	3	5	10	5	7
A_2	9	3	4	9	8	4	10	0	8	10	4	0	10	7	8	10	10	3	11	0	8
Gis_2	9	8	11	10	2	4	10	7	1	10	10	9	11	2	8	11	5	5	11	8	2
G_2	10	2	10	10	8	8	11	1	10	11	5	10	11	10	1	12	1	0	12	4	1
Fis_2	10	9	0	11	3	4	11	8	11	12	1	5	12	5	11	12	9	1	13	0	5
F_2	11	3	6	11	10	5	12	4	5	12	9	4	13	2	2	13	5	6	13	9	2
E_2	11	10	4	12	5	9	13	0	3	13	5	8	13	10	10	14	2	5	14	6	7
Dis_2	12	5	6	13	1	6	13	8	7	14	2	6	14	8	1	14	11	11	15	4	5
D_2	13	1	1	13	9	8	14	5	4	14	11	9	15	5	8	15	9	10	16	2	9
Cis_2	13	8	11	14	6	3	15	2	6	15	9	7	16	3	11	16	8	5	17	1	9
C_2	14	5	3	15	3	3	16	0	3	16	8	0	17	2	8	17	7	6	18	1	4

Figures 5.4–5.5 (continued)

§. XLI.

Hierzu findet man in der schon früher erwähnten Theorie des Herrn Prof. Weber eine Gleichung aufgestellt, nach welcher sich alle, bei den Zungenpfeifen dieser Art vorkommenden Größen für einzelne gegebene Fälle berechnen lassen. Es ist folgende:

$$n\,n = n'\,n' + \frac{2\,g\,\mu\,k\,p\,n'}{\pi\,\varrho\,c}\ \text{tang.}\ \frac{\pi\,l\,n'}{c}$$

Die Buchstaben haben in dieser Gleichung folgende Bedeutung:

n bezeichnet die Anzahl der Schwingungen, welche die isolirte Zunge in 1 Secunde macht.

n' bezeichnet die Anzahl der Schwingungen, welche die Zungenpfeife in 1 Secunde macht.

p bezeichnet das Gewicht eines Quecksilberprisma, das die Flächeneinheit zur Basis und die Barometerhöhe beim mittlern Druck der Luft in der Luftsäule zur Höhe hat. Drückt man alle Maaße in weimarischen Linien und die Gewichte in cöllnischen Richtpfennigstheilen aus, und setzt man die Barometerhöhe = 387‴,072 (= 28″ franz. Maaß), das Verhältniß des Quecksilbers zum Wasser = 13,593, die Schwere einer Kubiklinie Wasser = 2,10274 Richtpfennige, so ergiebt sich

p = 387,072 . 13,593 . 2,10274 = 11063 Richtpfennige.

k bezeichnet das Verhältniß der Druckszunahme zur Dichtigkeitszunahme in einer Schallwelle und wird nach Gay-Lussac und Welter = 1,375 gesetzt.

l bezeichnet die Länge der Zungenpfeife (vom obern Theile des Aufsatzes bis zum untern Ende der Zunge gemessen.)

μ bezeichnet das Verhältniß der Oberfläche des schwingenden Theils der Platte oder Zunge zum Querschnitt der schwingenden Luftsäule. (Man findet diese Verhältnißzahl durch Division des Querschnitts der Luftsäule in die Fläche der schwingenden Zunge.)

g bezeichnet die Fallhöhe in der ersten Secunde und kann nach hiesigem Maaß = 2504‴,448 gesetzt werden.

π bezeichnet das Kreisverhältniß = 3,14159.

c bezeichnet die Geschwindigkeit des Schalles in einer Secunde. Sie wird in Webers Theorie bei 28° C nach Laplace und in franz. Maaß

$$= \sqrt{\left[2\,g\,k\ 12\ .\ 28\ .\ 10494{,}8\ \frac{1 + 0{,}00375\ .\ 28}{1 - 0{,}375\ .\ 0{,}0322}\right]}$$ angenommen. Reduzirt man diese Angabe auf hiesiges Maaß, so erhält man 176886‴ = c.

ϱ bezeichnet das Gewicht eines Stückes der Zunge, von der Größe der Flächeneinheit. Man findet dieses Gewicht, wenn man mit dem Flächeninhalt der ganzen Zunge, (welcher in □‴ angegeben seyn muß) in das Gewicht der ganzen Zunge, (welches in Richtpfennigstheilen ausgedrückt wird) dividirt.

Figures 5.6–5.7

Töpfer's use and explanation of Weber's equation for the compensation of organ reed pipes in 1834. Source: Töpfer 1834, pp. 71–72.

eloquence as Beethoven's and Chopin's speaker."[138] And von Bülow was not the only one to comment on the lack of expression in German organs. In 1882 the British musicologist Dudley Buck praised English organs over their German counterparts.

> The lack of a swell is the weakest point of the great majority of German Organs. Even Dr. Burney, fifty years after swells had become common in England, expresses, in his famous work entitled "Continental Tours," his great surprise to find them utterly unknown upon the Continent. His remarks would hold true at the present day with but little modification, so far as Germany is concerned, but few instruments, outside the larger ones of recent date, possessing this great improvement. The reason of this is to be found partly in the extremely conservative character of their organ builders, almost a national trait, and still more in the fact that but little use would be found for a Swell Organ outside Solo playing.[139]

Free-vibrating reed pipes were common among German-built organs during the 1820s through the 1850s. Johann Friedrich Schulze of Paulinzella, Thuringia, and Eberhard Friedrich Walcher of Ludwigsburg, Wuerttemberg, employed them in their renowned organs. The British, on the other hand, preferred pulsating reeds, which, recall from the previous chapter, are pipes whose reeds are slightly longer than the pipe's aperture, and hence the reeds actually beat (or pulsate) against the pipe. The London organ builder Willis never used free reeds, while J. M. Walker and Sons claimed that they stopped using them some time back in the 1840s. The problem was that these pipes would not tolerate the requisite power, given the vastly increased airflow. Likewise, free reeds were not popular in France. Cavaillé-Coll in Paris decided to abandon them. When E. F. Walcher left Germany for London in 1851, he quickly switched to pulsating reeds. Similarly, when Schulze built the great organ in Doncaster, Britain, the only free reeds of the ninety-four stops were the trombone and its octave.[140] Hence, Weber's work on the organ reed pipe was relevant to organ manufacture throughout the German territories for about thirty years. As aesthetic tastes changed with time and country, Weber's work became irrelevant to organ builders.

Weber's Influence on Acoustics and Physiology

Weber's work on reed pipes, however, was relevant not only to organ building and adiabatic phenomena; it was also used by physiologists and

316

genpfeifen lösen, in welcher die Gleichung aufgestellt wird

$$n\,n = n'\,n' + \frac{2\,g\,\mu\,k\,p\,n'}{\pi\,\varrho\,c}\,\mathrm{tg}\,\frac{\pi\,l\,n'}{c}$$

die Buchstaben haben hier folgende Bedeutung.

n bezeichnet die Anzahl der Schwingungen, welche die isolirte Zunge in 1 Secunde macht.

n' bezeichnet die Anzahl der Schwingungen, welche die Zungenpfeife in 1 Secunde macht.

p bezeichnet das Gewicht eines Quecksilberprisma, das die Flächeneinheit zur Basis und die Barometerhöhe beim mittlern Druck der Luft in der Luftsäule zur Höhe hat. Drückt man alle Maße in weimarischen Linien und die Gewichte in cölnischen Richtpfennigstheilen aus, und setzt man die Barometerhöhe = 387,072 Linien (= 28 Zoll franz. Maaß), das Verhältniß des Quecksilbers zum Wasser = 13,593, die Schwere einer Cubiklinie Wasser = 2,10274 Richtpfennige, so ergiebt sich

p = 387,072 . 13,593 . 2,10274 = 11063 Richtpfennige.

k bezeichnet das Verhältniß der Druckszunahme zur Dichtigkeitszunahme in einer Schallwelle und wird nach Gay-Lussac und Welter = 1,375 gesetzt.

l bezeichnet die Länge der Zungenpfeife (vom obern Theile des Aufsatzes bis zum untern Ende der Zunge gemessen.)

μ bezeichnet das Verhältniß der Oberfläche des schwingenden Theils der Platte oder Zunge zum Querschnitt der schwingenden Luftsäule. (Man findet diese Verhältnißzahl durch Division des Querschnitts der Luftsäule in die Fläche der schwingenden Zunge.)

Figure 5.8

Töpfer's use and explanation of Weber's equation for the compensation of organ reed pipes in 1855. Source: Töpfer 1855, pp. 316–317.

317

g bezeichnet die Fallhöhe in der ersten Secunde und kann nach hiesigem Maß = 2504,448 Linien gesetzt werden.

π bezeichnet das Kreisverhältniß = 3,14159.

c bezeichnet die Geschwindigkeit des Schalles in einer Secunde. Sie wird in Weber's Theorie bei 28° C nach Laplace und in franz. Maß =

$$\sqrt{\left[2\,g\,k\,12\,.\,28\,.\,10494{,}8\,\frac{1+0{,}00375\,.\,28}{1-0{,}375\,.\,0{,}0322}\right]}$$

angenommen. Reducirt man diese Angabe auf hiesiges, Maß, so erhält man 176886 Linien = c.

ϱ bezeichnet das Gewicht eines Stückes der Zunge, von der Größe der Flächeneinheit. Man findet dieses Gewicht, wenn man mit dem Flächeninhalt der ganzen Zunge, (welcher in □ Linien angegeben sein muß) in das Gewicht der ganzen Zunge, (welches in Richtpfennigstheilen ausgedrückt wird) dividirt.

Um diese Gleichung richtig auf meine oben gegebenen Mensuren der Zungen anwenden zu können, wog ich eine Zunge, welche bei 27 Linien Länge den Ton c^0 und also bei 38,184 Linien Länge den Ton C_0 gab, und deren Dicke ich $\frac{1}{3}$ Linie fand.

Die Länge der ganzen Zunge betrug 48,7 Linien,
die Breite war 4,7 Linien,
der Flächeninhalt also 228,89 Linien □,
die Zunge wog 1376 Richtpfennige,

eine □ Linie derselben Zunge also $\frac{1376}{228{,}89} = 6{,}0116$ Richtpfennige.

Als Probe der Richtigkeit berechnete ich das specifische Gewicht des zur Zunge angewendeten Messings auf zweierlei Art; nämlich:

1) Da ich die Dicke der Zunge = $\frac{1}{3}$ Linie gefunden hatte, so mußte eine Cubiklinie dieses Mes-

Figure 5.8 (continued)

anatomists interested in understanding the physical organs involved in generating the human voice. By the mid-nineteenth century, the relationship between the mechanical and the organic had changed. Whereas nearly a century earlier reed pipes were being constructed by mechanicians to mimic the human voice, they were now primary to the physiological and anatomical understanding of how the voice worked. Weber's research was of particular interest to the renowned "Romantic" physiologist Johannes Müller, who applied the physicist's equations to simple membranous reeds. Müller noted that those reeds slowly increased in pitch after being agitated by a brief, strong blowing, drawing upon Weber's explanation for the increase of a string's pitch to explain this analogous phenomenon.[141] Bindseil's textbook *Die Akustik* summarized Weber's work on reed pipes, labeling it "fundamental."[142] And he proffered detailed experiments, much along the same lines as Weber, on the relationship between the fundamental tone of a single-lipped reed and the length of its connecting tube and discussed the importance of this work to anatomy.[143] Müller and other physiologists such as Heinrich Adolf Rinne worked on the differences and similarities between metal and membranous reed pipes.[144] And Weber's work marked the starting point for the acoustical research of the leading physicist and physiologist of the second half of the century, Hermann von Helmholtz (figure 5.9).

Helmholtz's earliest work on acoustics dates back to 1848–1849, when he authored a two-part review of recent acoustical research published several years later in *Die Fortschritte der Physik* (The Advances of Physics).[145] He criticized Guillaume Wertheim's 1848 attempts to measure the velocity of sound in organ pipes. The assumed initial stationary condition of the air failed when it was substituted by water. Helmholtz argued that the boundary conditions were incorrect, paying particular attention to the motion of the fluid at the tube's opening.[146]

In order to correct for earlier physicists' oversimplifications, Helmholtz first concentrated his efforts on the motion of the air near the organ pipe's opening by experimentally determining the position of the nodal points, the intensity of the sound, and the phase relations of the organ pipe.[147] Then he corrected the theoretical value of the resonance frequency by developing a method of diffraction based on an application of Green's theorem, which relates the surface integral over some bounded region to a line integral over the boundary curve of that same region.[148] He

Figure 5.9
Portrait of Hermann von Helmholtz (1821–1894). Photo: Deutsches Museum, Munich.

compared his theoretical calculations for organ pipes with small lengths with the results obtained by Wertheim and Friedrich Zamminer. His expressions worked well for wide tubes, but were rather inaccurate when dealing with narrower pipes. He reckoned that friction due to air's viscosity increased with the narrowing of the tube, resulting in a shift of the resonance frequency of the organ pipe.[149] Helmholtz's work on organ pipes, unlike the earlier work of Weber, was not directly linked to adiabatic phenomena, but rather led to important work initially on vortex motion and subsequently on internal friction.[150]

In 1861 Helmholtz demonstrated mathematically in his "Theorie der Luftschwingungen in Röhren mit offenen Enden" (Theory of the Vibrations of Air in Tubes with Open Ends) that free reeds must fit their frames exactly in order to produce a powerful tone. Weber and his mechanicians had experimented with metal reeds that could generate only one pitch.[151] Reeds of clarinets, oboes, and bassoons, of course, are not made of metal, but of softer materials, such as cane wood. Each of these instruments possesses a single reed that is responsible for generating all of its pitches. By alternating the pressure of the vibrating column, these instruments produce high pitches, which are close to the pitch of the isolated reed, as well as the lower pitches, because the waves of air in the column excite an alternate pressure of air adjacent to the reed that is strong enough to make it vibrate. In a vibrating column, the change in pressure is at a maximum when the velocity of air particles is at a minimum. And since the velocity is always at a minimum at the end of a closed air column, and the change in pressure at that point is always at a maximum, the pitch of a reed pipe is equivalent to the pitch produced by the air column alone, without the reed.

The proper tones of a clarinet correspond to the third, fifth, seventh, etc. partial tone of the prime. Skilled performers can alter how they blow into the mouthpiece, changing the resulting note from the prime tone to the twelfth of the higher major third. By opening the side holes of the clarinet, performers can change the length of the air column, thereby altering the note's pitch. The oboe and bassoon possess conical air columns and produce proper tones that are therefore equivalent to open tubes of the same length. As a result, the tones of oboes and bassoons come close to those of open pipes. Performers blowing harder into these instruments can produce the octave, twelfth, second octave, and so on.

Helmholtz argued that the moment of greatest pressure within the pipe must occur precisely between a maximum of placement of the reed outward and a maximum of the reed's displacement inward. The phase-angle ε by which the maximum of pressure precedes the passage of the reed through its position of equilibrium is expressed as

$$\tan\varepsilon = \frac{L^2 - \lambda^2}{\beta^2 L\lambda}$$

where L is the length of the wave of the reed in the air without the pipe, λ is the wavelength of the pitch produced, and β^2 a constant depending on the mass, friction, and elasticity of the reed.[152] The maximum velocity of air through the opening precedes the maximum pressure within the pipe by a phase-angle δ:

$$\tan\delta = \frac{-\lambda^2}{4\pi S} \quad \sin\frac{4\pi(l+a)}{\lambda}$$

where S is the transverse section of the tube; l is the length of the tube, and a is a constant depending on the type of opening, which is 45°. For organ pipes, clarinets, oboes, and bassoons, the reed strikes inward and the maximum velocity of the air going outward coincides with the maximum displacement of the reed inward, so that $-\varepsilon - (\pi/2) = \delta$.[153] Much like Weber, Helmholtz was committed to improving the design of musical instruments by eliminating, wherever possible, the wasteful tinkering technique of instrument makers.

This chapter has illustrated how entwined the histories of music and physics really were during the early-to-mid-nineteenth century. Wilhelm Weber plied his craft as an experimental physicist interested in adiabatic phenomena, in collaboration with skilled artisans, so critical to both the scientific enterprise and national economy. Helmholtz's interest in reed pipes led to his influential work on vortex motion and friction. Looking at a piece of material culture, the reed pipe, this chapter has traced the contours of and points of intersection among musical-instrument manufacture, physical theory, and experimentation. This chapter has shed light on the kinds of experimental objects with which both mechanicians and physicists worked during the period before laboratories such as the Physikalisch-Technische Reichsanstalt or the Cavendish Laboratory existed in the modern sense. Ever since the eighteenth century, the organ had been used for experimental inspiration, as evidenced by the works of Euler, D. Bernoulli, Chladni, Dulong, Weber, and Helmholtz. Like the submarine telegraph cable in the 1850s and '60s, organs provided additional resources and techniques for the nineteenth-century physicist. By exploring the sites of experimentation and their intersections, one can see that the works of acoustics scholars were being tested in the dark shops of Berlin mechanicians in the late 1820s and '30s. Indeed, analysis of these shops elsewhere as well—Leipzig, Paris, and London—reveals how extensive experiments that proved to be critical for the study of adiabatic processes and thermodynamics were carried out by musical and scientific instrument makers. And conversely, the cultural history presented here offers new insights into developments in music itself, which are not usually considered to be related to physics—developments such as the improvement of musical instruments—and, as discussed in the ensuing chapters, the standardization of pitch and musical virtuosity.

6 The Fetish of Precision I: Scheibler's Tonometer and Tuning Technique

While Wilhelm Weber personified nineteenth-century precision measurement in physics, Johann Heinrich Scheibler was his counterpart in manufacturing. Whereas the previous chapters dealt with mechanical musical and scientific instruments, here "mechanical" has an economic and industrial resonance, namely mechanization. And, as discussed in this chapter, the mechanical precision so emblematic of the factory resonated with German physicists. The 1830s and '40s witnessed the rise of factories in Prussia, particularly throughout Berlin and the Ruhr area. Scheibler came from a family of very wealthy towel and cloth manufactures. His grandfather, also named Johann Heinrich Scheibler (1705–1765), had been one of the first textile manufacturers in the region. He was born in the "Red House" (now a historic landmark) in Monschau (referred to as Montjoie until 1918) on 11 November 1777, the eighth of fifteen children. After attending school in Monschau, he embarked on numerous *Bildungsreisen* (educational journeys) throughout Europe, particularly Italy.[1] On returning to Germany he moved to Krefeld, the capital of silk manufacturing in Germany, second in Europe only to Lyon—and according to both Chladni and later Johannes Brahms, a city of high musical taste—where he married into the renowned silk-baron family of Heydweiller in 1806. From that point on, Scheibler managed the family business, and in 1834 founded his own silk and velvet firm, Scheibler und Cie. Such a commitment to his profession, however, belies his devotion to music and acoustical studies. Typical of a child in an upper-middle-class German household, he received a thorough musical training from the renowned musician Johann Nikolaus Wolff, who was recruited to Krefeld by Johann Heinrich's family.[2] Scheibler played numerous instruments, including the guitar, altering its fingerboard in order to tune it with greater precision in equal temperament.[3] And in 1816, after four years of attempts, he finally perfected the aura, an instrument composed of twenty well-tuned mouth harmonicas.[4]

This chapter links Scheibler's attempts to deskill organ and piano tuners to his deskilling of factory workers. By deskilling, I mean the process by which factory managers and owners set out to replace laborers with more reliable and efficient machines. In 1798 William Cockerill brought the wool-spinning frame to nearby Eupen, and Johann Heinrich's uncle, Bernhard Georg Scheibler, was the first to employ the machine in his Eupen factory. In 1806 the shearing machine was introduced to the area as well.[5] From 1809 to 1835 the number of silk factories in Krefeld increased from 11 employing 6,264 workers to 28 employing approximately 10,000, as silk weaving moved away from a cottage industry toward factory production. Those 28 factories in 1835 were powered by more than 1,600 Jacquard looms, while factories manufacturing velvet could boast 1,280 weaving looms.[6] By 1840 there were 6,700 weaving looms with 13,675 workers.[7] Although the number of workers doubled from 1809 to 1840, their wages certainly did not. Skilled weavers were being replaced by various machines during the 1820s and '30s, forcing them to accept lower wages.[8] Monschau and Krefeld factories in particular were filled with Jacquard looms. And Scheibler's particular commitment to mechanization is evinced by his being the first in the area to power his factory by a steam engine.[9]

Not surprisingly, the workers responded with violence. On 3 November 1828 the news of a new tax levied on the workers, which contributed to the lowering of their salaries, was met with fighting in the streets with policemen and the smashing of windows of the homes of the factory owners, who, in a fit of panic, established a security watch to prevent further disturbances.[10] Throughout the 1840s, weaving looms were being shut down, and over 4,500 workers were forced into unemployment.[11] The workers' discontent culminated in civil unrest on 21 March 1848, when many ransacked their factories.[12]

During this period of mechanization and civil unrest, Scheibler worked to revolutionize the tuning of keyboard instruments in equal temperament. Just as he drew upon technology to deskill the region's laborers, Scheibler turned to mathematics and (ironically) a highly skilled mechanician to deskill keyboard tuners. Scheibler often stated that he wished to free the practice of music from the grasps of capricious tuners. However, his desire to replace both factory workers and keyboard tuners, though upsetting to workers, was not entirely irrational, as it was in the case of deskilling laborers with machine tools described by David Noble, whereby owners actually lost money by mechanizing.[13] Scheibler saw the worlds of music and manufacturing as based on precision, accuracy, and efficiency. And the elimination of what he saw as wasteful laborers was part of those worlds. With the assistance of a Krefeld mechanician, Hermann Kämmerling, he invented a technique for tuning keyboard instruments in equal temperament with unprecedented accuracy using his tonometer, an improved metronome, and the physical phenomenon of beats, of which he had a rather limited physical understanding.

As we shall see, Scheibler never truly achieved either goal. Although he had hoped to replace all skilled labor with the Jacquard looms, the German silk industry was the most resistant among textile industries to thorough mechanization during that period, because silk was in the end too delicate for the harsh treatment of the early looms.[14] Neither did his technique for tuning keyboard instruments ever replace tuners, as in some instances tuners actually drew upon it to aid them in plying their craft. This chapter illustrates how the notion of precision can be used to illuminate the contours among physics, manufacturing, and musical practice.

A Brief History of Musical Temperament

Before one can appreciate Scheibler's method for tuning keyboard instruments in equal temperament, a brief foray into musical temperaments and intervals is in order. A musical interval is simply a step up or down in pitch defined by the ratio of the frequencies. An octave is defined as a musical interval whose ratio is 2:1. For example, the a′ key above middle C on the piano sounds at 440 vibrations per second (Hz); it is one octave higher than a cello's open A string, which sounds at 220 Hz. Certain musical intervals are consonant, or pleasant, to the human ear. In Western music any combination of the triad, which is a chord of three tones, the root tone, its major or minor third, and its perfect fifth, is consonant. In just intonation (discussed below), a major third represents two pitches with the integer ratio 5:4, for example, C to E, while a minor third represents the integer ratio 6:5, for example from C to E♭. A perfect fifth is a musical interval in the ratio of 3:2, for example, the interval from C to G. Also, all inversions[15] of the triad are consonant. All other intervals, such as sevenths and seconds, are considered to be dissonant, or unpleasant to the ear.

Debates on the musical temperaments have existed for millennia.[16] Up to the sixteenth century, instruments were generally tuned to the Pythagorean system, which is based on the both the octave and the fifth.[17] All fifths and fourths are pure, or based on the integer ratios of their respective pitches.[18] The opposite of pure is tempered. Temperaments are necessary for all keyboard instruments (such as the piano and organ), since these instruments produce fixed tones.[19] Whereas a skilled cellist or violinist can slightly alter the pitch of a note by shifting the position of his or her finger on the string, organists and pianists cannot modify the pitches of their instruments' notes. The phenomenon of musical temperament arises because the concords of triadic music (octaves, fifths, and thirds) are incongruous in their pure forms.[20] For example, three pure major thirds (say, the interval between A♭ and C, C and E, and E and G#) fall short (i.e., are flatter in pitch) of a pure octave (2:1) by one fifth of a whole tone,[21] while the four pure minor thirds (representing the integer ratio 6:5, or say the intervals between G# and B, B and D, D and F, F and A♭) are slightly greater (sharper in pitch) than the pure octave.[22] In order to equate three pure major thirds and four pure minor thirds with an octave, equal temperament (discussed below) slightly increases the interval of a major third and slightly decreases the interval of a minor third. The series of twelve perfect fifths need to be tempered by an overlap of 1/12 of a so-called Pythagorean comma, or 1/12 of the ratio 531441:524288.[23] Fourths need to be enlarged beyond their pure integer ratio of 4:3.

From the sixteenth through eighteenth centuries, a plethora of temperaments were devised in order to compensate for these inequalities, the most important of which were the various meantone temperaments and equal temperament. Meantone temperament is a system of tuning with flattened fifths and pure major thirds.[24] It is referred to as "meantone," since the whole tone (for example C–D) is exactly half of the pure third, C–E.[25] There are numerous variations of meantone temperaments with flattened fifths and sharpened thirds.[26] These systems of temperament are named after the amount that the size of the perfect fifths is flattened: 2/7 comma,[27] 2/9 comma,[28] 1/4 comma,[29] 1/3 comma,[30] and 1/5 comma.[31] By the late eighteenth century the most common version of meantone temperament flattened all fifths and increased all fourths by 1/4 of a comma and kept all major thirds pure, or what keyboard tuners often referred to as "sweet." Meantone temperament, in essence, rendered all whole tones the mean of fourths and fifths, 1/2 of a comma smaller than the whole tone, and 1/2 of a comma larger than the semitone, leaving the diatonic semitones E–F and B–C 1/4 of a comma sharp.[32]

Unfortunately, there was a price to pay for keeping perfect thirds. Meantone temperament resulted in some intervals being unpleasantly dissonant. These tones, referred to as "wolves" for their howling sounds, plagued keyboard instruments tuned in meantone temperament. Composers and performers did their best to avoid evoking wolf tones. For example, the seventeenth-century German-born London organ builder Father Bernard Smith (Schmidt) actually split the black keys for E♭ and G# down the middle, and added extra pipes in order to differentiate between D# and E♭ and G# and A♭. Meantone temperament was the preferred tuning method for organs and pianos throughout the German territories until about the third quarter of the eighteenth century.

Just (or pure) intonation was another type of temperament. It tuned harmonic intervals so purely that no beats could be detected. Each justly intoned interval can be represented by a numerical ratio, based on the measurements of a monochord. Pythagorean intonation shared with just intonation the ratios for the octave, fifth, and fourth. But Pythagorean intonation excluded any multiple of five or any higher prime number. Just intonation used multiples of five to provide for pure thirds and sixths. On keyboard instruments, however, fifths needed to be significantly smaller than pure so that fifths and thirds will not beat. Only extremely complicated keyboards could ameliorate such a problem.[33]

Another type of temperament, well temperament, not to be confused with equal temperament discussed below, refers to the temperament whereby (unlike Pythagorean or meantone) fifths are of different sizes, usually smaller around the home keys to get sweet (i.e., pure) thirds, and larger in the distant keys. In well temperament one can modulate through all the keys (meaning that one can change keys throughout a piece without fearing wolf tones), but all the keys have slightly different tuning, hence distinct characters, thereby preserving the uniqueness of the various keys, or what is referred to as "key color." J. S. Bach's compositions for the well-tempered clavier were the first extended keyboard works to utilize the possibilities of well-tempered tuning.

German theorists had begun discussing the merits and disadvantages of another type of temperament, equal temperament, during the late seventeenth and early eighteenth centuries. Equal temperament required each interval between the semitones to be equal (i.e., one semitone is the twelfth root of two vibrations per second times greater than the preceding semitone), while octaves remained in the ratio of 2:1. Hence, neither thirds, fourths, nor fifths were pure, or kept in their integer ratios. Purists, of course, defended the pure thirds against the revolutionary scheme, arguing that perfect thirds must be kept at all costs. Perhaps the most influential theoretician supporting equal temperament was Friedrich Wilhelm Marpurg, as detailed in both his *Principes du clavecin* (Principles of the Harpsichord)[34] and his *Versuch über die musikalische Temperatur* (Essay on Musical Temperament),[35] the latter of which offered a spirited defense of equal temperament. This work was particularly influential on both Chladni and Scheibler.[36] Although meantone temperament was still preferred by one of Germany's leading eighteenth-century theorists, Johann Philipp Kirnberger,[37] by the end of that century, equal temperament was beginning to gain the upper hand among music theorists.[38] Daniel Gottlob Türk's *Clavierschule* (Piano School) of 1802 advocated equal temperament, because, he claimed, it contributed to the unity of a composition. During the early nineteenth century, keyboard instruments, pianos in particular, were being tuned in equal temperament on the Continent.[39] Britain followed suit some forty years later.[40] Johann Lehmann advocated equal temperament in 1827, and indeed cited Chladni on the topic.[41] Johann Nepomuk Hummel's *Ausführlich theoretisch-practische Anweisung zum Piano-Fortespiel* (Detailed Theoretical-Practical Instruction on Piano Playing)[42] concluded with a lengthy discussion of tuning that argued for equal temperament since it was easier for novice tuners, which turns out not to be true.[43] Jean Jousse, however, argued against equal temperament tuning for pianos and for a subtle unequal temperament in his *Essay on Temperament* published in London in 1832. According to Jousse, the major drawback of equal temperament was that "it cannot be obtained in a strict sense, as may be proved, not only mathematically, but also by daily experience; therefore the best equally tempered instruments are still unequally tempered, and, what is worse, oftentimes in wrong places."[44] In 1836 Claude Montal published his defense of equal temperament for the piano.[45] Alfred James Hipkins, Britain's leading piano technician, introduced equal-temperament tuning to the Broadwood firm in 1846.[46] At the same time, German organ builders were using equal temperament for tuning with increasing frequency.

Mathematicians seemed far less enamored of equal temperament. The English mathematician Wesley Stoker Barker Woolhouse serves as an example. Although he proposed equal temperament for organs in 1835, he conceded that "no tuner can preserve this exact quality."[47] In 1857 Augustus De Morgan, professor of mathematics of University College, London, discussed equal temperament tuning for keyboard instruments:

In equal temperament, for example, the tuner gets one octave into tune, with its adjacent parts so far as successions of fifths up and octaves down require him to go out of it; and the notes thus tuned are called *the bearings*: all the rest is then tuned by octaves from the bearings. The method of tuning the bearings, after taking a standard note from the tuning-fork, consists merely in tuning the successive fifths a little flat, by the estimation of the ear, making corrections from time to time, as complete chords come into the part which is supposed to be in tune, by the judgment of the ear upon those chords. Proceeding thus, if the twelfth fifth appear to the ear about as flat as the rest, the bearings are finished: if not, the tuner must try back.[48]

De Morgan argued that this method of tuning constituted both "a loss of time and a loss of accuracy."[49] Two organ tuners tuning in equal temperament would tune registers of the same organ differently. And he informs the reader that "an old professional tuner" with whom he had spoken at length about equal temperament claimed that "equal temperament was equal nonsense."[50] Unaware of Scheibler's work, De Morgan recommended the manufacturing of tuning forks for all of the semitones of the octave. One could tune by counting beats using a half-minute sand glass filled with emery powder, rather than a regulated metronome, which he thought would be too troublesome to use.[51] He felt that "[t]he system of equal temperament is to my ear the worst I know of. . . . A newly-tuned pianoforte [to equal temperament] is to me insipid and uninteresting. . . ."[52]

Scheibler's Theory of Beats and Tuning Technique

While perfecting the aura some twenty years earlier, Scheibler was to recall in 1834, he had noted how challenging it actually was to tune the various portions of the instrument.[53] During that period, instruments were typically tuned by comparing their pitches to a pitch generated by a monochord. Scheibler first determined two neighboring positions on the monochord's string that produced four beats per second with an a tuning fork, the same pitch as a cellist's open A string (220 Hz).[54] By means of this method, he could transfer the monochord's perfectly tuned a to other instruments. In order to determine the other semitones of the scale, two new neighboring points needed to be found, which would produce four beats when struck simultaneously. Scheibler then determined these positions for all semitones of the equally tempered scale, from a to a′.[55] This procedure, however, required much time and effort. The constant use of the monochord caused the string to warm up (much more quickly than a tuning fork), resulting in a higher pitch. Scheibler concluded that tuning the twelve semitones (a to a′) of an instrument via a monochord was not very reliable for the accuracy he required. Scheibler then decided that tuning forks themselves should be the basis for tuning all the semitones of the octave, and he embarked on his invention of the tonometer.

Scheibler's method of tuning was based on the beats generated by the sounding of two pitches with different frequencies. For every difference of two simple vibrations in pitch, a beat is produced. For example, a pitch of 440 simple vibrations per second (svps, or 220 vps or cycles per second, now called Hertz) will produce 5 beats per second with a pitch of 450 svps. Two tones of equal pitch will not produce any beats, merely a change in loudness, depending on the phase difference between the two. In modern terms, the beat frequency is just the difference between the frequencies of the tones. Hence, Scheibler reckoned that beats could provide an easy and extremely accurate way of testing to see whether two pitches were the same. With his technique, "Mr. H. Scheibler was enabled to obtain the great *desideratum* of a perfectly equalized temperament, and to benefit the musical world by an invention, which is as beautiful as it is ingenious, and as simple as it is complete."[56]

Scheibler had fifty-six tuning forks[57] made by the Krefeld watchmaker and precision mechanician Hermann Kämmerling,[58] each fork producing four beats a second with the one before it, or being eight simple vibrations per second higher than the previous fork (see figures 6.1 and 6.2). His a′ was 880 svps, calibrated with a monochord.[59] This a′ was the average pitch of Viennese pianos, based on measurement that their pitches rose and fell

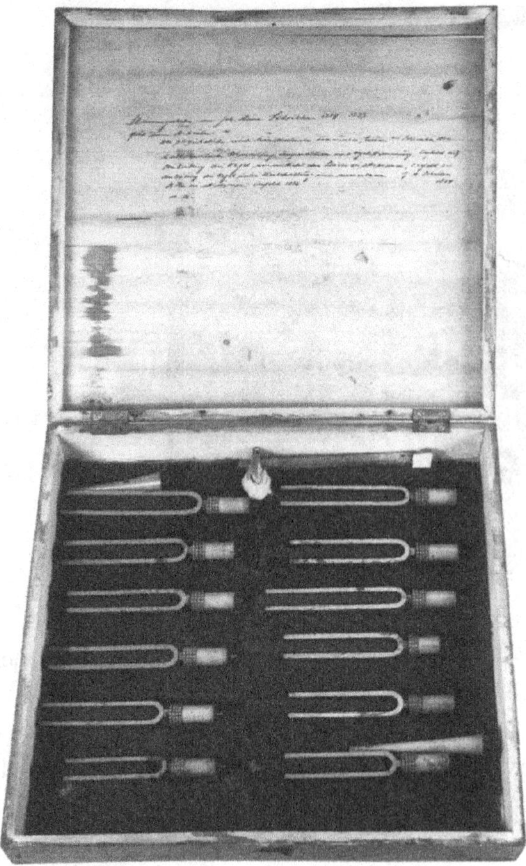

Figure 6.1
Kämmerling's tuning forks made for Scheibler. Source: Medienzentrum Rheinland Landesbildstelle, Düsseldorf LBR 35/59/1.

by 1.5 simple vibrations per second due to temperature changes (although he never informed us what the precise change was). As discussed in the next chapter, this is the origin of the German concert pitch a′ being 440 vps (also referred to as the Stuttgart pitch or the Scheibler pitch). Scheibler wished to confirm that pitch experimentally. The difference between fork 1 (a) and fork 56 (a′) must be 220 beats per second, since 55 × 4 = 220. His lower a had to be 440 svps. Scheibler then determined the number of beats that two successive semitones of the scale make when sounding with each other. He had the two pitches representing the octave with forks 1 and 56,

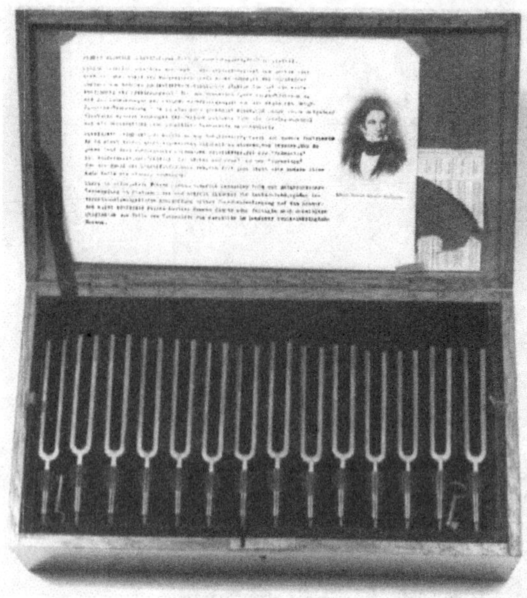

Figure 6.2
Kämmerling's tuning forks made for Scheibler. Source: Medienzentrum Rheinland Landesbildstelle, Düsseldorf LBR 35/59/183.

but the other musical intervals within the octave needed to be determined as well. As we saw, the pitch ratios of these intervals depend on the temperament. For example, the interval C to E (a major third) has a pitch ratio of 5 : 4 in just temperament (or intonation), but would have a different ratio depending upon which type of meantone temperament was chosen. In equal temperament, the pitch ratio of a major third is $1:2^{(4/12)}$, or approximately 1.259.

Scheibler busied himself with different types of musical temperaments. Reflecting upon his earlier interest in unequal temperament, he wrote: "I have been often requested for a manual on a good, unequal temperament, which could be easily and precisely achieved by my method. Because no one could ever provide me with the numbers, and my own troubles to find such a temperament were fruitless, I came across Marpurg's *Versuch über die musikalische Temperatur* [Essay on Musical Temperament] and quickly realized that this author was quite correct in only recommending equal temperament."[60] Tuners, however, still had problems tuning accurately in equal temperament with the unaided ear.[61] He continued to refine his tuning method with the singular purpose of enabling tuners to tune keyboard instruments in equal temperament.

Before Scheibler's efforts of the 1820s and '30s, no means existed by which a tolerable approximation to equal temperament could be obtained on keyboard instruments.[62] The mathematics of equal temperament had been established theoretically since the work of the seventeenth-century Dutch mathematician and engineer Simon Steven: each semitone of the equalized scale needed to be the twelfth root of 2 (approximately 1.059) times higher than the previous semitone. The calculation was rather straightforward; the practice of tuning keyboards accurately and reliably, however, was not. "[W]hen it came to practice—when a musical instrument had actually to be tuned—then all the calculations of the theorists proved so much worthless rubbish, because practice knew of no other means or criterion to regulate the pitch of the different sounds and their ratios to each other, than the ear."[63]

Before Scheibler's technique, piano and organ tuners had used the intervals of fourths and fifths for tuning, since impurities and dissonance of those intervals are most easily detected by the human ear. Tuners had employed twelve ascending fifths and descending fourths, tuned according to their mathematical ratio, for example, C–g–d–a–e–b–f#–c#–g#–d#–a#–e#(f)–C. All the semitones of an octave were now tuned, and the higher and lower notes on the piano were tuned by octaves.[64] Two problems with this technique, however, quickly become apparent. First, series of perfect fourths and fifths can lead in succession to different semitones of an octave, and no ear can detect whether or not an interval is in its pure mathematical ratio. Second, in equal temperament, fourths and fifths are not pure, but tempered. Fifths must be decreased by 4/5 of a vibration twelve times in a row in order for the octaves to be in tune and equal temperament realized. Unfortunately, no ear can discern 4/5 of a vibration per second.

Whereas tuners had previously relied on the ear to detect differences in pitch, Scheibler wished to circumvent the imperfections of the ear and put more of a burden on the eye, as the eye is a "much quicker and more acute judge of small differences."[65] With the mathematics of equal temperament in hand, Scheibler could now quickly ascertain the number of beats that two successive semitones of the equalized scale needed to make when sounding with each other.

Semitone Interval (in svps)				Diff. in Simple Vibs.	No. of Beats
a	440	b♭	466.16	26.16	13.08
b♭	466.16	b	493.88	27.72	13.87
b	493.88	c	523.25	29.37	14.68
c	523.25	c#	554.36	31.11	15.55
c#	554.36	d	587.33	32.97	16.48
d	587.33	e♭	622.25	34.92	17.46
e♭	622.25	e	659.26	37.01	18.50
e	659.26	f	698.46	39.20	19.60
f	698.46	f#	739.99	41.53	20.76
f#	739.99	g	783.99	44.00	22.00
g	783.99	g#	830.61	46.62	23.31
g#	830.61	a′	880.00	49.39	24.69

Looking at the vibrations of these semitones, a slight problem becomes evident. Only a and its octave a′ correspond to Scheibler's tonometer forks (1 and 56). For example, the semitone b♭ has a value of 466.16. Scheibler's forks would sound at 440 svps, 448, 456, 464, and so on. He therefore needed forks that would produce the other semitones of the scale. Since a and b♭, when sounded together, will produce 13.08 beats, fork 3 with 456 svps would need to make 13.08 – 8 = 5.08 beats with b♭, since fork 3 makes 8 beats per second with fork 1. But how can anyone measure 8/100 of a beat? No ear can detect such a minute phenomenon. Scheibler now needed to alter his technique to allow for the limitations of the ear. If fork 4 is to make 5.08 beats with fork 3 in the span of one second, there must be another time interval in which both forks produce four beats, which most humans can detect rather easily: 5.08 : 4 = 1 sec. : 0.78 sec. Scheibler now needed a device that could provide an accurate time interval.[66] This he had in the metronome, invented by Johann Nepomuk Mälzel in 1816. Scheibler used a modified metronome for his measurements, as the precision he needed surpassed anything that musicians would need to count the beats of a measure.

The metronome consists of a pendulum set in motion by a spring. The pendulum oscillates in front of a graduated index that shows the number of vibrations the pendulum will make in one minute, dividing the minute into as many as 280 equal increments. Scheibler had Kämmerling engrave Mälzel's graduated index on the pendulum itself, rendering measurements more precise. And he had Kämmerling make ten subdivisions between each number on Mälzel's scale.[67] Scheibler simply needed to determine the metronome marking, which he could easily achieve by using proportions: 4 beats: 5.80 beats :: 60 sec: 76.2 sec. He proceeded to adjust his fork 4 by filing down the prongs until it made 4 beats with fork 3 during each oscillation of the pendulum. He thus knew that his two forks made 5.08 beats in a second, since 4 beats in 1/76.2 of a minute are equivalent to 304.8 (76.2×4) beats in one minute, which equals 5.08 beats per second. Fork 4 now made 13.08 beats with fork 1, or 26.16 vibrations, and therefore was $440 + 26.16 = 466.16$ svps, which is the pitch of b♭ in equal temperament when a is 440 svps. He continued this procedure until he had a tuning fork for each of the semitones of the scale, and the octave of the tonic, or thirteen forks in total.[68]

As time went on, Scheibler perfected his technique. One problem that he encountered was related to the fallibility of the human ear. When comparing the sound of a piano string and a tuning fork, the ear tends to hear a unison of the consonance produced by both tones, rather than hearing beats, signifying that the two pitches are not the same. This phenomenon is particularly acute in tuning pianofortes due to the small mass of the oscillating string. To overcome this potential problem Scheibler had Kämmerling produce pulsating forks (*Nebengabeln*, literally "adjacent forks"), or forks that are not in tune with the pitch they represent but are perceptively lower. The ear will not be inclined to hear unity, but will tend, in this case, to concentrate on the beats. Kämmerling produced twelve of these types of forks from a to g#, with the a fork corresponding to 432 svps, rather than 440, thereby producing four beats per second with a.[69] Every other pulsating fork was 4 beats (or 8 svps) lower than its corresponding semitone.[70] Indeed so much more accurate was the piano tuned with pulsating, rather than unison, forks (save those forks used to tune the remaining octaves of the piano after all of the semitones of an octave had been ascertained) that Scheibler wrote to the German violinist, conductor, and composer Louis Spohr in early 1832 that a piano tuned with pulsating forks was "infinitely, infinitely purer than the best tuner had ever achieved."[71]

The tuner, armed with both pulsating and unison forks and a metronome, could now tune the piano with unprecedented reliability using Scheibler's technique. "Thus he [Scheibler] succeeded in providing for the musical world a measure of sound which, in regard to accuracy, surpasses the most rigorous demands that can be made upon it, whilst its application is as easy as it is free from a possibility of a mistake. According to such a fixed scale only, as contained in Scheibler's 12 forks, can an instrument be tuned with any hope of success, the perfection of intonation is such as cannot be obtained by the *finest musical ear,* although with the assistance of Scheibler's apparatus no musical ear whatsoever is necessary to arrive at it."[72] The forks were placed in a hole drilled in a wooden box, which served as a resonance board, and were struck by a mallet. The forks had wooden handles, so that they would not change their pitch due to the warmth transferred from the tuner's hand. The tuner was instructed to screw the a fork in the board, and to place the board on or near the piano. He was then simultaneously to strike the fork with the mallet and to depress the a key (with one of the two strings dampened), preferably with equal force. The string was then tuned until no beats were detected by the ear (if unison forks were used, or 4 beats per second if pulsating forks were used). The tuner next tuned the second string of the a key, dampening the first, in the same manner. The two strings were then sounded simultaneously. If the tuner detected beats, then one of the strings had become flat, and the procedure needed to be repeated. If, however, no beats were noticed, both strings were in tune with 440 svps. The tuner now moved on to b♭, above a and tuned that note in the same fashion, and continued this procedure for all the semitones of a to a′, completing the octave. Finally, the remaining keys were tuned by octaves, ensuring that no beats were produced.

The experienced piano tuners did not need all twelve tuning forks (for the tonic and the eleven semitones). He normally used only six: b♭, c#, d#, f, g and a.[73] After adjusting these keys (preferably with pulsating forks), he would tune the lower octaves of d#, f, g, and a, and the upper octaves of b♭, c#, and d#. The remaining keys (b, c, d, e, f#, and g#) could be tuned by comparing them with their ascending and descending fifths, ensuring that all ascending fifths were flattened, while descending fifths were increased for equal temperament. Scheibler suggested using the method of combination tones to tune the musical intervals to increase the precision of the measurement.[74] A combination tone is a pitch that is generated when two pitches are sounded together. It is lower than the original two pitches. According to Scheibler, the first combination tone sounded at a frequency equivalent to the difference of the original two pitches. The second combination tone[75] sounded at a frequency equal to the difference between the lower of the two initial pitches minus the pitch of the first combination tone. For example, two pitches, say 659 and 440 svps, sounding together will produce the first combination tone at 659 – 440 = 219 svps, and the second combination tone will sound at 440 – 219 = 221 svps. Since the difference in simple vibrations per second of the two combination tones is two, they will produce one beat per second. The interval between 659 and 440 actually represents an ascending fifth from a to e. The tuner would have already tuned a, and would tune the e key until the combination tones of both pitches made one beat per second with each other. The d key, fourth above a with 587 dvps in equal temperament, will share its first combination tone with a at 587 – 440 = 147 svps, and its second at 440 – 147 = 293, and its third at 293 – 47 = 146. The second and third combination tones differ by only one svps (or one beat every two seconds). Combination tones were particularly useful in tuning organs. The organ tuner needed only to choose the a note, which was called the *Hülfston* ("assisting tone") that produced measurable combination tones with one of the keys tuned with a tuning fork.[76]

Tuning an organ had always been a more difficult task than tuning a piano, but Scheibler nevertheless provided organ tuners with a much easier and more accurate method to tune their instruments in equal temperament as well. "This process is easy and most expeditious, and has like that of tuning pianofortes the great advantage, that it requires no knowledge of music whatsoever, not even a cultivated musical ear. All the tuner has to do, is to look to his table [provided by Scheibler] and watch the beats."[77] Only tuning forks were necessary for determining the organ's pitch for a′. The organ tuner needed three forks: fork 1 needed to resonate at 880 svps, fork 2 at 860, and fork 3 at 840. The tuner drew a soft stop[78] (one in unison with the tuning stop, open diapason 8 feet) and stepped back from the pipe until its sound and the sound of the tuning fork were of equal volume. If the a′ was in tune with one of the three forks, the organ tuner was instructed to use the appropriate tuning chart provided by Scheibler.[79] If a′ did not match any of the three forks, then some adjustment was required. If the organ's a′ was higher than fork 3, but lower than fork 2, then the turner needed to ascertain the number of beats produced by the pipe with the forks per second, with the metronome adjusted to 60, or one oscillation per second. After determining the pitch based on the number of beats, the corresponding table was to be consulted. If the organ's a′ was higher than fork 1, then Scheibler recommended replacing a′ with g#, one semitone lower, and each note of the tuning table would apply to the semitone below.[80]

After determining the pitch of the organ's a′ and selecting the appropriate tuning table, the tuner needed to select two organ stops, one principal and one auxiliary, both being an open diapason pipe of eight feet. The auxiliary stop had the sole function of providing a convenient number of beats with the desired note of the principal stop. Scheibler's technique necessitated that the turner (1) stand as close to the pipes as was possible during the tuning, (2) ensure that the flow of air through the pipes was equal and constant, (3) tune each pipe pure (i.e., with no beats), and then raise or lower the pitch as the tuning table required, and (4) situate the metronome's regulator on the required line as precisely as possible.[81] To change the pitch of an organ pipe, Scheibler had tuning rings made. The previous method of widening or compressing the top portion of an open metal pipe in order to raise or lower its pitch had been clumsy and often inaccurate. Scheibler's tuning rings, which were composed of wood from a walnut tree, were permanently fixed to the open end of the pipes. By turning a brass peg connected to the ring, the diameter of the pipe's aperture would be changed, thereby altering the pipe's pitch.[82]

Scheibler's tuning technique was an attempt to replace the well-trained, organic ear with mechanical and mathematical precision. By providing an algorithmic procedure for tuning keyboard instruments, he hoped to render the mystical skills of the artisto-genius tuners superfluous.

Musicians' Reception of Scheibler's Technique

Scheibler's tuning technique, based on the counting of beats occurring between pitches and their generated combination tones, offered unprecedented precision for tuning in equal temperament. Musicians with the revised metronome in hand now had control over tuning their own instruments. The earliest response we have to Scheibler's technique was published by the Krefeld physician, Johann Joseph Loehr. The editor of Loehr's work called it an "incomparably beautiful discovery in the area of acoustics, namely the determination of the pitch of tones."[83] It had not been put to use, he continued, because tuners were unfamiliar with the essence and character of equal temperament, and most did not possess enough knowledge of the mathematics involved to apply Scheibler's technique. Loehr accordingly set out to explain at a basic level the relationship between mathematical and equal temperaments as well as to analyze the most important theoretical aspects of Scheibler's discovery.[84] Loehr proceeded to discuss the limitations of tuning pianos and organs in mathematical temperament, and the drawbacks of tuning in meantone temperament, and concluded that "not only is equal temperament necessary, but particularly with those instruments, whose pitches cannot be changed, . . . it is solely applicable."[85] Loehr, unlike Scheibler, carefully and clearly discussed Scheibler's tuning technique for both pianos and organs, and concluded his pamphlet by asserting, "[w]hen Scheibler claims that it is not possible to tune a pitch flawlessly in unison one directly after the other, and even less so to continue the desired tuning of other intervals in this manner, he is indeed correct. Everyone who undertakes the experiment will clearly be convinced and will accept Scheibler's standard of measure. How wonderful does this new invention once again appear to us from this vantage point! This technique, itself perfect in the utmost degree, hitherto unachieved, indeed to a level of perfection never thought possible."[86]

Six years after the publication of Loehr's essay on Scheibler's tuning method, Johann Gottlob Töpfer, who, recall from the previous chapter, was one of Germany's leading organ builders and organists of the period, penned his own essay aiming to convince piano and organ tuners to draw upon Scheibler's system. By 1842, the publisher of Töpfer's work pointed out, Scheibler's technique had generally become known throughout the German territories, but unfortunately, it still was not used as much as its merit warranted, particularly for organs.[87] Apparently, the tuners avoided the process for two reasons. First, it necessitated the purchase of a modified metronome and tuning forks, which were often beyond the modest means of a tuner, and the implementation of these devices proved quite wearisome. Second, tuning by means of *Hülfstöne* was not only a waste of time, but also deleterious to the principal pipes of the organ. Töpfer hoped to alleviate these two setbacks by offering his own modified version of Scheibler's technique, a method that rendered tuning forks and the modified metronome superfluous and tuned an organ in equal temperament without *Hülfstöne*. Töpfer's modification was put to the test by his colleague, the organ builder Johann Friedrich Schulze, who tuned organs in

Thuringia based on Töpfer's improvements on Scheibler's technique. The new technique was a success: Schulze claimed to be able to tune organs much more quickly and accurately than ever before.[88]

As Töpfer correctly noted, Scheibler's tuning method could not be reliably implemented unless the pitch of at least one tone was known precisely. One could determine such a pitch rather easily with a tuning fork, but Töpfer wished to free tuners from the expense of purchasing such devices. He, therefore, suggested a technique based on the tuning of three major thirds.[89]

Although an instrument was still needed to provide small intervals of time, Töpfer argued that it did not need to be an expensive, modified metronome. He produced a table with the lengths of an oscillating pendulum, which could reproduce the metronome's precision.[90] Töpfer then discussed an alternative to using *Hülfstöne* in order to tune the semitones of an octave on an organ. Since major thirds and major sixths generate more beats, which can be counted with greater certainty than those produced by *Hülfstöne*, than fourths or fifths produce, Töpfer recommended employing them, but an octave lower than usual, or with A = 220 svps.[91]

The organ and piano builder Otto Kraushaar suggested another method for tuning keyboard instruments to equal temperament, also drawing upon Scheibler's technique but without using so many tuning forks.[92] He accepted Scheibler's recommendation to use combination tones; however, he only used ones that could be heard between chords, rather than relying upon the ratio of vibrations and the number of the degree of combination tones—both of which Scheibler had utilized. In addition, Kraushaar used only one tuning fork to tune the instrument. He tuned the note corresponding to the pitch of the tuning fork and then tuned that fundamental tone with its octave. He then proceeded to tune using the circle of fifths, or a circle of all twelve pitches arranged such that any pair of adjacent pitches represents an interval of a perfect fifth.[93]

The limited success that Scheibler's technique did achieve owed much to Töpfer's and Loehr's elucidations and elaborations, although other early support was forthcoming. Not surprisingly, the first verdicts were rendered in Scheibler's hometown of Krefeld. J. N. Wolff, music director in Krefeld and Scheibler's music instructor, argued that a piano tuned with Scheibler's technique in equal temperament "was so perfectly pure and equal, even by the strictest tests, that nothing else could be wished for."[94] He added that the tuning method was quick and reliable. F. A. Wortmann, another music instructor in Krefeld, lauded Scheibler's precision measurement. He tuned his own piano using Scheibler's method, and claimed that it possessed a purity that it never had before. He was also impressed with how quickly and effortlessly he tuned the semitones of the scale from a to a′, in fifteen to twenty minutes. He concluded his letter of support: "It is to be wished that this new method of tuning, whose great advantages I have learned through experience, be generally put to use."[95] The Cassel pianist Hauptmann reported his joy of owning Kämmerling's tuning forks. He tuned his own piano himself, and claimed that after three weeks of rigorous playing it was still "purer" and still sounded better than it had ever sounded after being in the hands of the best piano tuners.[96]

A small, yet significant, fraction of the musical community throughout Germany quickly took an interest in Scheibler's efforts. His works were reviewed in Germany's leading musical periodicals, such as the *Allgemeine musikalische Zeitung* and *Caecilia* as well as prestigious scientific journals such as Poggendorff's *Annalen der Physik und Chemie*. An article in the *Allgemeine musikalische Zeitung* recommended Scheibler's technique for tuning organs to equal temperament.[97] The musician C. W. Finck favorably reviewed Scheibler's *Ueber mathematische Stimmung, Temperaturen und Orgelstimmung nach Vibrations-Differenzen oder Stöße* (On Mathematical Tuning, Temperament and Organ Tuning according to Differences in Vibrations or Beats)[98] as well as J. J. Loehr's pamphlet, which rendered Scheibler's technique intelligible to piano and organ tuners.[99]

The composer and pianist Ignaz Moscheles read of the new method in the London newspapers. Having met Wortmann, who had traveled to London, Moscheles commented on Scheibler's tuning technique, "The method of tuning is so reliable and flawless by means of mathematical calculations that the finest ear can wish for nothing more. I consider it a duty to bring this man to the attention of the musical public."[100] Moscheles succeeded in bringing Scheibler's method to the attention of London's musical elite, including the director of the Royal Academy of Music, the director of the Royal Opera, and the director of the Italian Opera.[101]

The renowned composer Sigismund Ritter Neukomm had also heard of the new technique while in England. Upon his return to Germany, he played on both a piano and organ tuned according to Scheibler's requirements. His praise was unreserved.

One tuned an organ and a piano under my watchful eye according to this theory, and you would be amazed to see how the beats are regulated with such mathematical precision based on the metronome's pendulum. It is so easy to tune with this procedure that one can easily do it by oneself, because one proceeds according to simple and definite laws, and nothing depends upon the immediate disposition of the tuner. The result is that one can tune more quickly than by using the ear alone, and one is always sure what needs to be done. Only on such an instrument can one modulate and improvise with perfect freedom: this is not the case with tuning by the ear alone.[102]

Neukomm's remarks brilliantly summarized the major benefits of Scheibler's invention. First, the performer no longer need fear that the capricious tuners would mistune his or her instrument. Scheibler's technique, much like a machine in his factory, was flawless and efficient. Second, with meantone temperament, the further away from C major one deviated (i.e., the greater the number of sharps and flats in the signature), the more dissonant the intervals sounded. Modulation, or the shifting of keys throughout a piece, was rendered much less difficult on an equally tempered piano or organ. One of Europe's leading composers of the period was now one of Scheibler's leading supporters. Other musicians and composers followed suit, including Cherubini, the violinist and singer Eduard Rieß, and Louis Spohr.[103] The Parisian piano tuner Henri Pape, the first to put felt on piano hammers, paid a tuner 3,000 francs a year to tune his piano, using only six tuning forks, according to Scheibler's method.[104]

By far Scheibler's greatest support came from the virtuoso violinst, conductor, and Romantic composer, Louis Spohr. From 10 November 1826 until 30 May 1836, the two corresponded.[105] Scheibler's second letter, written in February 1831, more than four years after his initial letter, discussed his tuning forks: "I think I can make a product [*Handlungsartikel*], which will be found in every boutique where light snuffers and knives are sold."[106] Scheibler's entrepreneurial interests are manifestly evident here. His desire to include his tuning forks in such shops also illustrates the status of high-quality music among German bourgeois households in the 1830s. Nevertheless, one year later, on 3 February 1832, Scheibler complained to Spohr about the relative obscurity of his musical invention: "Neither in Leipzig nor in Berlin nor Paris have I been able to break through. It appears that my assurances are taken as hypotheses, and I am considered a very distinguished scholar of acoustics, and that is it."[107] Scheibler then proceeded to detail his first tonometer and explain to Spohr how he employed both unison and pulsating forks to tune a piano so precisely that no tuners had even come close to his achievement. Indeed he even claimed that virtuosi of string instruments could not match the purity of the protracted notes of his piano.[108]

Scheibler penned his fourth letter to Spohr on 3 May 1833. He briefly spoke of how *Naturwissenschaftler* were becoming interested in his work (as discussed in greater detail below) and boasted about how the accuracy of his tuning technique had continued to improve: "In this manner, by means of a pendulum, I can prove that my greatest error in one chord is 1/17,000, and that my second largest error is 1/50,000."[109] In his next letter, dated 20 June 1835, Scheibler once again complained how physicists were taking notice of his work much more so than musicians: "The physicists in Stuttgard have appreciated the total correctness and absolute necessity of my laws, but Euterpe [the Muse of Music] does not seem to care."[110] Six weeks later, Scheibler sent Spohr two tuning forks for tuning an organ with the promise that six more for tuning the piano were on their way.[111] And we are told that he had heard that Spohr actually "feared that the ear would become too accustomed to the purity of instruments tuned with my method that no one will want to hear orchestral music any more."[112] On 30 May 1836, Scheibler wrote his last letter to Spohr. Once again he reported how his work had been reviewed and recognized without opposition, thanks to the Versammlung deutscher Naturforscher und Aerzte meetings in Stuttgart and, in the following year, Bonn.[113]

Although German organs were all generally being tuned in equal temperament by the 1880s, apparently Scheibler's tuning method was now barely used. In 1886 the Lucerne organ builder Friedrich Haas argued, "I cannot understand why Scheibler's reliable tuning method is not being used, considering the present advances and progress of organ builders."[114] After discussing how Neukomm had, back in the 1830s, convinced organ builders in the Krefeld and Düsseldorf areas to tune in equal temperament using Scheibler's method, Haas added, "I am convinced that when one carefully employs Scheibler's tuning method, one will obtain a purity with total certainty, which can never be achieved by the ear alone. It is hoped that all organ inspectors . . . demand that new organs be tuned by Scheibler's method."[115] Haas then detailed a revised procedure of Scheibler's technique, using a′ as 870 svps—the new French *diapason normal* (or standard tuning fork)—rather than 880 svps.[116] Haas, too, employed

Hülfstöne in the tuning of organs.[117] According to the late-nineteenth-century British acoustician Alexander J. Ellis, Scheibler's "counting seems to have been wonderfully perfect."[118] Indeed, Ellis, nearly a half century after Scheibler's death, used the silk manufacturer's 56-fork tonometer for the measurements in his classic 1880 paper on the history of pitch.[119] When discussing the techniques of others in determining the pitch of organ pipes, Ellis declared, "in the first rank of these stands Scheibler himself," above the later work of French physicists Lissajous and Delezenne.[120] In the end, however, Scheibler's tuning method was never universally adopted because it was too complicated (even after simplification) and deemed unnecessary for tuners.

The Scientific Response: Combination Tones

As the letters from Scheibler to Spohr strongly suggest, German *Naturwissenschaftler,* particularly physicists, greeted Scheibler's tonometer and tuning technique with enthusiasm. On 22 September 1832, Scheibler was invited to present his tonometer and relevant findings to a meeting of the Physikalischer Verein (Physics Society) in Frankfurt. The physicists were particularly interested in the use of vibrations of two sounding tones (and their resulting beats) as a new measurement of pitch and the application of that measurement in tuning keyboard instruments.[121] Present at that meeting was Georg Wilhelm Muncke, professor of physics at Heidelberg, who spoke of Scheibler's invention at both the Royal Society of London in 1834 and at the Royal Academy of Sciences at St. Petersburg in the following year.[122] Muncke also was the first to publicize Scheibler's technique in J. C. Poggendorff's *Annalen der Physik und Chemie.*[123] Muncke carried out experiments with Scheibler, using a modified Mälzel metronome calibrated with Muncke's astronomical clock. The greatest error detected using Scheibler's technique was a mere 0.13 vibrations, rendering it more exact than Cagniard de la Tour's siren, which, according to Muncke, "had been discarded for good reason due to its unreliability."[124] In an addendum, however, Muncke admitted that Pierre Louis Dulong convinced him that the number of a pitch's vibrations could be determined with great precision by using a small organ and a siren. Indeed, he conceded that the resulting tones produced by the tuning forks are slightly sharper than those produced by organ pipes.[125]

Of particular interest to physicists was Scheibler's work on using combination tones to tune keyboard instruments. Combination tones had been first detected by the renowned German organist and music theoretician Georg Andreas Sorge in his *Anweisung der Stimmung der Orgelwerke und des Claviers* (Instruction on the Tuning of Organs and Pianos) of 1744.[126] At a public session of the Sociéte Royale des Sciences of Montepellier in December 1751, J.-B. Romieu announced that when tuning an organ he could detect a third, rather low pitch when two organ pipes sounded in unison.[127] A more thorough investigation of combination tones, however, was carried out by the renowned Italian violin virtuoso, Giuseppe Tartini. Tartini encouraged his students to improve their intonation by listening to these tones, which even today are sometimes referred to as Tartini tones.[128] He neither proffered a physical explanation for these phenomena, nor he did correctly determine their intervals.[129]

Starting in the early nineteenth century, combination tones migrated from the realm of music into the sphere of physics, challenging experimental natural philosophers to provide an explanation for this interesting phenomenon. During the first years of the century, Thomas Young offered a physical explanation based on his theory of coalescence, arguing that combination tones were "beat tones." Hence, beats and combination tones were actually the same physical phenomenon.[130] He reckoned that vibrations of different sounds are able to cross each other in all directions without affecting the particles of air which transmit them. Surely, he argued, the total displacement of the air particles is equal to the algebraic sum of the displacement of the particles of each of the waves acting alone.[131] Indeed, with such an explanation, he was able to point out that multiple combination tones are possible when two pitches are generated simultaneously. Two pitches possessing the same amplitude but different frequencies "communicate the idea of a continual sound; and this is the fundamental harmonic described by Tartini."[132] For Young, the musical phenomenon of combination tones generated by two pitches sounding together were proof positive of his theory of coalescence.[133] They were "a product of combination which possesses properties totally different from those of its constituent parts; its pitch is always much lower, its quality of tone is perfectly singular, and the hearer is sometimes almost inclined to imagine its direction to be different from that of the original sounds; a circumstance which he sometimes expresses by saying, that the sound 'rings in his ears.'"[134]

In 1827 François Ane Alexandre, Baron Blein, an officer in the French army corps of engineers, published a treatise on combination tones.[135] He took two stretched strings, each tuned to middle C (or c′ at 256 vps), and methodically raised one of the strings in twenty small increments, until that string was an octave higher than the other.[136] Blein noted that the resulting combination tones changed slightly as the second string was tightened to a higher pitch. The pitch of the combination tone then increased until it was at the same pitch as the pitch of the unaltered string. When the second string was tuned to the midpoint of the octave, Blein observed two combination tones, only one of which followed the aforementioned pitch pattern.[137]

Alexander von Humboldt brought Blein's work to the attention of Wilhelm Weber, most likely while both were at the Berlin meeting of the Versammlung deutscher Naturforscher und Aerzte in 1828. According to Weber, combination tones were rapid beats. Drawing upon Young's theory of coalescence, Weber claimed that these beats were a result of the rapid constructive interference of sound waves.[138] He argued that Tartini tones arise in the following manner. Two wave trains of, for example, four and five waves arrive at the eardrum at approximately the same time. The trains are followed again by two more trains, again of four and five waves, which strike the eardrum nearly successively, and so on. A combination tone will be heard that is two octaves lower than the lower of the two original pitches. If this process is repeated with wave trains of five and six waves, the Tartini tone will be a major third and two octaves lower than the lower of the two pitches. If two wave trains hit the eardrum in succession, again nearly simultaneously, and if the frequencies are equidistant between 5 : 6 and 4 : 5, then it would be possible to hear two Tartini tones, one two octaves lower than the lower of the original two pitches, and the other a major third and two octaves lower than that lower pitch.[139] Weber then provided an algebraic expression for the relationship between the wave trains and Tartini tones.

> If "a" waves of a wave train reach the eardrum in the same time as "b" waves from another wave train, one can deduce from this relationship a : b a break in the chain: $a/b = 1/[(a + 1/b + 1/\gamma)]$, etc., such that the relationship a : b is little different from $b/(ab + 1)$, $b\gamma/(ab\gamma + a + \gamma)$, etc. The resulting Tartini tone can behave in this case as the lower of the two sounding pitches, either as $1 : b$ or as $1 : b\gamma + 1$, etc. Which one of these two Tartini tones will be received by the ear depends on how many beats

need to follow on the eardrum and on whether the ear is sensitive to such a low pitch. The smallest of the given relationships: b/(ab + 1), bγ/(abγ + a + γ), etc., which differs so little from the true relationship a : b—such that so many beats in the necessary time would reveal no interfering deviation, will determine the Tartini tone, as if it itself were the true relationship, with the only difference that the Tartini tone will be barely audible.[140]

Weber concluded his short article by chiding Blein for being unaware of the mode of origin of these Tartini tones, and hence for being unable to provide the physical law his observations follow.[141]

Weber's paper was not as important for its content as it was for the response it sparked from the Swedish physicist living in Finland, Gustav Gabriel Hällström. In 1819 Hällström had written his dissertation, "De tonis combinationis," deciding to translate an abridged version of it in German, which was published in *Annalen der Physik und Chemie* in 1832.[142] Although Hällström concurred with Weber and Young that combination tones arose from positive interference of the sound waves, thereby championing the beat-tone theory of combination tones, he demonstrated experimentally that combination tones occurred at frequencies not predicted by earlier theories. He started his investigations by arguing that beats commence at a slow rate when two strings sound in unison, and then gradually increase as the pitch of one string is sharpened.[143] If x is the beat frequency, then $1/x$ is the time between beats, and the pitches of two tones r and s will produce r/x and s/x vibrations. Hällström proceeded to argue that since the beat is a result of the constructive interference (as we now call it) of waves, one pitch had to be one vibration greater than the lower pitch in the time it took to generate the beat, or mathematically expressed: $r/x = s/x - 1$, or $x = s - r$.[144] Using that same formula, Hällström calculated the frequencies at which subsequent combination tones would appear.

Generating Tones	Combination Tones
r, s	$s - r$ (first order)
r, $s - r$	$2r - s$ (second order)
s, $s - r$	r (second order)
$2r - s$, r	$r - s$ (third order)
$2r - s$, s	$2r - 2s$ (third order)
etc.	etc.[145]

As Turner has argued, Hällström was "the foremost champion of the 'beat theory' of combination tones."[146]

August Roeber, instructor of mathematics and physics in Krefeld, detailed Scheibler's work in *Annalen der Physik und Chemie* and in his article on acoustics in *Repertorium der Physik*.[147] Roeber praised Scheibler's precision measurement as well as his idea to use the number of beats to tune instruments with an unprecedented degree of exactness and certainty.[148] And he noted that Scheibler's research confirmed Hällström's "theoretical experiments" and derived "law."[149] Although he confessed that he was originally unaware of Young's theory of coalescence until told of it by Poggendorff, he, too, was a staunch supported of the beat theory of combination tones.[150]

Roeber wished to diagram the vibrations of two pitches, in a ratio of 2:1 (i.e., pitches representing an octave), which resulted in the formation of a combination tone. The graph depicts a "system" of two sinusoidal frequencies, and the resulting sinusoidal combination tone, whose maximum recurs with a frequency equivalent to the difference of the two original pitches. Referring to figure 6.3, the point of compression of wave 1 (*abc*) is the point of rarefaction of wave 2 (*a'b'c'*). The line *pq* is the abscissas axis of those wave trains; *ap* represents the maximum point of compression, and *a'p* the maximum point of rarefaction. As these wave trains reach the ear, each pitch retains its original characteristic features, because the system of the one wave is not altered by crisscrossing with the other.[151] In addition to these two unaltered pitches, one can perceive a third pitch, contingent upon the reciprocal reinforcement and cancellation of the colliding compressions and rarefactions. The curve *pαBq* designates the resulting combination tone from the two original wave trains as well as the alternating successions of compressions and rarefactions. At *p* two maxima of two opposing simple vibrations meet. A weak compression then occurs, *pAα*, followed by a strong rarefaction *αBβ*. The compression reaches a

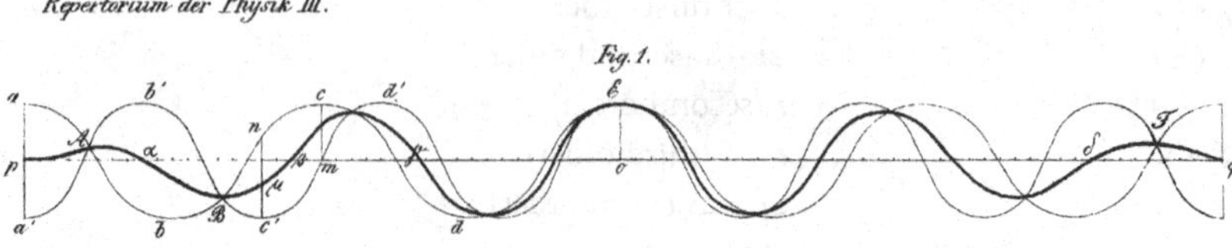

Figure 6.3
The graph of the formation of a combination tone. Source: Roeber 1839, vol. 3, table 1.

maximum at *Eo*, the point of strongest combined effect and the middle of the wave system. The combined effect of the two original pitches decreases throughout the second half of the system, resulting in ever weakening compressions and rarefactions, just as it had increased during the first half of the system. The system concludes with a weak combined compression of the combination tone *δFq*, and then starts over again.[152] Roeber concluded, "with regard to the combination tone according to this representation, it is likewise the regular return of one and the same impulse, which causes the generation of this tone, whose pitch is determined by the number of successive pulses in one second. It differentiates itself from the other tones in that this impulse is neither given by the appearance of a simple rarefaction, as is the case with sirens, . . . nor through the combination of a compression and rarefaction, as is usually the case, but rather by the union of a large number of dissimilar compressions and rarefactions into a large whole, which we have called the system."[153]

Roeber pointed out that Scheibler's work on the beats resulting from the coincidence of the compressions was "a convenient relationship of a rule for the calculation of beats," a rule governing "the actual method of the formation of two simple tones."[154] Critically, Roeber believed, as nearly all acousticians believed in 1839, "that the generation of [a combination] tone is occasioned merely through the regular repetition in the hearing of any kind of impulse, in connection with which there occurs in all cases the same dependence of the pitch of the tone upon the number of impulses occurring per second."[155]

Roeber's detailed article encouraged an intervention from the editor himself, Poggendorff.[156] Poggendorff was quick to point out that Scheibler had certainly not discovered a new physical phenomenon, as the notion of beats and combination tones had already been explained by Young's pattern of coalescence and superposition of sound waves earlier that century.[157] Poggendorff did, however, argue that Scheibler's experiments gave reason to discuss Young's theory once again. Poggendorff noted that although he agreed with Scheibler and Roeber's explanation of first-order combination tones based on Young's theory of interference, he was far less pleased with both their explanation of second-order combination tones and their inability to predict, in advance, which order of combination tones could be heard, and which could not. According to Young, the first-order combination tone results from the waves of both generated tones traveling in the same direction. The second combination tone must, therefore, arise from the first combination tone and the reflected waves of one of the two generated tones. Poggendorf argued that Scheibler and Roeber needed either to proffer experimental evidence of the creation of the second combination tone, or to show that this assumption of the explanation of beats and the simultaneous appearance of second-order combination tones, which Hällström used, was superfluous.[158] The editor wished that Scheibler's research would encourage physicists to explain why, if two tones did indeed coalesce, they preserved their own identities in the process.[159] Finally, Poggendorff expressed his hope that the interference of waves be explained mathematically, simply by determining the maxima of the sum of two or more sine and cosine terms.[160] If Y is the distance of an affected air particle from its point of equilibrium caused by several wave trains traveling in the same direction at time t, with t, t', t'' etc. the duration of an entire oscillation of the individual wave trains, and a, a', a'' etc. the maxima of deflection, which the particle of air would experience in the individual wave trains, then $Y = a \cos 2\pi\ [t - c/t + a' \cos 2\pi\ [t - c'/t'] + a'' \cos 2\pi\ [t - c''/t''] + \ldots$ [161]

By 1840 nearly all acousticians subscribed to the beat theory of combination tones: Hällström, August Seebeck, Heinrich Wilhelm Dove, Poggendorff, Roeber, and Wilhelm Weber.[162] One physicist stood alone in his opposition, Georg Simon Ohm. In 1839 Ohm published a short essay on combination tones, rejecting Hällström's account of the formation of higher-order combination tones.[163] Ohm asserted that for two pitches to give rise to first-order combination tones, their forms needed to be similar.[164] This claim, however, meant that the combination tone would not form higher-order combination tones with the original pitches.[165] There is one way to determine experimentally if Ohm was correct, a method he himself suggested. Since acoustical theory shows that the upper partials of oscillating rods are not harmonics of their fundamental, if he was right, then higher-order combination tones could not be generated by tuning forks. According to Hällström's beat theory, they could.[166]

In 1856 Hermann von Helmholtz picked up where Ohm left off. He realized that if Ohm was correct, then the beat theory of combination tones must be fallacious. He wished to devise a theory of combination tones that would be consistent with Ohm's law of combination tones. Ohm believed, yet never proved, that when a combination tone is detected, the pendular vibrations that the ear hears are not the product of the two original pitches, but are a result of the interaction of those pitches with the ear.[167] According to Helmholtz, contrary to the prevailing theory that superposition does not disturb the waves in any way, there must be some physical distortion of the original wave that results in the creation of new pendular vibrations, which the auditory nerve detects.[168] In his "Ueber Combinationstöne" (On Combination Tones) Helmholtz explicated his "transformation theory" of combination tones.[169] He went on to differentiate between combination tones, or what had been previously called first-order combination tones, and summation tones, which had been referred to as higher-order combination tones. Summation and combination tones taken together were called difference tones. Summation tones, which were much more difficult to detect than combination tones, and which were not predicted by the earlier beat theory, confirmed Helmholtz's transformation theory.[170] Summation tones originated at the junction between the eardrum and the short process of the malleus; they were a product of human hearing. Combination tones, on the other hand, arose as physical vibrations, which the cochlea nerve endings perceived. Hence, they have both a "subjective" and "objective" existence. "According to this interpretation, combination tones would have not merely a subjective existence, but rather also could exist objectively, even if for the most part only in the oscillating parts of the ear."[171] In order to prove the existence of combination tones independent of our perception of them, Helmholtz sprinkled grains of sand on a membrane suspended over his tuned resonators. The tones generated Chladni figures.[172] During the mid-1870s, however, the renowned Parisian scientific-instrument maker Rudolph Koenig revived the beat-tone theory by offering experimental evidence for which Helmholtz's transformation theory of combination tones could not account.[173] Debates over whether Helmholtz or Koenig was correct lasted well into the twentieth century.

In France, Scheibler's work was not taken up in the combination-tone debate. Rather, the French scientists praised his experimental acumen. In October of 1855 Auguste François Lecomte, a corresponding member of the Société Impériale des Sciences, de l'Agriculture et des Arts of Lille brought Scheibler's work to the attention of his French colleagues.[174] Lecomte echoed sentiments that Scheibler offered the musical world a procedure that could determine the number of vibrations of a musical sound with unprecendented accuracy. This "practical method" was "new, easy and precise for the tuning of instruments."[175] Both Felix Savart and Cagniard de la Tour commented on how unintelligible Scheibler's work was at first glance. And Aristide Cavaillé-Coll noted both the "advantages" of Scheibler's works as well as the "difficulty" in understanding them.[176] The mathematician and "musical archaeologist" A. J. H. Vincent had wanted to render Scheibler's ideas and methods intelligible to everyone back in 1849.[177] He had been assisted in this task by an artisan named Wolfel, who after two years of labor and reading Scheibler's works was able to construct a tonometer of twelve forks in order to generate beats of all the chords of the scale. Like Kämmerling, Wolfel was able to adjust the pitch of each fork by attaching a small slider (*curseur*) on the prong. In so doing, he was able to divide the scale into its twelve semitones.[178] Wolfel tuned a piano "perfectly" using Scheibler's technique and his own tuning forks.[179] Lecomte detailed Scheibler's use of pitches generated by a series of tuning forks, each tuned eight simple vibrations per second higher than the previous fork. He explained how Scheibler used beats and combination tones to determine precise musical intervals.[180] He concluded his memoire by praising Scheibler's "service to the art of music," which was both "ingenious and precise, justified by exact theory."[181]

In short, the method of tuning keyboard instruments in equal temperament was provided neither by a physicist nor by a professional musician, but by a silk and velvet manufacturer who was a musical dilettante. Acousticians greeted Scheibler's work on combination tones with enthusiasm, hoping to bolster the beat theory of the superposition of waves. Certainly the experimental accuracy afforded by his tonometer was unmatched. The best experimental evidence for combination tones and their composition in the 1830s, then, was generated by the Krefeld mechanician Kämmerling, far removed from the physics halls of Königsberg, Heidelberg, Göttingen, and Berlin. Combination tones were important to Scheibler not because of his knowledge of or interest in the beat theory, but because they could be used by tuners to tune keyboard instruments more precisely. Yet they successfully migrated into the world of the acoustician, who was naturally more interested in proffering physical accounts of wave superposition than in tuning in equal temperament.

This chapter began with a brief remark on Scheibler's dedication to deskilling. As mentioned earlier, he failed at deskilling both silk weavers and tuners. Tuners for their part were not, and today still are not, armed with twelve tuning forks and a metronome, as Scheibler required. Tunings are nowadays carried out by registered craftsmen of the Piano Technician Guild. They tune the middle section of the piano. Pianos of the same make, condition, and age are generally tuned to the same high standard; their pitches match each other very closely. The single mathematical basis for this procedure is now well known and is part and parcel of all tuners' education. Hence, pianists need not have their personal tuners by their side as they tour.[182] But such standardization of the distances between semitones came with a cost. As a number of musicians lamented, pure thirds and fifths were sacrificed. As William Polet remarked in his *The Philosophy of Music* of 1879: "The modern practice of tuning all organs to equal temperament has been a fearful detriment to their quality of tone. Under the old tuning an organ made harmonies and attractive music, which it was a pleasure to listen to. . . . Now, the harsh [i.e., as opposed to pure, or sweet] thirds, applied to the whole instrument indiscriminately, give it a cacophonous and repulsive effect."[183] One is reminded of De Morgan's claim of 1857 that pianos tuned to equal temperament were "insipid" and "uninteresting." There was a complete loss of key coloration; all keys now sounded alike, the uniqueness of key gone forever. They, in a sense, were the musical equivalent of the silk patterns produced in Scheibler's factory.

7 The Fetish of Precision II: Standardizing Music

The best compositions did not produce the desired effect because the appropriate tempo was missing. For ages one sought to counter this evil by seeking refuge in a remedy, which could only be found in mechanics.—Anon.[1]

Why can I not remember that the good, the beautiful, the true, the false, the ugly are not the same to everyone?—Hector Berlioz

Why is it that when I learned to play the cello, I tuned my A string to a tuning fork sounding at 440 vps and not say 435 vps, or 432 vps?[2] Was there something special or natural about 440? And why did I need to count meticulously the beats of a measure by following the tempo generated by a metronome? Did I not have the right to play with whatever speed I chose? Surely, metronome markings were merely suggestions, not ironclad rules to be followed blindly. In both cases, I always wanted to rebel against these mechanical constraints, much to the outrage of my teachers. This chapter helps to answer my childhood (childish?) questions by analyzing two historical episodes where mechanicians and acousticians played a critical role in the standardization of two musical characteristics: tempo and pitch. Both episodes illustrate the interplay and various negotiations among musicians, physicists, and instrument makers. Whereas previous chapters underscored the harmonious relationships among these three groups, this chapter sheds light on examples of tension and disputes. Who possessed the power to dictate the tempo of a piece or the concert pitch of an orchestra, city, or nation: composers, performers, instrument makers, or physicists?

The late eighteenth and nineteenth centuries, as many scholars have argued, were a period of scientific standardization throughout Europe and in the United States.[3] This chapter will focus on standardization of music in the German territories. Scientific standardization of commercial units were of crucial importance to early nineteenth-century Prussia. Although the Prussian government first commissioned members of Berlin's Akademie der Wissenschaften in 1816 to assist with the reform of weights and measures, as Olesko has demonstrated, it was not until the work of Friedrich Wilhelm Bessel with the mechanician Thomas Baumann in the 1830s that it became clear to the Prussian civil servants that physics could play a critical role in such standardization. Throughout that decade, German physicists expressed their anger at the lack of such standardization throughout the German territories. In 1834 the German Zollverein was established with a view to improve upon the accuracy of locally determined weights and measures and their conversions.[4] A year later, the Berlin physicist Hermann Wilhelm Dove bemoaned the lack of progress of standard determinations throughout the German territories. And in 1836 the Hanoverian ministry asked Carl Friedrich Gauss to determine a precise value for the Hanoverian pound in relation to measures used in other territories.[5]

Why was standardization such a desideratum for German physicists in the nineteenth century?[6] As Olesko has argued, by the 1860s the establishment of standards throughout the German territories was a "complex social, political and economic procedure."[7] Because each state had its own standard measures, German territories, lacking national uniformity, desperately needed conversions of these measures for trade. Reforms in weight and measures had been passed in Wuerttemberg in 1806, Bavaria in 1809, and Prussia in 1816. The need for standardized measures became painfully clear at the international exhibitions throughout the 1850s.[8] Cries for such measures echoed throughout the lands from such diverse occupations as foresters, engineers, machinists, architects, and representatives from industry, trade, and transportation. In 1860 a German commission comprising physicists, technicians, civil engineers, civil servants, and machinists met in Frankfurt to select a unitary system of weights and measures. Prussian delegates were conspicuous in their absence.[9]

The standardization of musical pitch and tempo shares with these other accounts the inextricable links among science, technology, economics, and politics. But I would suggest that the standardization discussed here is rather unique. There is clearly an aesthetic issue at stake, as the lead-in quotation by Berlioz suggests.[10] Ironically, as discussed below, he played a prominent role in establishing the French *diapason normal*. And the

Leipzig Kantian philosopher C. F. Michaelis and the Romantic music critic G. L. P. Sievers were well aware of the aesthetic ramifications of both ever-increasing pitches and ever-fluctuating tempi as well as the adoption of mechanical devices as guidelines in the organic art of musical performance. They were repulsed by the thought that machines would dictate something as uniquely human as musical taste.

Standardizing the Tempo

Calls for a standard measurement of beats date back to the late sixteenth century. In 1592 Zacconi, an Augustan monk and music theorist of the Venetian School, had argued in his *Prattica di musica* that the human pulse should serve as a basis for calibrating tempi, a view that was to last well into the eighteenth century.[11] Nearly three-quarters of a century later, Christopher Simpson penned his *Principles of Practical Musick*, suggesting that one count time with the now-familiar up-and-down motion of the hand, or by tapping one's foot.[12] In 1696 the English composer Henry Purcell recommended the use of a mechanical instrument, the pendulum clock, which had been invented by Christiaan Huygens in 1657, to keep time, tapping one's foot, or moving one's hand up and down, to the beat.[13]

Also in 1696, the French musician Étienne Loulie described a chronometer, a pendulum measuring approximately six feet, divided into seventy-two sections, in his *Éléments ou principes de musique, mis dans un nouvel ordre* (Elements or Principles of Music Newly Arranged).[14] A piece of wood was attached at the top, from which was suspended a ball of lead tied to a string. The length of the cord could be changed by means of a peg inserted into one of the holes on the graduated scale: the longer the string, the slower the beat.[15] Loulie's compatriot, the natural philosopher Joseph Sauveur, constructed a *métromètre*, a chronometer modeled after Loulie's device, which itself served as a prototype for Michel l'Affillard's chronometer described in his *Principes tres-faciles pour bien apprendre la musique* (Principles Made Easy for Teaching Music Well).[16] This work supplied detailed instructions on building and using Sauveur's chronometer for various musical tempi. In 1732 Louis-Léon Pajot, Chevalier, comte D'Ons-en-Bray announced his invention of a *métromètre,* which had the distinct advantages over previous chronometers of possessing a dial and pointer to indicate the speed.[17] Throughout the second half of the eighteenth century, pendulums had often been used to count beats throughout Britain and the Continent.[18] The renowned English maker of nautical clocks John Harrison invented a musical timekeeper that apparently kept perfect time, but no musician could afford to purchase it.[19]

What started as isolated calls for mechanical instruments to count more accurately the beats of a composition culminated in a deafening crescendo by the century's end. In response musicians and mechanicians attempted to construct portable chronometers. The major drawback of eighteenth-century chronometers was their size; pendulums for the slowest tempi were huge, measuring up to seven feet in length and being rather heavy and cumbersome. In 1782, a Parisian clockmaker named Duclos constructed a machine, which he called a *rythmomètre*, that could be used by the composer to note the exact tempo at which his piece should be played. Two years later the composer Jean-Baptiste Davaux published an article detailing a pendulum clock, "Sur un instrument ou pendule nouveu," in the *Journal encyclopédique ou universal*. The clockmaker A. L. Breguet constructed a pendulum using Davaux's plans.[20] Around that time the Parisian clockmakers and mechanicians Pelletier and Renaudin introduced their own chronometers. In 1790, Professor of Mathematics at the Royal Military Academy of Berlin Abel Bürja and Cantor Weiske of Meissen also attempted to market variations of the chronometer. But, we are told, their products did not succeed, perhaps because of the cost, perhaps because of the awkwardness in use.[21]

In 1798 the harpsichordist and mechanician August Heinrich Wenk published his *Beschreibung eines Chronometers oder musikalischen Taktmessers und seines vortheilhaften Gebrauchs für das musikliebende Publikum* (Description of a Chronometer or Musical Measure of the Beat and Its Advantageous Use for the Music-Adoring Public).[22] The Magdeburg organist and Cantor G. E. Stöckel argued that chronometers were necessary for beginners interested in learning to play music, as well as those *Liebhaber* ("amateurs") who do not play professionally, but simply for their enjoyment.[23] In addition, Stöckel underscored the importance of a correct standard measure of time for compositions: "the advantage of an as yet unknown, absent, generally accepted standard for musical time, or a chronometer, according to which this time is precisely determined . . . will be clearly recognizable enough."[24] The chronometer was, for Stöckel, the ally of the composer, ensuring that his "spirit" would be preserved.[25] And he hoped that chronometrical markings might ensure a beat based on the universal standard of time, replacing the constant referral to vague terms, such as andante, allegro, and presto. "If this chronometer were to be accepted as a firm standard for musical time, one could make oneself understood by writing the correct tempo of already well known, and particularly important, musical pieces. One could say, for example, that this piece is performed in Paris with this tempo, in Vienna and in Berlin etc., which is until now not impossible, but will continue to be difficult, since a firm, determined standard of time has not been accepted."[26]

Stöckel recommended that his chronometer, measuring between four and four-and-a-half feet and resembling a clock, be set next to the piano in order to guide the performer most efficiently.[27] The time interval was regulated by a hand, which was moved between numbers etched on the face from 0 to 84. The pendulum was wound up, set into motion, and the time interval was signaled by a hammer striking a small bell. Composer and leading music critic Johann Friedrich Reichardt, music director, keyboard instructor, and composer Daniel Gottlob Türk, cantor, music director, and keyboard instructor A. E. Müller, and critic, writer, and music editor Johann Friedrich Rochlitz all added their support to Stöckel's contraption.[28] Stöckel argued that the chronometer was just the device to assist pupils "who by nature do not at first possess a sensitive and sure feeling of the beat [*Taktgefühl*] to sustain the most exact, rhythmical practice of the pieces."[29] The chronometer could provide mechanically what the performer lacked organically. In 1806 the piano instructor F. Guthmann penned an article describing his idea for a *Taktmesser*. He expressed his concerns about using a chronometer as a teaching tool that students would follow throughout the practicing of the entire piece, as this was anathema to "the spirit of true music."[30] He added, "[the] greatest artistes have also proven that they could not play following such a mechanical instrument. It repulsed their sensitivity, and they deviate from it involuntarily."[31] For Guthmann the proper use of the chronometer was for beginners to gain a sense of the tempi at the beginning of the piece. Unfortunately, the majority of chronometers were ill suited for this purpose, as they were too large to be moved from place to place. A small pocket watch, which could be set to a particular rate, whereby a bell was rung was also impractical, since the bell could be heard throughout the performance of a piece, thereby disrupting it. Guthmann suggested a small thermometer-like pocketwatch that could rest next to a musician's ear at the beginning of the performance, and he challenged clockmakers to construct such a device.[32]

The early nineteenth century, not coincidentally the period that witnessed the increase of individual interpretation and improvisation of the performers—particularly the virtuosi—gave rise to what some would see as the mechanical tyranny of precision. By the second decade of the nineteenth century, debates began to rage within German musical circles as to whether or not chronometers or timepieces should be used. Critics noted that the mathematical tempo corresponding to chronometer markings might simply be too mechanical. Precision just might not be all that beneficial. By not divulging the exact tempo markings for the piece, the composer allowed for individual interpretation. Mechanical control was diametrically opposed to the art of music—a theme echoed by a number of performers.

Perhaps the most informed piece on the subject was written by Gottfried Weber. Born in 1779, Weber had studied law at Göttingen, subsequently taking up a position in the Imperial Court of Law in Wetzlar. He later was a tribunal judge of Mainz until 1818, when he was elected court legal advisor of Darmstadt, where he was named to a legal commission in 1825 that drafted the civil laws of the region. His civil and legal accomplishments, however, belie his undying passion for music. He was an accomplished cellist and flautist, and learned much from his friend, the acclaimed composer Carl Maria von Weber. Gottfried, who founded *Caecilia*, an important journal for musicians and friends of music, authored various books and essays on a myriad of topics, including critiques of Mozart's compositions, a theory of tonality, the acoustics of wind instruments, the improvement of musical-instrument design, and the physiological properties of the human larynx.[33]

Weber set up his essay on the chronometer in the form of a dialogue between a composer and music director.[34] The composer wishes to convince the music director that chronometers and timepieces are necessary for the proper performance of a composition. The music director, sensing that he is losing his authority to interpret the piece, protests staunchly. After the music director argues speciously against using new instruments to determine the beat by drawing upon such classic authorities as Sébastien de Brossard's *Dictionnaire de musique*, the crux of the argument ensues. He first condemns the use of mechanical instruments that make noise while keeping the beat. The composer, however, wholeheartedly agrees. Then the music director openly disputes the use of a chronometer to replace his conducting duties. "Now, so you want to invent a new, expensive, mechanical work of art, through which an artificial arm shows the choir and orchestra on command all tempi, just as a live music director is accustomed to do with the baton? Yes, of course, then you have a director, which surpasses all living directors on rigidity, pedantry, lack of pity, and inability to lend a hand, which around the world assists no soloist, or covers no mistake by giving way, which takes no notice with a stringendo or ritardando, and which thrashes out a piece from beginning to end according to rule in an unchanging tempo. That is of use to you? Well, fine. So replace me with the machine and send me off to retire."[35] In a sense, the machine would be too perfect for the imperfect art of music, and those who perform it. Once the composer allays the music director's fear of replacement, he continues the conversation: "I was never of the opinion that a chronometer would be suitable for directing music. No lifeless or insensitive machine would ever be suitable for that."[36] For the composer (who represents Weber's own views here), the chronometer had the unique function of serving as the interpreter between the composer and the performers of the piece. The composer adds: "Please do not misunderstand me: I do not want the chronometer to appear and be used in small or large musical performances: no, not at all! It should remain banned from performances and rehearsals as well. It is merely a standard of measure for the composer to be able to show exactly the performer, or the director of large musical pieces, in which tempo he wishes to have his work performed."[37] The composer merely needed to provide the music director and performers with the length of the pendulum, which produced the required time interval. The music director, by the dialogue's conclusion, is convinced of the chronometer's importance.

The early chronometers, as ingenious as they might have been, never proved to be nearly as successful as Johann Nepomuk Mälzel's metronome. He had decided at an early age that while the concept of a chronometer was indeed a very good one for musicians, the previous models needed vast improvements. In 1813 he announced that his chronometer (he called it both a *Chronometer* and a *Taktmesser*) in the *Allgemeine musikalische Zeitung*. Salieri and Beethoven immediately offered their support for Mälzel's device: "As a proof of the high appreciation of this invention by local composers, we may say that in addition to their promise to avail themselves of it as a welcome means of safeguarding their compositions everywhere, and for as long as they exist, from all mutilation due to the use of the wrong tempo, the Court Chapel-Master Salieri, on account of his deep veneration of the late musicians Gluck, Handel, Haydn and Mozart, has obliged himself to mark their masterpieces in accordance with *Mälzel's chronometer* to make sure that, in future their compositions will be performed in the spirit of the composers, with which Mr. Salieri is more familiar than anybody else."[38] Mälzel's earliest chronometer consisted of a small lever connected to a toothed wheel. This toothed wheel caused the lever to oscillate, which in turn resulted in a wooden anvil beating out the time in the form of regular beats. The duration of intervals between beats was determined by the position of a vertical rod, which was divided into degrees with the number of beats the lever makes in 60 seconds. Mälzel's chronometer could beat from 48 times a minute at the largo end of the range to 160 beats a minute at the presto end.[39]

A consummate businessman, Mälzel traveled from late 1812 throughout Europe advertising his chronometer, panharmonicon, and automatic trumpeter. Passing through Amsterdam in the summer of 1815 in route to London, he met up with the renowned Dutch mechanician, Diederich Nicolaus Winkel. Winkel, the son of a clockmaker from Lippstadt, showed Mälzel his own chronometer, which had received the highest praise from the Royal Dutch Institute of the Arts and Sciences earlier that year.[40] Winkel had made a critical improvement on the chronometer. Up until that point, as previously stated, the tremendous length of the pendulum required for the slowest tempi often thwarted its use as a standardizing instrument. Winkel, however, overcame the length problem by inventing a pendulum in which a brass bar BC pivots on the horizontal axis A (see figure 7.1). The weight G could slide up or down along a graduated scale between A and B. There was a counterweight G′ affixed to the end at C. The sliding weight G determined the speed at which the pendulum moved. This construction enjoyed a major advantage over other chronometers; it was small, since this double pendulum would oscillate as slowly as a singular one five to six times longer in length, a characteristic crucial for slower movements.[41] It was, therefore, portable.

Mälzel was impressed by Winkel's mechanism, which he immediately realized was superior to his own. He attempted to purchase the rights from

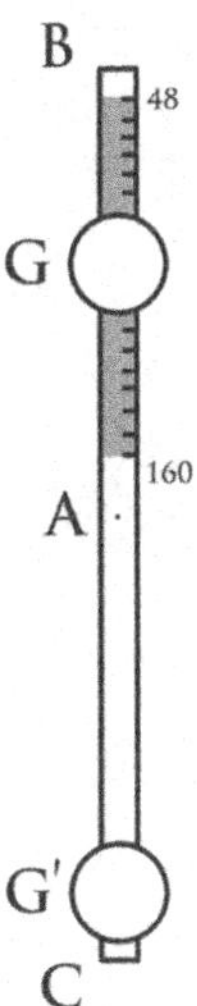

Figure 7.1
Winkel's mechanism for his chronometer based on Tiggelen 1985, p. 145.

the Dutchman. After Winkel refused, Mälzel submitted his revised version of his chronometer, which he now called a metronome, with Winkel's mechanism for patent review to the Institut de France on 15 December 1815 (see figure 7.2). He then continued on to the goal of his journeys, London, where he quickly took out another patent for his metronome.[42] The *Allgemeine musikalische Zeitung* felt obliged to publish a brief article questioning whether Mälzel's invention indeed belonged to Winkel. The author cited a report of the Royal Dutch Institute, dated 14 August 1815, claiming that Mälzel's metronome was clearly modeled on Winkel's device.[43] By taking patents out in London and Paris, Mälzel ensured that his metronome was manufactured and distributed throughout Europe with great efficiency. Although controversy filled the musical halls and periodicals, he rapidly became, in the eyes of contemporary major musicians and composers, the inventor of the metronome.[44]

Mälzel's new metronome possessed three advantages over the earlier chronometers (save of course Winkel's!). First, it was much smaller, portable, and easier to use. Second, it was based on the number of beats per minute, a measure of time (unlike a measure of length) that every country recognized, and hence, the best measure possible for standardization. Third, it was relatively inexpensive. Indeed, Mälzel himself made 200

Figure 7.2

Johann Nepomuk Mälzel's metronome, circa 1817. It was ordered by the Physikalische Sammlung des Königlichen Lyzeum in Regensburg from England. Reproduced with the kind permission of the Historische Instrumentensammlung der Universtität Regensburg.

metronomes and sent them free of charge to the leading musicians and composers of the period: a brilliant business tactic to ensure that his device was the only one used for counting beats at various tempi.[45]

The Italian composer Gaspare Spontini reported that he attended a meeting of composers in the spring of 1816, which included Antonio Salieri. These composers, all of whom were impressed by Mälzel's metronome, decided to pen a letter addressed to composers at large. This letter was their attempt to exercise greater control over the performance of their works, by convincing other composers of the importance of tempo standards. Other composers joined the ever-increasing lists of supporters of Mälzel's mechanical contraption. They included Muzio Clementi,

Giovanni Battista Viotti, Luigi Cherubini, J. N. Hummel, Conradin Kreutzer, Louis Spohr, and Ignaz Moscheles. More important, we are told that "the aforementioned masters have promised to note all their future compositions according to the scale of Mälzel's metronome in order to confront every argument that can arise over the speed of the movement."[46] Salieri continued: "Italy, Germany, France, and England will cheer with joy in one voice the inventor and producer of a machine, which is the true interpretor of the ideas and feelings of every composer, and by which the performance of their future works will not be distorted by an incorrect measure of time, which unfortunately happens not all that rarely now."[47]

Mälzel's metronome, however, did have its critics. Gottfried Weber critiqued Mälzel's earlier chronometer in the *Allgemeine musikalische Zeitung.* As we have seen, Weber was fundamentally interested in the standardization of musical tempi, proclaiming that "the composer may note [the tempo] at the beginning of a work in order to inform the entire world in the most definite way how quickly or slowly it should be performed."[48] For Weber the chronometer needed to fulfill two criteria to be successful: "it must be 1) a sure, definite and reliable standard measure every place at any time, and 2) generally intelligible and known everywhere."[49] Weber concluded that Mälzel's earlier chronometer possessed the first criterion but not the second. In Weber's view, one could only make sense of Mälzel's chronometric markings printed on the score of a composition if one had Mälzel's chronometer in hand, which, of course, was an ingenious move by the entrepreneurial Mälzel.

Weber certainly was not attacking the importance of standardizing the beat; quite the contrary. Rather, he objected to Mälzel's method of standardization, since it necessitated the purchase of his device. Weber proposed the pendulum for standardizing tempi. "This simplest instrument, from nature itself, is the pendulum. And already widely disseminated, the Rhineland inch is known in all civilized places. The composer, who announces his tempi by means of the beats of a pendulum in Rhineland inches, can now be understood everywhere."[50] Although Mälzel purposely (and, as it turns out, wisely) sacrificed the length of the pendulum based on regional dimensions for determining tempi for the universality of the minute and second, Weber wanted the Rhineland inch to become the universal standard of length, despite its obvious local flavor. Constructing a pendulum with a thread and a piece of lead was rather straightforward;

one did not need to rely on a commercial product: "every composer can easily mark his piece so that every reader understands this universal language by means of a similar, simple pedulum and can inform himself of the design of the piece."[51] In short, Weber argued that Mälzel's chronometer (and therefore his later version, the metronome) could never, *contra* the wishes of its inventor, become a universal standard. "Every attempt that adopts a tempo notation only according to the degrees of a certain machine will never find general use, and as a result will only serve to complicate and delay the notation of the inches of the pendulum's length, and perhaps in the end even foster confusion . . . by means of language, much like the Tower of Babel."[52] Weber hoped that Mälzel would use his "mechanical genius" to build a device that "lets us Germans just this once seriously start with our Rhineland inch standard."[53] His chauvinistic fascination with the Rhineland inch is noteworthy, since it informed much of his writings on music. He continued, "I wager that our pendulum with our divisions according to Rhineland measurements will be accepted, just as all nations accepted Italian musical expressions [of tempi] with Italian music."[54]

Apparently, Mälzel visited Weber on his way to London from Vienna, where the *Hofmechanikus* demonstrated his chronometer. Weber informs the reader that he was able to convince Mälzel "*to display the notation according to the inches of the pendulum,* and either keep the notation according to divisions of a minute or remove it altogether. He had not quite ruled out the last possibility."[55] However, in a letter to Breitkopf and Härtel Publishers dated 9 August 1816 while in London, Mälzel quickly dispelled such an impression for his newly invented metronome.[56] He sarcastically wrote in October 1817 after his metronome became widely available, that he was, according to Weber, "forgetting the land where music is, as it were, at home and where, as everyone knows, nearly every city has its own measure and weight. Trieste and Naples, Rome and Milan use their own, different shoes. It must be amusing to collect all of the different inches and measures, to calculate and search how the musicians, who are not always the greatest calculators, rack their brains in order to find out the standard from all the others and to calculate the duration of their pieces according to the yard."[57]

In 1817 Mälzel's metronome received strong support from an anonymous author in the *Allgemeine musikalische Zeitung* who expressed the hope that Mälzel's machine would free musicians and composers from "such multi-headed anarchy, which rules over theses areas."[58] The metronome was easy to use, as it did not presuppose any mathematical or scientific knowledge on the part of the musician. The *Quarterly Musical Magazine and Review* of 1821 perhaps best summed up the device, describing it as being

> one foot high, the decorated exterior of which renders it an ornamental piece of furniture. Its interior contains a simple mechanical apparatus, with a scale resembling that of a thermometer. According to the number on this scale the index is set to, the audible beats produced will be found to embrace the whole gradation of musical time, from the slowest *adagio* to the quickest *presto.*
>
> The metronomic scale is not borrowed from the measures of length peculiar to any one country, but is founded on the division of time into minutes. The minute being thus, as it were, the element of the metronomic scale, its divisions are thereby rendered intelligible and applicable in every country: an universal standard measure for musical time is thus obtained, and its correctness may be proved at all times by comparison with a stop-watch.[59]

By 1824 the metronome had become part of a musician's and composer's life in Britain: "[a] metronome must be referred to in order to ascertain the movement of the adagio or the allegro. . . ."[60]

The musical publishing company, Breitkopf and Härtel, was overwhelmed with joy. They sold Mälzel's metronome as well as most of the new scores being composed in Germany. Since they now strongly requested that composers put the metronomic markings on the scores, they guaranteed the market's fidelity. Beethoven's first works with Mälzel's metronome markings were his string quartets op. 18, 59, 74, and 95, and the piano sonata op. 106. He then added his markings to all the movements of his symphonies.[61] In 1818 and 1819 Carl Czerny affixed tempo markings to Beethoven's piano works. These markings, however, did not guarantee that future generations would play at these tempi. In 1837, for example, Moscheles too added metronome markings to Beethoven's piano works; these tempi were a bit faster than the originals. And in 1850 when Czerny published a new edition of Beethoven's piano pieces, he altered the tempi by one or two metronome markings faster than those used in the earlier edition.[62] Hence, the metronome did not necessarily safeguard the wishes of the composer, but rather reflected the tempi of the culture where and when the works were (re)published and performed. That said, Beethoven did do his best to force his will on the performers, writing to Ignaz von Mosel:

I am delighted to know that you share my opinion of those headings, inherited from times of musical barbarism, by which we describe the tempo of a movement. For example, what can be more absurd than Allegro, which once and for all means cheerful, and how far removed we are often from the true meaning of this description, so that the piece of music itself expresses the very opposite of the heading! As far as these four main connotations are concerned [allegro, andante, adagio, and presto], which, however, are far from being as right or true as the four main winds, we would do well to dispense with them. Those words which describe the character of the piece [e.g., *Con brio, Maestoso,* and *Scherzando*] are a very different matter: these we could not give up, as the tempo is really no more than the body, while these refer rather to the spirit of the piece. For myself, I have often thought of giving up these absurd terms Allegro, Andante, Adagio, Presto. Maelzel's metronome gives us an excellent opportunity to do so. I give you my word for it here, in my further compositions I shall not use those terms. . . . That one will cry out "tyrants," I do not doubt. . . . It would still be better than us being accused of feudalism.[63]

Beethoven realized the issue was quintessentially one of control. He was particularly angered by the virtuosi, who constantly took liberties with the tempi of his piano pieces.[64] A rather infamously whimsical creature, he withdrew his support once he and Mälzel became embroiled in a lawsuit over the ownership of the "Battle of Vittoria," calling the metronome a "dumb thing; one must feel the tempi."[65] This view was the major objection to using the metronome. Rhythm should somehow be organic; one needed to feel it from within, not copy it from an external source. Artificiality, in a sense, is failed genius. This sentiment was echoed by the Romantic music critic G. L. P. Sievers, who wrote in September 1819 that both the metronome and the chiroplast (discussed in the ensuing chapter) were "error[s] of the human spirit, because music cannot be improved by mechanical inventions."[66] He felt that the performer could use a different tempo than that intended by the composer; hence the metronome was merely a triviality (*Spielerei*).[67] And he thought it strange that Mozart and other renowned composers would compose their pieces without realizing that the tempi of their compositions would be altered by performers.[68] According to Sievers, a "philosophical" point was at issue. "A piece of music stops being a product of its composer in the moment when it is performed by an artiste, then this artiste is its second creator, or something much more: an animator of the work, transferring his living spirit into the dead form of the composition. . . ."[69]

Sievers's opinion notwithstanding, Mälzel continued to tinker, subsequently improving the design of his metronome in 1818 and 1832.[70] By

1836 an anonymous pamphlet could boast two important attributes of the device. First, it gave the composer control over the performance of his or her piece, since "whoever wishes to occupy oneself with music will find in recent times with most pieces of music, instead of the previous typical superscript of words [adagio, allegro, andante, etc.], the number of Mälzel's metronome, which specifies the tempo in which the author wants his piece to be performed."[71] Second, it helped beginners learn the correct "feeling of the beat" (*Taktgefühl*).[72] The mechanical metronome helped discipline both the youthful beginner as well as the capricious, unbridled performer learn an organic trait.

Standardizing Pitch

The standardization of pitch also enjoys an impressive history, with calls for it dating back to Michael Praetorius' *Syntagma musicum* of 1620. Just as the late eighteenth century witnessed the ever-increasing number of calls for composers to provide a standard measurement of beat, chaos apparently reigned in performance pitches as well. In 1776 R. F. Reichardt claimed that the pitch of the Berlin orchestra was relatively "low," particularly compared to the range of the Viennese pitch, between 430 and 440 vibrations per second.[73] In that same year, the music theoretician Friedrich Wilhelm Marpurg estimated Berlin's pitch to be 414.[74] During the ensuing decade Saxon instrument makers apparently made their wind instruments sharper than their Prussian colleagues.[75] In 1791 the flautist J. G. Tromlitz remarked that "the pitch of all places is not the same, but sometimes varies up to a semitone higher or lower. . . ."[76] The lowest Viennese tuning fork was still sharper than the Leipzig pitch by a half tone.[77]

Not only did different countries possess different pitches for a', orchestras in the same city sharing the same pitch was a rarity. For example, in Paris, the Grand Opera's pitch for a' ranged between 427 in 1811 to 434 in 1829, while the Italian Opera preferred 424 in 1823.[78] The Dresden Opera's pitch began to rise, albeit rather slowly in 1821, reaching 435 some time between 1825 and 1830, while the organ of the Roman Catholic Church in Dresden sounded at 415 in 1824.[79]

In addition to the myriad of concert pitches throughout Europe, a gradual sharpening of the various concert pitches was well underway by the 1820s. As early as 1802 Koch had noted in his seminal *Musikalisches*

Lexikon that *Kammertöne* ("chamber pitches") were gradually rising.[80] Generally, with the formation of large orchestras and musical scores written to accommodate such orchestras, tonal color catered to the higher instrumental range rather than the human voice, or the *vox humana*, as had been the case during the Middle Ages.[81] The rapid rise in pitch was perhaps greatest in the German territories. Whereas *Kammertöne* throughout Prussia and Saxony had been consistently lower than those in Vienna and Paris during the early years of the nineteenth century, by the 1830s German pitches were slightly higher than most French cities and Vienna.[82] According to Sievers, by 1817 "the pitch of the three great Parisian orchestras is more than a semitone higher than the highest in Germany and Italy. The purely instrumental groups, where no singing is involved . . . tune even higher."[83] By the early 1830s, however, J. H. Scheibler took several tuning forks from Berlin, Dresden, and Paris, with the French capital possessing the lowest pitch. The pitches of these forks varied as much as three-quarters of a full tone. A concerned reader lamented in 1835 in the *Allgemeine musikalische Zeitung* the "excessively high pitch" of a number of German orchestras.[84]

Mid-nineteenth-century scholars pointed to a more dramatic rise in pitch after the Congress of Vienna in 1816.[85] Tsar Alexander of Russia, who was also the colonel of a regiment during the Napoleonic Wars, presented his military band "Tsar Alexander" with a set of musical instruments, newly made by Austrian instrument maker Stephan Koch, for a performance at the Congress. In June of that year, Vienna's Musik-Verein invited his band to play Abbot Stadler's *Die Befreiung von Jerusalem*. However, the new instruments were so sharp that the collaboration was canceled.[86] The band's conductor, Hieronymus Payer, commented on the "brilliance of its [the band's] tones," arguing that the sharper pitches were well suited to the horns.[87] Four years later, members of a Viennese regimental band, the *Hoch- und Deutschmeister-Infanterie*, led by Freiherr von Ertmann, were also given a new set of instruments, which were tuned even sharper.[88] Horns in particular, and military musical instruments in general, tend to possess a favorable tonal color with higher pitches. By war's end, these military musicians returned to play at local Viennese theaters. As a result, the mean pitches in theaters began to rise as well. As Koch built instruments for numerous virtuosi, who desired higher-pitched instruments for their performances, many European cities witnessed the sharpening of pitch.[89] And the invention of a horn with valves and the valve trumpet in the ensuing years also contributed to the rise in performance pitch.[90] During the late 1810s, then, concert pitch began its inexorable rise that was to last well into the mid-1880s. Although the phenomenon originated in Vienna, Dresden rose slightly during the 1820s, but eventually succumbed to their neighbors to the southeast.[91] Leipzig's pitch, although sharpened over time, was much more consistent than Dresden's. Berlin pitches rose from approximately 430 from 1806 to 1814, to 437 in 1822, 440 in 1830, and 442 in 1834. By 1859 the average Berlin pitch for a′ had soared to 452.[92]

Variation in pitch did not merely alter the aesthetic effect of pieces; it had an anatomical consequence as well. Vocalists in particular were distressed by the lack of pitch standardization. Throughout the first four decades of the nineteenth century, the *Allgemeine musikalische Zeitung* often reported various diatribes against the rising pitch.[93] Sopranos in particular protested most ferociously, the prima donnas carrying along a tuning fork to performances, insisting (with limited success) host orchestras tune to them.[94] In 1814 the philosophers C. F. Michaelis[95] and J. G. Schicht[96] had pleaded for a common pitch for singers and orchestras across Europe. They argued that the necessity of singers to adjust to orchestral pitches, which differed as much as a full tone, was both exhausting and deleterious to their health.[97] The Parisian prima donna Mme. Branchu forced the opera to lower its pitch in the early 1820s, as she feared the high pitch would result in a premature loss of her voice.[98] Arias requiring the mastery of very high pitches as well as pieces containing falsettos were particularly difficult, as the singers touring Europe were forced to readjust their ears and vocal cords to the particular a′ which that orchestra chose. The problem became particularly acute by the mid-nineteenth century, when vocalists and instrumental virtuosi touring Europe reached a feverish pitch.

Chladni, Weber, and Scheibler, and the Standardization of Pitch

To bring order to this cacophony of pitches, musicians and musical-instrument makers joined forces with physicists. Marin Mersenne repeated Praetorius' call for a common pitch in his *L'harmonie universelle* of 1636. In 1700 the French physicist Joseph Sauveur suggested that the proper eight-foot organ pipe C sound at 132 svps (or 61 Hz), which corresponds to a tempered a′ of 410 vps.[99] The eighteenth and early nineteenth centuries witnessed in increase in appeals for the standardization of pitch from acousticians as well. In 1800 Chladni recommended 128 svps (64 Hz) for the eight-foot organ pipe C, which corresponds to a tempered a′ of 430 vps.[100] The C pitch was often used as the fundamental tone in tuning other pitches, as 128 could be divided by two for the lower octaves and multiplied by two for the higher ones. Granted this number was within the range of pitches for the eight-foot C, Chladni was more interested in the mathematical ease of determining the absolute number of vibrations.[101] He sought a universal standard pitch "since the nearly ubiquitous increase in pitch has done nothing for performance, it would be most advisable to accept a standard pitch where the number of vibrations in one second is a factor of two for every octave of C."[102] Such a pitch could be measured accurately and easily with a tonometer.[103] Despite Chladni's valiant efforts, his advice fell on deaf ears.

Wilhelm Weber, in addition to his work on improving musical instruments, also addressed the problem of a lack of standard pitch. His well-known commitment to the standardization of scientific phenomena (particularly in electrodynamics) dates back to his reed-pipe research.[104] And Weber certainly played a pivotal role in the standardization of physical constants. Hence, it should come as no surprise that one of Weber's greatest contributions to music was in the rigorous application of experimental physics to musical-instrument design, in particular with the view to standardize pitch. He was especially interested in the effects standardization could have on trade. On 11 April 1830 he wrote to the Prussian Ministry of Arts and Trade complaining of the lack of agreement on various standards relevant to commerce.[105]

In his article "Ueber die Construktion und den Gebrauch der Zungenpfeifen,"[106] Weber discussed the suitability of reed pipes for establishing a standard pitch (*Normalton*). He argued that "[b]ecause there is no instrument known, other than the reed pipe, which possesses the advantage that it keeps its pitch constant with the various strengths of its tone, I believe that reed pipes can be recommended as the most suitable apparatus for standard pitch [*Normalton*], in order to achieve an even greater precision [*Genauigkeit*], where the previously used tuning forks do not suffice."[107] Just as the determination of physical standards garnered him fame as Germany's leading experimental physicist of the nineteenth century, Weber was interested in establishing a standard pitch for musical instruments and voices upon which everyone could agree. He lamented,

"We are still far removed from possessing a generally accepted, scientifically established standard pitch. And it is now very difficult to come to an agreement on the tuning of instruments. It is worthy of such an extensive art as music, and of such an important branch of industry as instrument making is, to test the object in question from many sides."[108] Weber argued that tuning forks were not suited for all musical purposes, and that they were even less suitable for acoustical research.[109] He recommended his compensated reed pipes for tuning instruments. Although Cagniard de la Tour's 1819 invention of the siren was celebrated by French savants for providing a constant pitch, sirens were not immediately accepted by German acousticians as critical instruments to study pitch. As late as 1839, Bindseil's influential textbook *Die Akustik* argued that "only compensated reed pipes built according to W. Weber's laws can serve as an infallible means [for determining a standard pitch], protected through compensation not only against the influence of strength and weakness of blowing, but also against the influence of temperature on pitch."[110]

Reed pipes were not the only instruments that piqued Weber's curiosity. He was interested in the monochord as well, since it was simultaneously a physical and musical instrument. Indeed, this fascination—using the same mechanical device to investigate differing physical phenomena relevant to a plethora of disciplines—including music, informed Weber's work on the monochord.[111] Weber discussed his instrument, to which he referred as both a monochord and tonometer, which was built to his demanding specifications by J. August D. Oertling, the mechanician who built Weber's reed pipes.[112] The instrument was composed of one or more perpendicularly suspended iron strings between two pegs. By altering the length of the weights attached to the string(s), different pitches could be generated. Weber immediately commented on the various uses for which this instrument could be employed:

In the hands of an *experimental physicist,* this instrument can be used as a measure of the comparison of tones. And because the pitch of a tone is dependent on the velocity of the repeating vibrations or beats of the sounding body in the same periods, it can assist in measuring the duration of vibrations of the smallest intervals of time, which last from one beat to the next. This instrument can even measure smaller parcels of time than the *Tertienuhren* and all other instruments, which are at our disposal.

In the hands of *the instrument maker, the composer,* and *the practical musician,* the instrument is a standard measure, by which instruments can be tuned, and by which the composer can determine for all future occasions, for which voice his composition is intended. And by means of this instrument a general agreement can be established by all nations on the acceptance of an unchanging pitch.[113]

Clearly, Weber wished to achieve a standard pitch accepted equally by musicians, composers, and physicists.

Weber's monochord was composed of three elements (see figure 7.3). The first were the strings. He used metal strings (typically made of iron), since those strings, in contrast to strings of catgut or intestine, are not subject to changes in pitch due to dampness or dryness. The second was a device through which the strings could vibrate freely.[114] This apparatus, which surrounds the string but does not hinder its motion, is composed of two crafted steel clamps, *cd* and *ef* in fig. 1 of figure 7.3. Each clamp is composed of two cubes, 1/2 inch thick, *a* and *d*, and *e* and *f*, which must be parallel. The clamps limit the vibrating portion of the string and secure its endpoints without altering its tension.[115] To ensure that the string was not flattened by the two clamps, Oertling constructed two half-cylinders of steel, which were fastened to each other. Between these two half-cylinders (fig. 2 in figure 7.3) a groove was cut with the same diameter as the string. This cylinder was then fitted into a brass clamp, *defg* (fig. 3 in figure 7.3), and tightened with a screw.[116] The third part of the monochord was a wooden tray of a known weight that was suspended from the end of the string. Weights were added to the tray in order to alter the string's pitch.[117]

As we have seen, Weber recommended this device to both musical-instrument makers and performers for the precise measurement of the highest and lowest notes. In particular, he argued that his contraption could be used to tune the lowest keys of a piano. In order to tune the highest keys, the monochord needed to be modified, as the string(s) needed to be shortened. The resulting higher tones, however, would not be pure, since a short, thick string no longer behaves acoustically merely as a string, but also as a rod. Hence, in order to measure and tune very high pitches, one had to use a very long string and bow it with a violin bow. For the actual bowing practice, Weber borrowed the technique of his deceased friend and colleague, Chladni, in order to elicit longitudinal vibrations, rather than transverse vibrations, as discussed in chapter 2. These vibrations give rise to longitudinal tones as one gradually shortens the string. "In this manner one can track the series of tones to the highest, which a human ear can detect, and these high tones are fuller and better

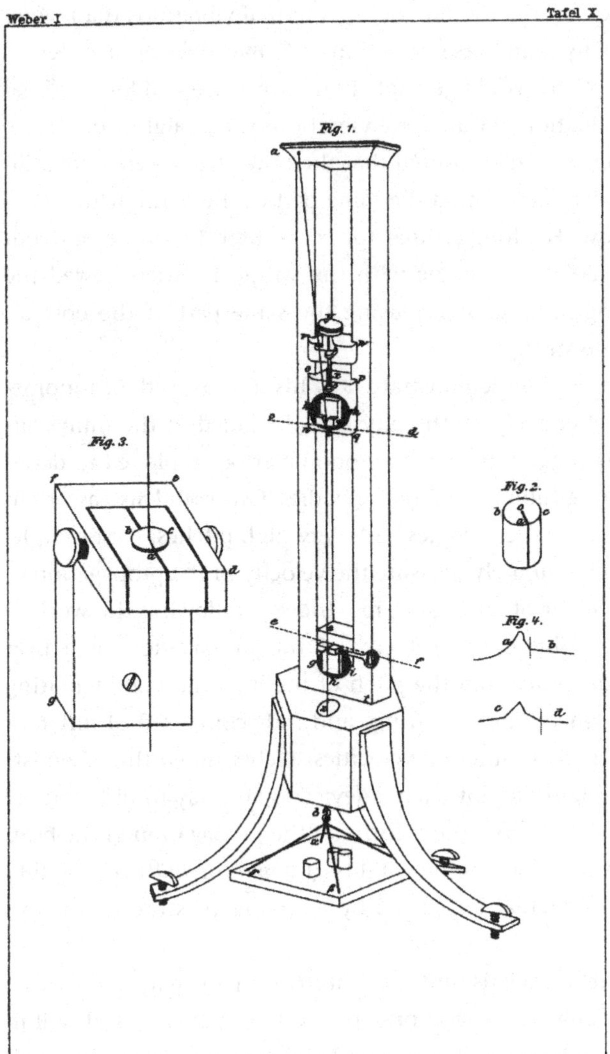

Figure 7.3
Wilhelm Weber's monochord (also called tonometer). Source: Wilhelm Weber, *Werke* (1892–1894), vol. 1 (1892), table X, pp. 358–359.

sounding when they are the product of longitudinal vibrations than when they are produced by transverse vibrations."[118] By applying a different bowing technique, Weber could generate both transverse and longitudinal vibrations. The production of transverse vibrations was straightforward. He simply used a violin bow and ensured that the bow's hairs were perpendicular to the string. He then bowed the same part of the string with different parts of the bow. For longitudinal tones, he used the same bow but tilted it to form a 20° to 30° angle with the string. He then bowed the string along its length, successively using the same part of the bow on different parts of the string.[119]

Weber then proceeded to demonstrate how his monochord, or tonometer, could be used effectively by the physicist. Included in the numerous possibilities was his suggestion that the tonometer be employed to determine the pitch of longitudinally vibrating bodies. One could also ascertain the various properties of these bodies with very high pitches. For example, one could easily and accurately measure the velocity of the propagation of the beat through the air of an organ pipe. Indeed, following the work of Chladni, one could obtain the velocity of the propagation of a beat through all solid materials from the pitch of the longitudinally vibrating string or rod. And since different strings and rods composed of different metals produce beats with differing velocities, Weber noted that chemists and mineralogists might employ such a device as an assay to differentiate between substances. And, from the velocity of the propagation of the beat, a physicist could study the compressibility (*Compressibilität*) or ductility (*Dilatabilität*) of a particular substance by applying pressure or tractive force.[120]

Weber feared that musicians and musical-instrument makers might be discouraged from using this device; one needed both patience and skill in weighing the strings. And few, if any, musicians possessed precise enough balances for the required accuracy. Weber, therefore, attempted to ameliorate the problem by using a tested tuning fork (a compensated reed pipe would be too expressive and its precision is not necessary in this instance) in conjunction with the monochord so that an instrument maker or musician need not burden him- or herself with the weighing of the strings. He or she need only stretch a string of a known length until it generated precisely the same pitch as the tuning fork.[121] One struck the tuning fork with a mallet and bowed the monochord's string. If one detected beats, then the two sounding bodies did not possess the same pitch. The string was then either tightened or loosened (depending on whether it was too sharp or too flat) until no beats were heard, and then weighed. Weber envisaged the instrument makers shipping a table of weights of the strings along with the monochord. He also hoped that instrument makers would provide a guaranteed tuning fork so that the musician "need not calculate anything, but can simply find the weight of the string for the observed tone."[122] He provided an example of such a table. If the length of the freely vibrating portion of a string between two clamps of the monochord is one Parisian foot (approximately 0.325 meters), and the tuning fork produces an a′ tone or 864 simple vibrations per second (or 432 Hz), then when the string is suspended with a weight of 1,000 grains, one foot of the string would weigh 40.45 mg:

1300 g	52.58 mg
1600 g	64.72 mg
1900 g	76.85 mg
2200 g	88.99 mg
2500 g	101.13 mg
2800 g	13.26 mg
3100 g	125.40 mg
3400 g	137.53 mg
3700 g	149.67 mg
4000 g	161.80 mg[123]

One last problem from which the tonometer/monochord suffered, which Weber could not easily remedy, was the instrument's cost. For it to be useful for musical and physical purposes, it needed to be extremely accurate; therefore, a mechanician needed to build it. Monochords made by Oertling, for example, ranged in cost from 60 to 200 *thaler*.[124] Weber appreciated that most musicians could not afford to purchase such an instrument; however, he did hope that musical institutes, which produced and tested tuning forks, would have such devices at their disposal for general use. He added, "These musical institutes from various lands and parts of the world would gradually establish a generally accepted pitch." Musicians, working hand-in-hand with mechanicians and physicists, could now enjoy the proper standardization they so desperately sought.

J. H. Scheibler, like Weber, yearned for a recognized standard pitch: "It is very much hoped that one will assume a′ to be the same every where."[125]

His reputation for the precision of his tonometer began to spread throughout the German scientific community. He also was renowned among scientific (and to a much lesser extent, musical) circles for his work on attempting to standardize performance pitch. In September of 1834 Scheibler presented his work on the determination of the pitch for a′ at 440 vibrations per second (Hz, or 880 simple vibrations per second). After conducting several experiments with various pitches used in Paris, Berlin, and Vienna, Scheibler decided to choose his a′ at 440 as "the middle of the extremes between which Viennese pianos rise and fall" due to change in temperature.[126] The pitches of these pianos were determined by a monochord, and the pitch 440 vps was checked by his tonometer, discussed in the previous chapter. This precise value was approved as the official, national German pitch by the physics section of the Versammlung deutscher Naturforscher und Aerzte in Stuttgart in September 1834.[127] Scheibler demonstrated the precision of his physical and musical tonometer to the sixty-nine *Naturforscher* of the chemistry and physics section. An error of 1 in 75 vibrations per second greatly impressed the *Naturforscher*, all of whom immediately appreciated the tonometer's potential use in tuning musical instruments.[128] In the following year, Scheibler presented a paper entitled "Mittheilung des Wesentliche musikalischen und Physikalischen Tonmessers" (Announcement of the Constituents of the Musical and Physical Tonometer) to the chemistry and physics section at the Versammlung deutscher Naturforscher und Aerzte meeting in Bonn.[129] Clearly, physicists were very interested in Scheibler's work. Bindseil's *Die Akustik* mentions Scheibler's work on pitch determination using the tonometer.[130]

Scheibler's interest in precision was extraordinary. Although musicians appreciated that his technique was far more precise than previous methods, his desire for ever greater degrees of precision became an obsession. He was pursuing precision for precision's sake, as no ear, not even that of an accomplished musician or keyboard tuner, could hear an error of 1 in 75 vibrations per second. The entrepreneur's fetish for this aesthetic of precision manufacture maps nicely onto his tuning feat. Such a desire resonated with physicists more so than it did with musicians, who initially seemed to ignore Scheibler's pitch. It had no immediate effect on performance pitch. But the Viennese piano maker Streicher tuned his pianos to 440 during the late 1830s. And in 1847 an anonymous critic recommended in the *Allgemeine musikalische Zeitung* that 440 be the German standard performance pitch, not because that was the pitch approved by the Association of German Investigators of Nature and Physicians, but rather because it served as a practical pitch for vocalists and an average pitch of those in use throughout the Continent.[131] In that same year, the renowned Bavarian flautist and flute maker Theobald Boehm based his fingering chart for flute playing on a pitch of 440.[132]

More Calls for Standardization

Not willing to succumb to a German pitch, the French were the first to establish a commission with a view to providing a national standard performance pitch. On 17 July 1858 the French Minister of State created the Commission to "[i]nvestigate the means of establishing in France a uniform musical diapason, to fix on a standard of sonority which might serve as an invariable type, and to point out the measures to be passed in order to secure its adoption and preservation."[133] The Commission comprised twelve members: six composers—Hector Berlioz, Gioacchino Rossini, Giacomo Meyerbeer, Daniel-François-Esprit Auber, Jacques Fromental Halévy, and the Perpetual Secretary of the Academy of Fine Arts and the Commission's Reporter, Ambroise Thomas; four civil servants of the State—J. Pelletier, Councillor of State, Secretary General in the Ministry of State, and the Commission's President; Camille Doucet—head of the theatrical department in the offices of the Minister of State; General Mellinet—in charge of the organization of the army bands; Eduard Monnais—Imperial Commissary to the Lyrical Theaters and the Conservatory; and two physicists, Jules Lissajous, Professor of Physics at the Lycée Saint-Louis, and Member of the Council of the Society for the Encouragement of National Industry, César-Mansuète Despretz. The relentless elevation of pitch, it was argued, was deleterious to musical performance.[134] In addition, the wide variety of pitches being used throughout Europe was "a constant source of embarrassment for concert music" and, just as important, resulted in "difficulties in commercial transactions."[135]

The Commission placed the blame for the ever-increasing performance pitch on those "who manufacture tuning-forks, or have them manufactured, [they] are the authors of the evil and masters of the situation. It is the musical-instrument makers, and we can understand that they have a legitimate and honourable interest in elevating the diapason. The more elevated the *tone*, the more brilliant will be the sound."[136] Traveling virtuosi exacerbated the problem by playing more sharply in order to impress fee-paying audiences. "Once introduced into the orchestra, they [the instrumentalists] sway and rule it, easily dragging it to the heights in which they delight. In fact, the orchestra belongs to them, or rather, they are the orchestra, and it is the instrumentalist who, by giving the *tone*, regulates, without desiring to do so, the studies, the efforts, and the destiny of the singer."[137] Military bands in particular were singled out as the culprits of increasing performance pitch. Both singers and composers, they argued, were innocent of the crime, as higher pitches damage the singers' vocal cords, and "the composer does not create the diapason; he submits to it."[138]

There were no instrument makers, instrumentalists, or vocalists on the Commission. Some of France's leading musical-instrument manufacturers did, however, express their views to the Minister of State in a letter, which he then passed on to the Commission, calling for his Excellency "to cause this kind of anarchy to cease, and to render the musical world as important a service as that formerly rendered to the industrial world by a uniform system of measures."[139] The letter was signed by some of the nation's leading musical-instrument manufacturers: Georges-Louis-Guillaume Triébert, Auguste Buffet, and Adolphe Sax (the inventor of the saxophone), makers of wind instruments; Aristide Cavaillé-Coll, the renowned organ builder; a representative of the Erard company, and Pleyel-Wolff and Henri Herz, all piano makers; Jacob Alexandre, a melodium-organ builder; and Willaume, a string-instrument maker.[140] The twelve-member Commission made it clear that the viewpoints of these artisans would be taken very seriously. "The Commission entertains the highest consideration for the interests of our great trade in the manufacture of instruments, which is one of the sources of riches in France, a brand of industry intelligent in its products and felicitous in its results. The clever men who direct, and have raised it to the first rank, cannot question our solicitude; they know we are friendly towards a trade which supplies some of the members of the Commission with valuable and charming auxiliaries."[141]

But some instrument makers feared "the inconveniences" that a *diapason normal* could bring about.[142] Smaller merchants, catering to a local market, would most likely be forced to make new instruments, and their livelihood would be threatened by larger manufacturers, who could more easily and efficiently serve a national clientele. To ameliorate fears that the contrasting views among the musical-instrument manufacturers would not be taken into consideration, the Commission gathered together France's leading artisans who garnished first prizes at the Universal Exposition of 1855 (including the authors of the aforementioned letter to the Minister of State) and orchestral conductors (including Narcisse Girard, conductor of the Imperial Academy of Music at the Société des Concerts du Conservatoire; Mohr, band master of the Imperial Guard, and Delaffre, conductor of the Théatre Lyrique). The Commission wished to hear these musicians' and instrument makers' view on establishing a relatively low pitch as the *diapason normal*.[143]

Although the French sought a national standard, they clearly were hoping to establish a pitch that would be accepted worldwide. As Alder has argued, this gesture toward internationalism bears a "Gallo-centric stamp."[144] The Commission requested tuning forks from various cities noted for their musical sophistication throughout France, the German territories, England, Belgium, Holland, the Italian states, Russia, "and even America."[145] Most cities complied, sending both current and antiquated tuning forks from their cities' operas, theaters and military bands. The Commission reported that an overwhelming majority of musical directors throughout Europe praised the French attempt to both standardize and lower the performance pitch.[146] For example, Karl Gottlieb Reissinger, the first *Kapellmeister* of Dresden's Court Theater, argued that the newer, sharper pitches destroyed the "effect and character of ancient music—of the masterpieces of Mozart, Gluck, and Beethoven."[147] And Joseph Abenheim, director of the chapel of the King of Wuerttemberg, claimed "this will be another act of service rendered by your nation to art and commerce."[148] Ferdinand David, director of the Leipzig Conservatory, assured the Commission that "we take a lively interest throughout musical Germany in the execution of your project."[149]

The physicists Lissajous and Despretz noted that an overwhelming majority of these forks were above the Scheibler/Stuttgart pitch of 440; indeed, the forks' mean pitch was in the vicinity of 449.[150] They verified the forks' pitches using a Cagniard de la Tour siren with a constant pressure bellows constructed by A. Cavaillé-Coll. This ensured the siren's constant pitch.[151] More important, however, was Lissajous' subsequent work for the

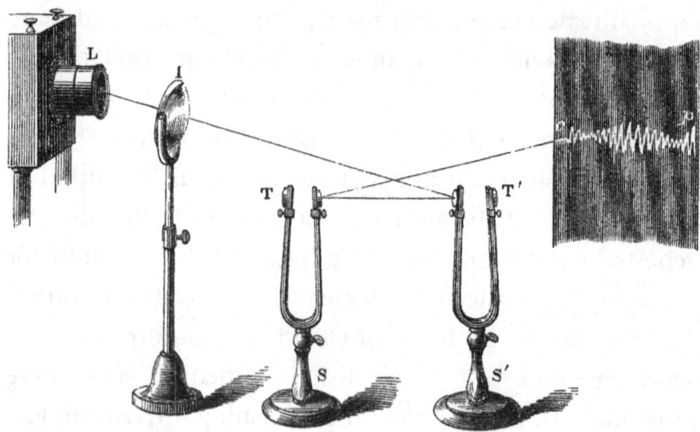

Figure 7.4
Lissajous' reflected light beam. Source: Tyndall 1867, p. 267.

Commission. In 1855 he had invented a method to visualize acoustical vibrations. Not surprisingly, his dissertation, written five years earlier, proffered a thesis on vibrating bars that employed Chladni's method of sprinkling sand on the sounding bodies in order to demarcate the nodal points. Much like Chladni's, Lissajous' work was committed to rendering the invisible visible: allowing the eye to scrutinize what the ear could not hear. He reflected a light beam from a small mirror attached to a vibrating tuning fork and then from a larger, rapidly rotating mirror onto a screen (see figure 7.4). After further study, he was able to generate his "Lissajous figures," by reflecting a beam of light from mirrors perched on top of two vibrating tuning forks positioned perpendicular to each other (see figure 7.5). These curves are determined by the relative frequency, phase, and amplitude of the tuning forks' vibrations as depicted on the screen (see figure 7.6). If one of the two tuning forks is of a standard pitch, one could calibrate the pitch of the second fork by analyzing the resulting Lissajous figures with his invention, the "phonoptomètre," a microscope with a tuning fork attached to the objective (see figure 7.7).[152] His work on the 1858 Commission resulted in the development of the procedure used by mechanicians for the manufacturing and testing of precision tuning forks.

Berlioz was less concerned with Lissajous figures than he was with the consequences of decreasing the concert pitch by too much. He pleaded with the other members of the Commission to establish a pitch that represented

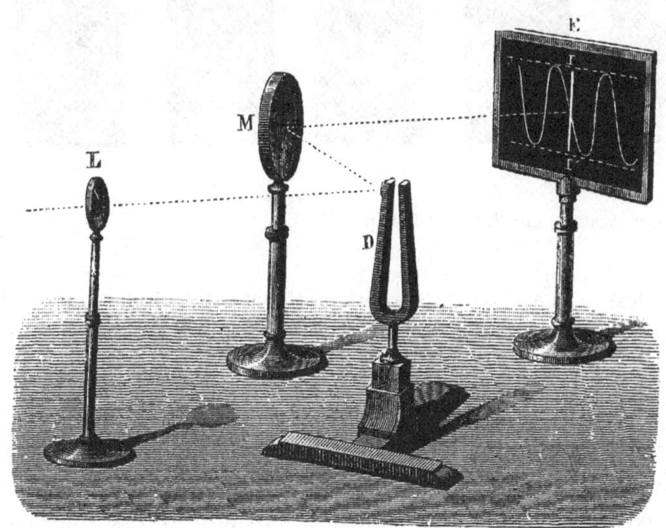

Figure 7.5
Lissajous' technique for generating his optical figures. Source: Guillemin 1881, pp. 716–717.

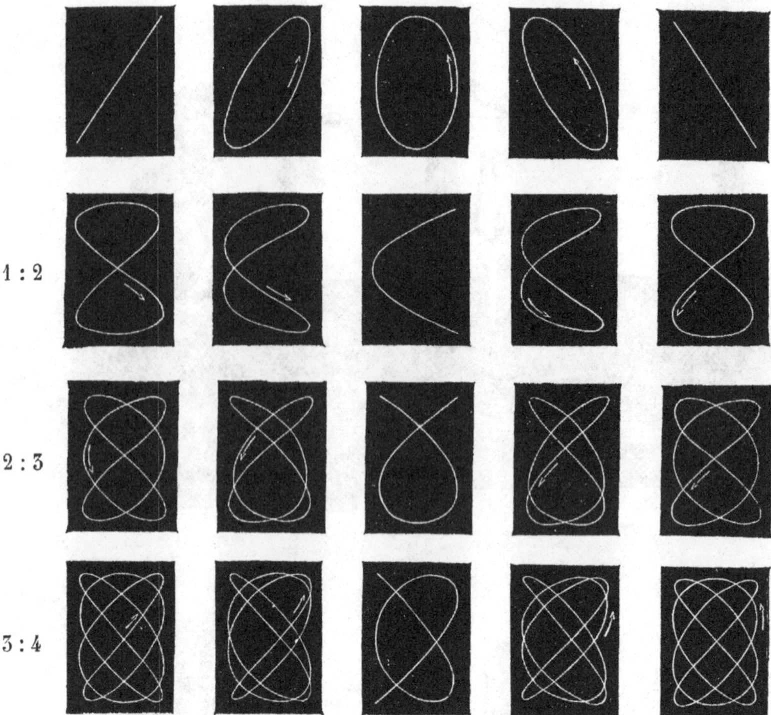

Figure 7.6
Lissajous figures. Source: Guillemin 1881, p. 719.

a decrease of less than an eighth of a tone of the average pitch of all tuning forks they gathered. He argued that lowering the pitch considerably would be impractical, confusing, and expensive, as all fixed-tone instruments would need to be remanufactured.[153] The other members remained unconvinced. A majority of the Commission did agree that the greatest practical reduction would be a semitone. Several members, particularly Berlioz, however, balked at such a drastic reduction. Wishing to seek a fair compromise between the higher tuning forks favored by musical-instrument makers and performers and the lower pitches insisted upon by vocalists, a majority of the members settled on a decrease in pitch of a quarter tone, or a standard pitch of 435.[154] "[T]his would sensibly moderate the trouble attending the studies and executions of singers, and thus insinuate itself, so to speak, incognito, into the presence of the public, without causing too great a perturbation in established habits; it would facilitate the execution

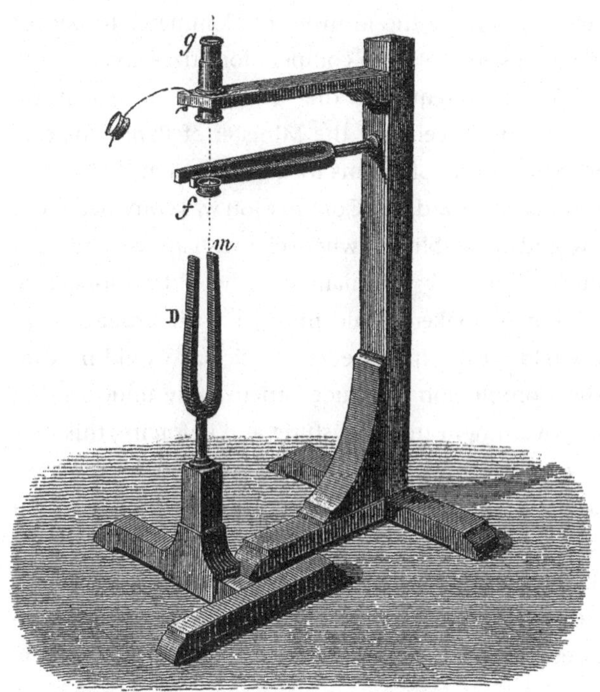

Figure 7.7
Lissajous' instrument for precise measurement of pitch. Source: Guillemin 1881, p. 721.

of ancient master pieces, and would bring us back to the diapason employed about thirty years ago, the period of the production of works of which most have remained on the repertory, and which would thus be in the position they occupied when composed and first represented."[155] And, the Commission added, the value of 435 was close enough to the 1834 Scheibler/Stuttgart pitch of 440 that German orchestras, operas, and theaters could accept the *diapason normal* without great adjustment.

The Commission advised the Monsieur le Ministre that (1) a model tuning fork producing 870 simple vibrations (435 Hz) at 15°C should be constructed by qualified individuals; (2) at some point the *diapason normal* should become obligatory throughout the country; and (3) officers should inspect tuning forks and musical instruments in all French theaters, schools, and other musical establishments. They also requested that the Minister persuade the Minister of War to accept the tuning fork for the military bands, as well as convince the Minister of Commerce to permit only musical instruments tuned to 435 to compete for prizes given at the industrial expositions. And they requested that the Monsieur le Ministre exercise his influence over his Excellency the Minister of Public Instruction and Worship for the retuning of organs to the new pitch.[156]

By accepting this national standard, the Commission was convinced that "order and regularity would be established where chance, caprice, and carelessness now sometimes reign. . . ."[157] Human voices would no longer be sacrificed. Finally, instrument makers could improve their products and sales as their market would greatly increase, for musicians would need to play at this pitch. The Commission concluded their study underscoring the importance of the government in establishing and enforcing this new standard. "It is not unworthy of the government of a great nation to busy itself with questions of this kind, which may appear futile, but which possess a real importance of their own. . . . By directing your attention to the dangers to which an excessive love of sonority may expose musical art, and by endeavoring to establish a rule, a measure, a principle, your Excellency has afforded a fresh proof of the enlightened interest you taken the fine arts generally."[158]

On 1 February 1859 the Minister of State resolved to establish a standard, prototype tuning fork at 435, to be called the *diapason normal*, which was to be stored at the Imperial Conservatory of Music and Education. All musical establishments throughout France were to use this new, official pitch. Public concerts authorized by the State needed to observe this pitch, while State-sanctioned musical establishments were to be provided with a verified, stamped tuning fork. Proper officials were ordered to test periodically the state of tuning forks and instruments. The new pitch was to come into effect in Paris on 1 July 1859, and throughout the rest of France on the following 1 December.[159] The secretary of the Commission did supply copies of the *diapason normal* to cities throughout Europe, with the hopes that the French standard would become the universal standard of performing pitch. They were sent, for example, to the Societies of Arts of Berlin, Baden, and Munich. And a fork was supplied to the French Exhibition of 1867.[160]

French musical-instrument makers now constructed instruments with the new pitch in mind. As these instruments, particularly the French woodwinds, were highly sought after across Europe, the French pitch had an indirect effect on pitch beyond France's borders. Indeed, the *diapason normal* temporarily slowed down the increase in pitch. In 1862 the Vienna *Hofkapelle* and court theaters accepted the French pitch by imperial decree.[161] The Vienna Opera used 435 throughout the 1860s, but the pitch sharpened to 447 some time during the ensuing decade.

Not to be outdone, the very day the French commission filed its report, the Society of Arts in London appointed a fifty-member committee to determine a standard pitch.[162] The British seemed more interested in mathematical pitch than their French counterparts, who viewed the history of pitch more seriously. Sir John Herschel spoke forcibly on behalf of the mathematical purity of tuning. He wanted c′ to be 512 cycles per second (Hz). Echoing Chladni's view over a half-century earlier, Herschel emphasized that this number could be divided by two, representing the octaves, all the way down to the number two. The committee, however, chose 528. The problem with the British selection, being based on the mathematical, "natural" intervals of the scale by whole numbers, is that it was based solely on just intonation for the major C scale only. It could not be applied to equally tempered scales.[163] According to Ellis, the British committee neglected the effects of temperament on tuning, which led to some embarrassing mistakes. For example, while they chose c′ 528, they also chose the Scheibler/Stuttgart pitch a′ 440. But with equal temperament, which (recall) was gradually being adopted by fixed-tone instruments by the time of the meeting, a′ 440 yields an equally tempered c′ 523, whereas c′ 528 requires an equally tempered a′ 444.[164] While the French Commission went for a compromise pitch, the British sacrificed practicality for a utopic, mathematical purity. As a result of this sharpening, augmented by measuring errors made by the maker of the tuning forks—(Griesbach), 444 became 446 and equally tempered c′ became 534—the British pitches were among the sharpest in Europe. The British Society of Arts' pitch became absorbed into the higher pitches in general, so that the *diapason normal* was the only recognized compromise pitch.[165]

A number of German court theaters (mostly those in the western territories, such as in Cologne and Stuttgart) adopted the *diapason normal*, but its acceptance certainly was not universal.[166] The Dresden singing master Karl Näke claimed in 1862 that there were areas throughout Germany that were strongly opposed to the pitch.[167] He personally launched a crusade against the French pitch, arguing that it was still too high for vocalists, causing harm to their voices. He accused the French Commission of pandering to instrument makers while ignoring *Kapellmeistern*, voice instructors and singers themselves.[168] He was one of the first to draw causal and conceptual links between the increase in pitch and the decrease in singers' ranges. Voices were strained so much hitting the higher notes that they lost their power to sing at the extreme ranges, both upper and lower.[169] He feared that vocal instructors would now spend all of their time teaching their students to read higher pitches, while "worrying little about the beauty, richness and uniformity of the [singing] organ."[170] For Näke, pitch should be set neither by physicists nor instrumentalists, nor instrument makers. Rather the *vox humana* and historically determined pristine pitches were the arbiters of beauty. The Berlin music critic Otto Gumprecht concurred, writing in the *Berliner Nationalzeitung* on 22 November 1861, "the petition for the lowering of the concert pitch, if possible lower than the French standard, should by the way be the 'ceterum censeo' of all opera departments. The way things presently stand, not only will all collective voices, particularly the tenors, reach an early grave, but most musical scores in our classic operas will lose their original character."[171]

According to Näke, "the only true utility of the Parisian Commission are the measurements of the various older and current pitches from various locations."[172] His goal as *Gesangmeister* was to recreate the pitch used during Mozart's life. With Scheibler's method of tuning in hand to ascertain this historical pitch, he claimed that it was the only true one for vocal works. His nostalgic reverence (and indeed worship) of Mozart was clear: ". . . the piety for Mozart's works, the wish to hear his works unadulterated, as he himself heard them" should be the aim of establishing performance pitch.[173] He did his utmost to lower Dresden's musical pitch. According to the late-nineteenth-century acoustician Alexander Ellis, Dresden was the only city in Europe whose city theater possessed two distinct sets of instruments tuned to two different pitches independent of the new French standard. The lower one was the mean pitch somewhere between 418 and 424. The higher pitch was the Scheibler/Stuttgart pitch of 440.[174]

In September 1862 a number of musicians, composers, and instrument makers gathered to test which pitch was better suited to opera. The conference was headed by von Koenneritz, the general director of the Dresden Theater. Attendees included Näke; Franz Abt, composer and *Kapellmeister* of Hanover; Bernhardt Scholz, also *Kapellmeister* of Hanover; Carl Reiss,

Kapellmeister of Cassel; Eduard Thiele, *Kapellmeister* of Dessau; the composer, singer, and music director of Stettin, (Johann) Karl (Gottfried) Loewe; Friedrich Kaufmann, the mechanician from Dresden; Lendel, also an instrument maker from Dresden; C. Krebs, *Kapellmeister* of Dresden; J. Lauterbach and Franz Schubert, concert masters of Dresden; *Kapellmeistern* (Karl Gottfried) Wilhelm Taubert of Berlin; and Ernst Lampert of Gotha.[175] The group listened intently to Gluck's "Iphigenia in Aulis" and Mozart's "Idomeneus" in a low mean pitch, and Rossini's "William Tell" and Carl Maria von Weber's "Euryanthe" in the higher Stuttgart pitch. They were then treated to Weber's "Oberon," Mozart's "O cara imagine" and the trio of the three ladies "Oh stelle! soccorso" from Die Zauberflöte, the *Scena* of Bois Guilbert from Heinrich Marschner's "Templar and Jewess," and "Durch die Wälder" from Weber's Der Freischütz, all in both pitches; followed by excerpts from Mozart's "Die Entführung aus dem Serail" and "Son vergin vezzosa" from Bellini's *I puritani*, both in the higher Dresden Theater pitch.[176] On 1 October 1862 a resolution was adopted to accept the French *diapason normal*, which was close to the higher Dresden Theater pitch, or the Stuttgart pitch:

> A lowering of pitch to the new Paris standard appears equally desirable and satisfactory for singers and for orchestra. That quality of tone that would be gained from the brilliancy of the band would not be lost, and the power of the singers would not be so severely taxed or strained. On the other hand, it was probable that, especially for recent operas and for instrumental pitch, a further reduction down to Mozart's pitch, as it was called, would ignore the brilliancy and effect of the music under the present, entirely different, conditions. Consequently, such a reduction must be rejected as impractical. Uniformity, especially in the orchestral pitch—which is to govern the future, must be looked upon as the principle object and chief gain. Consequently, under present circumstances, it was desirable that the Paris orchestral pitch should be generally adopted in Germany; a course that would be certainly welcome to singers and be gratefully accepted.[177]

The only dissenting voice to this resolution was Näke's. He argued that uniformity should not be the key concern, but rather beauty and suitability to the singers. Hence he preferred the lower mean pitch.[178]

Although apparently the French pitch held rather steadily for a quarter of a century, it ultimately failed to become accepted as a universal standard.[179] On 12 April 1862 the Belgian government set up a commission, similar to France's, headed by François Joseph Fétis. Their recommendation of 451 as concert pitch was officially instituted in Belgium starting on 22

June 1863.[180] Twelve years later, on 21 December 1873, a second commission was established, this time under the direction of Gevaerts, to review Belgium's stance on performance pitch. After much deliberation, it concluded on 22 May 1877 that Belgium should adopt the *diapason normal* for State-subsidized theaters, conservatories, schools and symphony orchestras.[181] On 19 March 1885, the Belgian government issued another law stating that the nation was committed to the *diapason normal*.[182]

1885

After German unification in 1871, calls rang out for a national pitch. Pitch was rising once again, eclipsing the French *diapason normal*. The fear of high pitches destroying voices prematurely was again raised, and many string-instrument players felt that the higher pitches favored woodwinds and horns.[183] Orchestras and bands needed to purchase fixed-tone instruments from manufacturers, which produced instruments at the requisite pitch. Again, although it might have been in the financial interest of larger musical-instrument manufacturers to unite on a national pitch with concert and theater bands, instrument makers satisfied with a smaller market, particularly a number of piano manufacturers, often built their instruments in two pitches, a higher and lower, and they were not about to make it three. And these manufacturers were often bitter rivals, not interested in collaborations. Apparently singers, directors, and band musicians still fought bitterly over what pitch would be most aesthetically pleasing.[184]

A plea addressed to *Reichskanzler* Otto von Bismarck for a national pitch was published in the relatively new *Zeitschrift für Instrumentenbau*. The petition, signed by hundreds of instrument makers (including Schiedmayer & Sons, renowned piano manufacturers of Stuttgart), music directors, composers (including Johannes Brahms), theater directors, opera singers, and musicians, formally requested that Bismarck create a Ministerial Commission comprising acousticians, conductors, and instrument makers.[185] The petition's endorsers hoped that this commission would either accept the *diapason normal*, or accept Scheibler's *Normalton* of 440, since many felt that the French pitch was too low.[186] The petition concluded by asserting that "the introduction of a general standard pitch in all of Germany is so easy, and it accords all of the artistes as well as instrument manufacturers the greatest utility and advantage as well as the courage to attain this goal. We must feel ashamed in the company of foreigners if we persist in the hitherto differences in pitch."[187] The Frankfurt conductor A. Kalkbrenner supported the notion of a standard pitch, but not by a prescribed date.[188] He queried how a standard pitch could be best achieved, wondering if the lower French pitch should be accepted. The problem, he pointed out, with establishing a new *Normalstimmung* was the cost in producing new fixed-tone instruments. An orchestra's woodwinds are notoriously the most resistant to pitch alteration. He estimated that new woodwinds alone would cost the nation 170,000 Reichmarks, while new brass instruments would add 286,000 Reichmarks to the tab, culminating in nearly a half a million Reichmarks. Who would pay the bill? Also, if Germany rushed to accept a national pitch, instrument makers would not be able to keep up with the demand for new instruments. Instrument manufacturers from Markneukirchen in Saxony joined the fray. Although they initially expressed "cool reserve," by October 1884 they were encouraging Bismarck to intervene to establish a national pitch.[189] Without a national standard, German instrument makers needed to build instruments in as many as "five, six, or more" pitches.[190] The French, in contrast, had only one official pitch, which was much easier for instrument makers. Germany was united; so too should be those who were determining performance pitch.

Although Bismarck never responded, from 16 to 19 November 1885 delegates from six nations met in Vienna to establish, for the first time, an international standard pitch. Austria, the host nation, had the largest contingent, including the President of Vienna's Gesellschaft der Musikfreunde (Society of the Friends of Music) Freiherr Josef von Bezecny; the former Minister of Culture and Education Karl Fidler; the aesthetician, leading music critic of the period, author of the influentical text *Vom musikalischen Schönen* (On Musical Beauty), and the University of Vienna's Professor of Musicology, Eduard Hanslick; the Court *Kapellmeister*, violinist, Director of the Gesellschaft der Musikfreunde, Professor of the Vienna Conservatory, and Concert Master of the Viennese Court Opera, Josef Hellmsberger; the Viennese Court Opera Director, Wilhelm Jahn; the Military *Kapellmeister* and composer of numerous military marches, Karl Komzak; University of Vienna's Professor of Physics and Director of the Institute for Experimental Physics, Josef Stefan, who served as the International Conference's President; the composer and member of the Ministry of Education and

Culture, Carl Zeller; and the composer, music and harmony instructor at the Conservatory of the Gesellschaft der Musikfreunde (where he served as General Secretary) and editor of the *Blätter für Theater, Musik und bildende Kunst* (Pages for Theater, Music, and the Visual Arts), Leopold Alexander Zellner. Max Schütz was Hungary's lone delegate, while Italy was represented by the University of Rome's Professor of Experimental Physics and Director of the Physics Institute, Pietro Blaserna, and the composer, novelist, poet and librettist for both Giuseppe Verdi and Amilcare Ponchielli, Arrigo Boito. Dr. Svedbom was sent by Sweden, while Russia's representatives were Military *Kapellmeister* Grushka and the renowned pianist, composer, and piano teacher Theodor Leschetitzky. Germany sent three delegations, the largest being from Prussia: the composer and Director of the Singakademie, Martin Blumner; theater director Hans von Brosart, renowned for his Wagner productions; Prof. Engel; Musical Director Joachim; Prof. Kolsch; and music director, pianist, and *Lieder* composer Franz Wüllner. Saxony sent three representatives, all of whom were from Dresden: the flautist and professor of music Moritz Fürstenau; the *Kapellmeister*, director, and composer Carl Reinecke; and *Kapellmeister* Ernst Edler von Schuch. The third German state, Wuerttemberg, sent two delegates: organist, pianist, composer, and Director of the Stuttgart Conservatory Immanuel Gottlob Friedrich Faist, and Court Music Director Seifitz.[191] Two countries were conspicuously absent: France, which would go on to use the *diapason normal* regardless of what the International Conference decided, and Great Britain, which—as stated earlier—notoriously had some of the highest pitches in Europe, particularly at their cherished opera house in Covent Garden.

Carl Zeller spoke at length on the first day on the history of attempts to standardize performance pitch, singling out J. H. Scheibler for his practical attempts to determine such a pitch as the first real success at standardization.[192] Zeller continued his history of pitch by briefly discussing the *diapason normal* of 1859 and Belgium's decision in March 1885 to promulgate the French pitch, a pitch significantly lower than the previous official pitch of 451. And he recalled Kaiser Franz Josef's decision of 1862 to adopt the *diapason normal* in all court theaters, and in court orchestras a year later. However, measurements in 1883 in both the Viennese Court Opera and the Hofburgkapelle revealed a slight sharpening of pitch to 443 and 441, respectively. And, measurements several weeks prior to the International

Conference noted a continual sharpening: the Viennese Court Opera was now at 449 and the Hofburgkapelle's performing pitch was 450.[193]

Italy did not as yet possess a national pitch, perhaps in part because—like Germany—it had become a unified nation only fourteen years earlier. There was, however, a national pitch for military music, which standardized b♭′ at 456. This placed concert a′ at 432.[194] This pitch quickly became known as the Italian pitch. Hence, the International Conference needed to choose between two pitches, the Italian or French. Scheibler's pitch of 440 was not under consideration.

Before the 1885 international meeting, the Austrian government gathered advice from a plethora of individuals relevant to music—music scholars, acousticians, and musical-instrument manufacturers. It assembled a national committee on pitch standardization. Some of those members attended the International Conference: Bezechny, Freiherr von Hofmann, Court Piano Manufacturer Ludwig Bösendorfer, composer Johannes Brahms, Nikolaus Dumba, Professor Gänsbacher, Eduard Hanslick, *Hofkapellmeister* Hellmsberger, Court Opera Director Jahn, Choral Master Kremser, Military *Kapellmeister* Komzak, the instrument manufactuer Stecher, the physicist Josef Stefan, Court Instrument Maker Uhlmann, Court Opera Singer Winkelmann, and Leopold Alexander Zellner. This Austrian commission decided in favor of the French *diapason normal.*

Zeller's explanation to the members of the International Conference of why the Austrian's chose the French pitch over the Italian one is rather illuminating. Apparently, physicists strongly supported the Italian a′ of 432 (corresponding to b♭′ 456). The Parisian instrument maker Rudolph Koenig felt that the scientific pitch of a′ 432 was the wiser choice, a standard that would stand the test of time, noting that the difference between 435 and 432 was negligible.[195] The most passionate advocate of this pitch had been the Belgian physicist Charles Meerens, who was fundamentally committed to basing music theory and practice on the solid ground of mathematics.[196] Meerens, the son of a flautist and himself an amateur cellist, had traveled to Brussels in 1855 to study music; however, he eventually decided to turn his attention to acoustics. His first criticism of the *diapason normal* had appeared in his *Instruction élémentaire du calcul musical et philosophie de la musique* (Elementary Instruction of Musical Calculation and the Philosophy of Music) where he called the *diapason normal* of 435 an "unpardonable consequence" of the French Commission.[197]

The argument Meerens employed against the chosen concert pitch owed much to Chladni's reasoning. He argued that 432 could be divided by two (representing the octaves of a′) for four octaves, or the lowest range of human hearing.[198] And his a′ of 432, which he called "the correct concert pitch," corresponded to a just tempered c′ of 256, which was two to the eighth power, a perfect geometrical ratio.[199] Mathematical calculations of the octaves were therefore easier, since fractional pitches were not necessary. Also, tuning forks of whole numbers were more accurate than those attempted to be produced with fractional pitches.[200] More important for Meerens, it was the goal of applied science "in the domain of the arts . . . to point to anomalies of blind and arbitrary routine everywhere it is present in teaching."[201] Music merited a mathematical basis, not an arbitrary one. He continued, "In order to remedy this actual, upsetting situation of the question of the prototype of universal sound, it would suffice to rectify [this situation] by means of a simple line of a pen, which would lead to no material consequences, replacing the abnormal number of 870 V. [435 Hz] that figures in a few governmental decrees with the same correct number of 864 V. [432 Hz] so that the true and standard tuning fork becomes accepted throughout the world. Honor would be bestowed on the country that would take the initiative."[202] Meerens felt that the difference between 435 and 432 was so small that it was nearly imperceptible and therefore insignificant.[203] Long after the 1885 International Conference had endorsed the *diapason normal*, Meerens continued his assault, referring to the pitch as "abnormal,"[204] "absurd,"[205] "arbitrary,"[206] "incorrect," and "a hazard."[207]

Although the Austrian contingent appreciated "the scientific basis" (*wissenschaftliche Basis*) of the Italian pitch, they considered music to be a "practical"(*praktisch*) art.[208] Because the musical scale had been tempered since the time of J. S. Bach, it was not mathematically perfect. And, a tempered scale with b♭′ at 456 would yield not a concert pitch a′ of 432, but rather one of 430.[209] The commission decided that "we are called to find a fair settlement and a correct balance between the purely scientific [*wissenschaftlichen*] and practical standpoints in the interests of our art. . . ."[210] If Austria were to view the standard pitch from only a scientific standpoint, Zeller continued, then clearly the Italian pitch would be more appropriate. The Austrians, however, decided that practicality should trump science. And since the French pitch was in more common use than the Italian, the Austrian contingent decided, "despite the principle steadfastness of their [the representatives of the physical sciences] viewpoint," to vote for the French pitch.[211]

The Italian representatives then responded to Zeller's report. Blaserna recounted how the Italians at the Milan Congress of 1881 chose 432 for performance pitch, following a purely mathematical model.[212] Defending his nation's decision, he countered the French and Austrians' claim that practicality needed to rule the day by asserting that "the question of a tuning fork is not merely a musical question, but also a theoretical question."[213] *Pace* Meerens, Blaserna argued that the difference between 435 and 432 was so small as to be nearly imperceptible. And, he added, the "scientific worth of the easier mathematical calculation and the resulting easier, more natural construction [of the tuning forks] and also the more straightforward teachings should be given preference."[214]

Blaserna's compatriot, Maestro Boito, echoed these sentiments. Drawing upon the theoretical works of Meerens, he wished to draw the delegates' attention to the dangers of ignoring the scientific relevance of musical pitch. "Our century is totally turned to the sun of science. The rays of this sun warm through and penetrate all disciplines of human knowledge, even also art. Were we to avoid the influences of this on us by not selecting the scientifically preferred tuning fork and choosing 435, we would be committing a type of anachronism."[215] Two Prussian delegates, Blumner and Wüllner, disagreed with the Italians, and spoke in support of the French pitch. Seifriz expressed concern of the cost of manufacturing new instruments to even lower pitches. Svedbom also chimed in supporting the *diapason normal*.[216]

The physicist Stefan proffered his own view. If science were to dictate the determination of performance pitch, he would be obliged as a physicist to choose 432: "The number 864 [simple vibrations, or 432 Hz] possesses the only advantage of an arithmetic aesthetic [*arithmetischen Ästhetik*]."[217] Since the *diapason normal* was being used in France, Belgium, Russia, and in a number of German and Austrian institutions, he supported the French pitch.

The meetings concluded for the day, resuming discussion on 17 November. General Secretary L. A. Zellner thought it best to draft a proposal arguing that while 432 had considerable advantages for the theory and practice of music, the French tuning fork should be accepted as an international standard because of its popularity throughout Europe.[218] The

Prussian Joachim, however, disagreed. Both he and the Austrian musicologist Hanslick feared that by stating one pitch is better than another for certain cases, the international pitch would most likely not be as readily accepted and enforced.[219] After further discussion and debate, Boito and Blaserna acquiesced, and the French *diapason normal* was chosen as the official international performing pitch.[220]

Although the "scientific" pitch did not win out, it would be a mistake to assume that physical science did not play an important role in the new standard. The members of the conference all agreed that the standard pitch needed to be constructed according to scientific rules. The fork needed to produce 435.0 vibrations per second at 15°C.[221] To guard against shifts in the standard pitch, each country was to select an office that would be responsible for generating tuning forks with the correct pitch. The fork was to be made from unhardened cast steel, with prongs that were parallel and at least one-half centimeter in width.[222] The fork clearly need to be rust-free and polished white hot or tempered blue.[223]

Blaserna and Stefan drew upon their expertise to advise the other members of the conference on the physics of tuning forks in order to ensure the fidelity of the original and replica forks. Blaserna noted how the effects of temperature, both room temperature and the heating of wind instruments from the constant breath of the performer, had not yet been sufficiently researched. Tuning forks were less influenced by heat than musical instruments. Blaserna argued that a difference in temperature of 30°C would change the pitch by three-quarters of a complete vibration per second. The pitch of an organ pipe, however, was much more dependent on the temperature. Earlier studies had indicated that a pipe that sounded at 435 vps at 15°C increased in pitch to 457.7 vps at 30°C, nearly a semitone higher.[224] Stefan, supporting Blaserna's comments, informed his colleagues that the number of vibrations of a metal rod, such as a tuning fork, will lose one half of a complete vibration with an increase in temperature of 10°C, whereas the number of vibrations of an air column, such as an organ pipe, will gain eight complete vibrations per second with the same increase in temperature.[225] This emphasis on precision was exactly what the conference delegates desired. They voted to use an electromagnetically powered tuning fork to tune orchestras. Tuning to the oboe, which had been the preferred method of tuning an orchestra, was now to be used only when no tuning fork was available.

Stefan offered his technical and scientific expertise by recommending how the standard tuning fork should be constructed. He enjoyed the reputation of being Austria's foremost experimental physicist. His most important work had been completed in 1879, when he determined that heat radiation is proportional to the fourth power of the absolute temperature. Boltzmann theoretically deduced this relationship, which is exact only for black bodies, known as the Stefan-Boltzmann law of radiation. Stefan also worked on heat conduction in gases, devising a diathermometer used to measure heat conduction of clothing. He also contributed works on the kinetic theory of heat, heat conduction in fluids, the relationship between surface tension and evaporation, and acoustics.[226]

Stefan argued that first and foremost the standard tuning fork needed to produce 435 vibrations per second at 15°C, as set by the conference. The number of the fork's vibrations should not change over time, and the fork should sound for a relatively long period after being struck. Finally, the tuning fork should be suited for the experimental optical method, an earlier version of which was pioneered by Lissajous.[227]

The constancy of the number of vibrations of a tuning fork was predicated on the stability of its mass, form, and elasticity. The correct mass and form could be assured by using a material of great hardness and elasticity, such as cast steel. The tuning fork needed to be gold plated in order to prevent changes in mass due to oxidation. Stefan postulated that heavy use of tuning forks would damage their elasticity, thereby affecting their pitch. He therefore suggested producing two forks, one standard fork to be used rarely and the other tuned in unison to be used for tuning musical instruments and conducting physical experiments. This second fork would be periodically compared to the standard fork to check its fidelity.[228]

To ensure that the tuning fork would continue to sound well after its being struck, Stefan argued, its prongs needed to be of the same length, width, and thickness. They also needed to be absolutely parallel and separated further apart than the tuning forks from earlier periods. Older tuning forks made during the 1820s and '30s—those, for example, used by Wilhelm Weber—had their prongs relatively close together so that a decrease in amplitude led to an increase in pitch. Stefan suggested drawing upon the work of the artisan Rudolph Koenig.[229]

Lissajous' optical method, used to compare the pitches of two tuning forks, was ideal for the standard tuning fork. The ends of the outside of the prongs needed to be totally flat and reflecting from at least one centimeter from the edge upward. Stefan recommended against using separate mirrors attached to the forks, as Lissajous had initially employed, since the extra weight could slightly change the fork's pitch. It was sufficient for the fork to have an extremely lustrous point that could be seen through a vibration microscope. This set up in combination with a tuning-fork clock constructed by Koenig, Stefan argued, was the simplest method known to determine the fork's number of vibrations.[230] Stefan recommended that one use Koenig's tuning forks, as they were constructed with "extreme precision."[231]

After relying on science for state-of-the-art precision measurement, the conference members took steps to guarantee that the standard pitch would indeed be internationally recognized and enforced. They strongly suggested that the pitch be used in public and private schools where music was taught, as well as musical associations and theaters. Military bands, particularly guilty of sharpening the performance pitch over the previous half century, were to adopt the pitch as soon as possible, or at the very latest, at the time of the next restoration of the woodwinds. Churches, too, were to restore their organs to the pitch as soon as was feasible, or at the next organ repair. Each country was to set its own time limit for these changes. Instrument makers were initially instructed to tune their musical instruments to the *diapason normal* at 24°C, but the temperature was then lowered to 20°C, closer to the officially tuned tuning fork of 435 Hz at 15°C.[232]

In Germany the governmental institution responsible for overseeing the production of new tuning forks was the Technical Section of the renowned Physikalisch-technische Reichsanstalt (henceforth PTR). Leopold Loewenherz, the section's director, reckoned that an agreement on a unified pitch was just as important to the practice of music and the production of musical instruments as the determination of a standard weight and mass was for commerce. Although the recommendations for a standard pitch had dated back to the seventeenth century, Loewenherz underscored that Scheibler's a′, to which Loewenherz referred as the German pitch, was approved by the Versammlung deutscher Naturforscher und Aerzte in Stuttgart in 1834.[233] Alas, the Versammlung lacked the political clout to implement its standard. Lamenting the Versammlung's impotence, he rather sarcastically remarked that "regrettably, France has taken this step [in determining its *diapason normal*] without coming to an understanding with the other major cultural nations."[234] He also criticized the Vienna Conference's decision to accept the pitch notation in dvps, or 870 svps, rather than 435 vps. One notes, once again, the rhetoric of anti-French sentiment emerging in Loewenherz's essay:

> Then not to mention that by far the most influential acoustical examinations have stemmed from these [German and British] researchers [who used vps, rather than svps], the French method of counting cannot be applied to vibrations of a sounding air column, such as those occurring in wind instruments and in singing. The pitch that a pipe generates when it executes 435 vibrations against the reed in one second is equally as high as the pitch of a tuning fork, which produces 435 vibrations in one second, using the German method of counting. In addition, the most straightforward method to compare two tuning forks possessing pitches close to each other is to determine the number of vibrations, i.e. the alternating rise and fall of the pitch, which is heard in one second with the simultaneous sounding of both forks. This number immediately gives the difference of the number of vibrations according to the German method of counting.[235]

He was also rather critical of the Conference's choice of the temperature for the *diapason normal*, 15°C, since concert halls and theaters were generally warmer, particularly during the summer months. Finally, the creation of a standard tuning fork, in Loewenherz's eyes, was "superfluous" and "unnecessary": superfluous, because the physical methods in connection with a good astronomical clock offered the means at any time to determine the number of vibrations a fork with sufficient precision, and unnecessary, because with the well-known nature of steel, a change in the number of vibrations with time was highly probable.[236]

These reservations notwithstanding, Loewenherz realized that German mechanics needed to produce tuning forks that followed the Vienna's Conference specifications. Forks needed to be produced from one piece of steel and had to be totally symmetric. Both prongs had to be prismatic in shape and possess precisely the same perpendicular cross sections. The inner portion of the yolk should be semicircular. The breadth of the prongs (i.e., the side of its cross section in the profile of the fork) needed to be at least 5 mm. The prongs could not be thinner than 2.5 mm. And, the greater the mass of the fork, the fuller the tone. The Vienna Conference recommended that the tuning fork be attached to a resonance box and stimulated by an electromagnetic current. This recommendation, however, was impractical, as the pitch of the fork would change with the strength of the current. And, concertmasters could not be expected to use such physical methods. Hence Loewenherz suggested that one simply bow the fork, which was attached to a resonating box, with a cello or violin bow several times. One repeats the bowing once the fork begins to weaken in volume, which usually happened after twenty seconds.[237] If bowing failed to produce a full, rich tone, Loewenherz recommended that one gently hit one of the prongs with the wooden portion of the bow. He also suggested that each orchestra's tuning fork be tested every two years. He urged German mechanics to pay more attention to the production of tuning forks than they had previously.[238]

The PTR recommended three different types of tuning forks. The first was the shape of Rudolph Koenig's tuning fork, in which all the cross sections of the prongs and the yolks were at right angles (fig. 1 in figure 7.8). The base was round and tapered off to a screw, which could be attached to a resonance box. Each prong was approximately 5.5 mm thick and 14 mm wide. The second type of the tuning fork was suggested by W. Wolter of Vienna (fig. 2 in figure 7.8). It had the same basic shape as Koenig's, but was a bit smaller, with the prongs 4 mm thick and 9 mm wide. The third type was the style chosen by the Berlin mechanic C. Reichel (fig. 3 in figure 7.8). It was particularly well suited for hand use, and its pattern was based on the tuning forks of German military bands. Both prongs were separated by merely 2 mm; as a result, the fork's pitch sounded for a much longer period of time.[239]

To tune the forks to the desired pitch, Loewenherz recommended the following method: one used two tuning forks, which were 1 vps sharper and flatter than the fork to be tuned. Strike the fork to be tuned together with the higher-pitched fork: they would produce one beat per second. Repeat the process with the lower-pitched fork and the fork to be tuned, and one should once again count one beat per second. If this was so, the fork was perfectly in tune. If the fork was too sharp, then one would file the inner side of the yolk. If the pitch was too low, then the prongs' ends needed to be filed down.[240] In 1889 Helmholtz summarized the PTR's procedure for testing tuning forks.[241]

The need for the manufacture of precise, reliable tuning forks for music institutes raised important questions for mechanicians and physicists on the accurate measurement of pitch. In 1883 Reichel had reported that the pitch of tuning forks could not be accurately determined by using a mono-

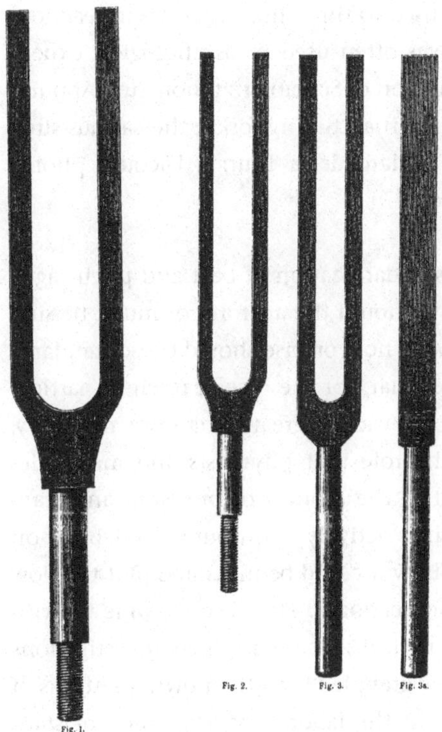

Figure 7.8
Three tuning forks recommended for determining concert pitch by the Physikalische-Technische Reichsanstalt in 1888. Source: Loewenherz 1888, pp. 263–264.

chord. He recommended an electromagnetically powered tuning fork with a small pin dipped in ink on the end of one of its prongs. That tuning fork would be struck by a mallet, and the pin would graph vibrations on a rapidly revolving cylinder and the waves would be counted.[242] This procedure was referred to as the graphic method of determining the frequency of a tuning fork. In 1890 Leman summarized the various methods the PTR was using for determining the pitch of sounding bodies, commenting on their advantages and disadvantages. He discussed the graphic method of calculating frequency, Koenig's tuning-fork clock, and V. von Lang's method of the Hipp chronoscope, a timing device powered by a weight hanging under the clockwork mechanism. Two dials indicated the time in seconds and tenths of seconds. The mechanism was controlled by a pair of electromagnets. The device was used to time short intervals in reaction-time experiments, and therefore was often used in psychological experiments. Leman also included a discussion of Scheibler's tonometer; Appun's sonometer, made of reed pipes (rather than tuning forks); the various siren designs of Seebeck, Savart, and Cagniard de la Tour; P. Lacour's phonic wheel; and Lord Rayleigh's method.[243]

In conclusion, debates about the standardization of beat and pitch raged throughout the nineteenth century. Should the art form of music be subjugated by mechanical standards? And how precise should those standards be? Although I concur with Haynes that, for the case of pitch in particular, the precision offered by physicists was far greater than was necessary, one must not therefore neglect the role that physicists and mechanics played, as musicians viewed them as the arbiters of precision and standardization. Physicists and mechanics actively contributed to debates on what concert pitch should be and how it could be measured. But the flow of knowledge and skill was not unidirectional. The need to come up with ways to produce and test new tuning forks resulted in new investigations by physicists and mechanics, as they grappled with their own questions of precision, accuracy, and reliability in the laboratory. The study of standardization also illustrates the role of science in determining aesthetic parameters. Whether pitch should be based on a mathematical-physical notion of aesthetics, or on a practical notion of performance, or on a nostalgic, historical basis was hotly contested throughout the nineteenth century. And the effects of its eventual resolution rippled throughout various European economies, nationalistic sentiments, musical-instrument manufacture, and acoustical research.

8 Physics, Machines, and Musical Pedagogy

Mechanical principles not only played a role in standardizing musical tempo and pitch, they were also summoned to assist young musicians learning playing techniques. And, some argued that they could be employed to explain musical virtuosity with a view to pass such information on to schools and conservatoires wishing to inculcate the skills of the great nineteenth-century virtuosi. This chapter concerns itself with two aspects of new aesthetics of music and expression discussed in chapter 4. The first half of this chapter discusses the role of mechanical devices in the musical mass education of the German *Bildungsbürgertum*. As seen in chapter 3, Carl Friedrich Zelter, director of Berlin's renowned Singakademie and founder of the *Liedertafel* (choral society), argued to King Friedrich Wilhelm III of Prussia that music was necessary to both the moral development of Prussian citizens (particularly school children) and the formation of Prussian nationalism. And Zelter supported the use of mechanical contraptions for the general musical education of the children of the *Bildungbürgertum*.

The second aspect of this new aesthetic analyzed in this chapter offers a rather different view on the purpose of music: one that takes the early Romantics' "surface allure . . . speaking directly to the listener"[1] to the extreme, namely the cult of the virtuoso. The 1820s, '30s and '40s witnessed the rebirth of the virtuoso.[2] Audiences in concert halls greeted the unprecedented skilled performances of Niccolò Paganini and Franz Liszt with unbridled enthusiasm. These virtuosi appeared to be the antithesis of mass musical production. Nevertheless, they can be seen as the inventors of mass entertainment.[3] The performances of the virtuosi astonished the public, who no longer identified with the performes, but rather was passively awestruck by them. Those who supported this trend did not seem interested in the role music played in the education and improvement of the individual, or in music's ability to fuse together a collective with a common goal.

These musical tastes did not merely reflect differing ideologies about music's purpose, but were indicative of the bifurcation of the middle class and indeed were in part responsible for such a divide. In his classic work *Music and the Middle Class*, William Weber has demonstrated that during the early nineteenth century, the middle class was not a monolithic group sharing common tastes. Rather, a split was well underway between the members of the upper class, comprising the aristocracy as well as bankers and merchants, and the liberal professions, such as academics, lawyers, physicians, and bureaucrats, who were far less wealthy. These two social groups formed around different ideas concerning musical taste; Weber refers to this social phenomenon as "taste publics":[4] "The people who went to the two major kinds of concerts made up distinct taste publics. The differences between them stemmed not only from their contrasting musical preferences, but also from a broad division in tastes and life-styles. The rise in the standard of living within the middle class made possible much more lavish and showy ways of living than before, and the propriety of self-display and adherence to fashion logically enough became an intense source of tension. . . ."[5] The members of the upper-middle-class and aristocracy favored the popular style of operas and virtuosic concerts. Popular operatic styles owed much to Gioacchino Rossini and the various alterations of his works by more recent composers, particularly Giacomo Meyerbeer. These concertgoers yearned for showmanship; they had a penchant for the virtuoso's unrivaled skilled performances. And salons served as the venues of these concerts.[6] As Weber claims: "The keynote of the popular-music public was novelty. The technical development of instruments had made possible far greater performing skills and more dazzling new effects than even [*sic*] before, and the virtuosi used them to their full advantage. As is usually true of popular culture, the concerts were concerned more with performance than with the music itself, for the virtuosi became famous less as composers than as skilled players who could use their instruments so dexterously in contexts of high showmanship."[7] Furthermore, this fetish for the lavish was not restricted to music: "Most business families accordingly had a self-confident, ostentatious life-style and valued cultural pursuits of the same order. They looked to the famous virtuosi for an exaggerated picture of the success and glamour they saw in themselves."[8]

The liberal professionals frowned upon such garish exhibitions of what they considered to be tastelessness. This "taste public," which favored the German classic style represented by the works of Joseph Haydn, Wolfgang Amadeus Mozart, Ludwig van Beethoven, and (later) Franz Schubert, was seen by the members of the upper classes as typical of intellectualism. It was staunchly opposed to the crass commercialism of the concerts promoting virtuosi. Indeed, a program of the London Musical Union, reeking of anti-Semitism, claimed that such commercialism only "fill[s] the pockets of shopkeepers and Jew speculators."[9] In 1846 a Viennese journalist lambasted the predilections of those fascinated by Franz Liszt as suffering from "a crude hyper-enthusiasm." He continued, "it is a truly deplorable state of the public when one receives such unedifying nourishment."[10]

The followers of the German style were generally amateurs who often saw music as a critical element of national identity. They were the ones who underscored the importance of music to the education of the self. And they often formed alliances with members of another distinct musical style and taste, namely choral societies, which were mostly made up of members of the lower-middle class, and in the German territories, members of the intelligentsia. As was seen in chapter 3, choral music was rather popular in the aftermath of the Napoleonic Wars, as it was an expression of nationalistic sentiment, seen to unify the German peoples in the absence of political unity.[11]

Although some music lovers did enjoy concerts of both styles, the distinctions in class and taste were obvious to contemporary columnists and reviewers. In 1835 the French writer, Emile Deschamps alluded to the "difference in their points of view toward art and . . . the rivalry between them." And he reckoned that "they are too different in this respect and too equally matched to be able to understand each other."[12] Similarly, the followers of "popular music"—the fans of virtuosi and operatic music—considered those advocating the classical style as arrogant and pretentious. Although they prized highly musical skill, they were far less enthusiastic about musical erudition.[13]

So how precisely does an account of physics and mechanical devices of the period fit into the story? First, musical instruction, like other forms of culture requiring manual skill, was undergoing mechanization. Publishers such as Breitkopf and Härtel of Leipzig had been mass-producing sheet music since the mid-eighteenth century, while musical-instrument manufacture slowly underwent mechanization later in the nineteenth century. Johann Bernhard Logier's chiroplast, a mechanical contraption patented in 1814 meant to discipline young pupil's hands, wrists, and fingers, and later Frédéric Kalkbrenner's hand-guide as well as Henri Herz's dactylion all assisted in the mass education of child pianists for two generations. Pupils using mechanical devices were all trained in the same fashion: everyone using these machines could achieve a similar level of skills. The universal principles of mechanics were seen as leveling the proverbial playing field.[14] The chiroplast was just the device to mass educate young Germans in piano instruction for the newly established seminaries cropping up throughout Prussia in the 1820s.

Second, physicists attempted to quantify virtuosity with a view to assisting musicians interested in forming teaching academies. Once again, musical skills and attempts to understand how they were transmitted to musicians were at issue. In this case, however, music for the bourgeois masses was no longer the goal, but rather the unparalleled skill of the virtuoso. Some musicians attempted to uncover the "mechanical parts" of the virtuoso's body, specifically his or her fingering techniques, by applying the laws of mechanics and dynamics to his or her performance. The key aspect of skill in all forms of culture—science included—is that it lends itself to quantification. Once a skill has been measured, inevitably the next step is to produce a machine that will mimic it. Some musicians turned to physicists to see if the mechanical principles of acoustics could communicate certain skills of the virtuoso to students. Interestingly, the virtuosi themselves were seen as machine-like, as they dazzled their audiences with machine-like rapidity and precision. Of particular interest to this chapter is the role that the so-called universal principles of mechanics, as elucidated in part by physicists, played when musicians wished to communicate their knowledge to their pupils. The story that unfolds touches on two interesting historical themes. First, this chapter explores the tension between contemporary accounts of mechanical skill on the one hand and the organic *Geist* and taste of the musician on the other. Could mechanical principles be used to replace some of the skills taught by a master? Many argued that they could not. Second, this chapter proffers a glimpse into how other forms of contemporary culture—in this instance, music—

perceived the roles of physics and mechanics in pedagogy. It illustrates how musicians used machines and mechanical principles to teach the skills normally taught by masters, and how some musicians, rather controversially, saw physicists as possible allies in pedagogical matters. This chapter does not take sides in the debate. Rather, it proffers a historical contextualization of skill in order to understand the complex relationship between music and physics.

Much of the argument centers on the theme of the organic individual versus the factory-like machine and its methods, a theme that runs throughout this book. And in part as a result of these aforementioned mechanical devices, critical questions concerning the true purpose of music rang out throughout musical circles. Was music about the ability of the individual to free him- or herself from the influence of the State in order to cultivate his or her own taste? Or was the purpose of music to assist civil servants in fostering camaraderie and nationalism? And could mechanical contraptions assist in these endeavors? Or should music simply reflect the extravagant lifestyles of the upper classes?

The Mechanics of Mass Musical Pedagogy

Johann Bernhard Logier[15] was born in Cassel on 9 February 1777. After his father taught him how to play the piano, he left for England in 1791 and later moved to Ireland, where he became the organist at Westport and director of the Kilkenny Militia Band in 1807. He moved to Dublin in 1809, becoming the musical director of the Royal Hibernian Theater for a year. A year later he opened up a musical shop specializing in sheet music and instruments. Logier's claim to fame was his invention of the chiroplast, or "hand-director," in 1814 (see figure 8.1). This mechanical contrivance, which he originally built to assist his seven-year-old daughter in playing the piano, was meant to discipline young pianists' fingers, hands, and wrists.[16] Logier surmised that the technique of proper piano playing was purely mechanical; therefore, he reckoned that a mechanical device could remedy any problems of technique.[17] Whereas piano instructors such as Johann Baptist Cramer, Johann Nepomuk Hummel, William Field, Frédéric Kalkbrenner, and Ignaz Moscheles were accused of training "dilettantes" to play simple waltzes and rondos, Logier hoped to increase the level of skill of his students.[18] He underscored that his mechanical contraption was

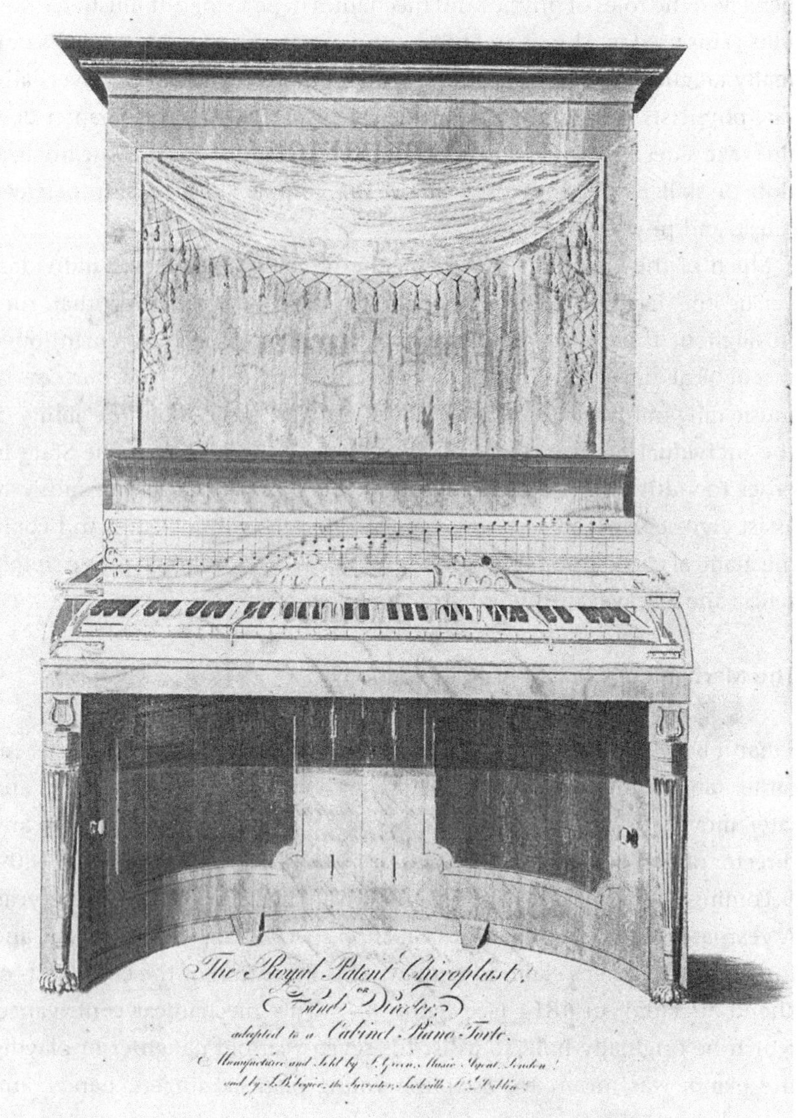

Figure 8.1
Johann Bernhard Logier's chiroplast. Source: Logier, circa 1818, frontispiece.

not meant to replace the master, but simply augment the process of instruction by alleviating his need to spend so much time teaching proper fingering technique and posture:

To the success of the plan of instruction here proposed, it has been objected, that the majority of the profession would oppose it, because it aimed to do that by an infallible mechanical means, which is now done by a long process of instruction, and would therefore tend to diminish the necessary call for their attendance. But no argument can be less founded than this; for, though this method greatly diminishes the most tedious part of professional labour, it must manifestly have the effect of enlarging the sphere of instruction, by giving so much greater inducement to the public to study the Piano Forte, when the acquirement of execution on that instrument is so materially facilitated.[19]

The chiroplast was made of two parallel rails extending from one side of the piano to the other on which a laterally sliding frame was attached above the keyboard. This frame contained two sets of brass frames, or "Finger-guides . . . through which thumb and four fingers are introduced. These divisions correspond perpendicularly with the keys of the instrument, and may be moved to any situation by means of the brass rod, or bracing bar, on which they are made to slide."[20] The wrist-guide, a brass wire, was attached to each finger-guide, "the use of which is to preserve the proper position of the wrist, and to prevent its being inclined outwards, which would necessarily withdraw the thumb from the finger command of its key."[21] The performer's chair should be elevated such that his or her shoulders are always kept down. "The hand must be a little elevated over the keys, but lying parallel to them, slightly turned outwards, so as to make a small angle with the outside of the wrist, thus preventing the thumb from being withdrawn from its key. The hand will be in its proper place, if in this position it is advanced upon the keys as far as the root of the thumb nail, the first joint of which is gently turned inwards, the fingers fall in graceful arches until they slightly touch their respective keys."[22] The students' bodies needed to be kept nearly erect, with a slight inclination toward the instrument. The arms had to remain as close to the body as possible, and from the elbows upward slightly tilted forward, except when the right hand was playing the upper octaves, and the left the lower ones. Clearly, in these extreme cases, the arms were to be extended.[23] In a very real sense, Logier was attempting to render the human body as machine-like as his chiroplast. The use of the chiroplast by pupils learning how to play the piano resulted in the proper positioning of the body, the graceful movement of the arms, and the acquisition of "independent motion" and "equality of power" by the fingers.[24]

This last point is rather crucial. Logier continually stressed "that all the *force* used in striking the key must proceed from the *finger* alone."[25] The influence of the body—including the arms and wrists—was to be minimized, if not totally eliminated. Hence, the chiroplast permitted the pupil's hand to move only horizontally, not vertically, enabling the fingers "to strike the keys with force and motion all their own, independent of any foreign assistance."[26] Logier insisted that only the fingers should generate the force striking the keys, and that "any additional force or motion that may receive from the hand or arm must be detrimental. . . ."[27] The weight of the arms and wrists should never be used to depress the keys, as this would thwart the ability of the pupils' fingers to "acquire a freedom, strength, and equality of power."[28] Each finger needed to be disciplined such that each conveyed equal strength to the keys. As we shall later see, Chopin and Liszt attacked such a technique, arguing that the entire body should be used while performing. Chopin, in particular, asserted that each finger should possess its own strength while depressing the keys, thereby enhancing the range of the pianist's touch. The last part of Logier's mechanical contraption was the "Gamut-board," an oblong board that contained two staves (one for treble and one for bass) with all the notes used in music. Pupils could now associate the key with its corresponding note, learn to read notes more quickly, and be able to differentiate between clefs, treble and bass, more readily.[29]

According to Logier, his "infallible mechanical means" could replace "the most tedious part of professional labour," thereby "enlarging the sphere of instruction, by giving so much greater inducement to the public to study the Piano-forte, when the acquirement of execution on that instrument is so materially facilitated."[30] Logier sharply delineated between the art and science of music. The mechanics involved in the use of the chiroplast "give[s] them [the instructors] that time for the formation of taste, and the study of the more difficult species of composition. . . ."[31] Taste and the art of composition could not be reduced to mechanical principles, whereas the handicraft of playing could.

Logier's chiroplast and piano-teaching method were rather successful for private instruction, or so he himself claimed. By 1818, he informs us,

40,000 copies of earlier editions of his *First Companion to the Royal Patent Chiroplast* had been sold in numerous languages.[32] These texts not only described the chiroplast and its use, but also included numerous musical passages for practicing with the device. By 1824 Logier's works had sold over 50,000 copies in Great Britain alone, while the London musical-instrument maker and sheet-music publisher I. Green had sold over 1,600 of his chiroplasts.[33] Three years later, he claimed that the number of copies of his *Companion* had reached 70,000.[34]

The quintessential entrepreneur, Logier published letters written by renowned musicians and piano instructions in support for his chiroplast, including Muzio Clementi, Hummel, Moscheles, and Kalkbrenner, the French pianist, legendary instructor, and composer of the first half of the nineteenth century.[35] Kalkbrenner argued that the chiroplast was a boon not only for beginners, but also "for every performer who has contracted bad habits in the position of his hands."[36] In 1831 he modified Logier's chiroplast, which he called a *guide-mains*, or "hand-guide," by allowing the turning under of the thumb. And he published his *Méthode pour apprendre le piano-forte à l'aide du guide-mains* (Method of Teaching the Piano with the Hand-Guide), which offered a detailed account of how to play the piano with his device.[37] In the preface to this work, he underscored his reliance on the mechanical device: "After a few days I understood all the advantage that this new working method [the hand-guide] gave me, my hand-position could no longer be incorrect, I had nothing more to occupy me, playing only five-finger exercises. Soon I decided to try reading while feeding my fingers their daily nourishment. For the first few hours it seemed difficult, by the next day I was already accustomed to it. Since then I have always read while practising."[38]

Logier's method was music for the masses brought about by means of mass production. The mechanical processes of playing the piano, according to Logier, were universal and could best be communicated to a large number of students by machines. Twenty or more students could receive a common training at the same time. And if a student moved from one instructor to another employing Logier's chiroplast and method, he or she would not miss a beat "as if the teacher had never been changed."[39] One often reads that the major advantage of this system was its ability to teach many students at once: "The most remarkable feature of this new system is that the pupils, who frequently amount to thirty or forty in number, all practice their lessons at the same time."[40] The students and teachers alike, in a sense, were interchangeable parts within the machine-like structure of Logier's system of piano playing. The household could serve as the site of this algorithmic instruction, "And so mothers should be able to provide for the musical education of their children with equal self assurance without assistance from teachers. And even children from 9 to 10 (Logier assures us) are able to assist younger siblings in many cases; so simple and certain is the basis of the method."[41] As Logier himself admitted in a letter dated 12 November 1817 to Mr. Bishop of the London Philharmonic Society, which had been criticizing Logier's chiroplast, "I pursue to bring my pupils into that state of advancement in the science of music."[42] The universal principles of mechanics were called upon to train an army of young pianists. Logier informs us that Louis Spohr commented on this mass instruction:

> . . . as all who practise the same lesson are [ar]ranged close to each other, the master, when near them, is capable of judging of their performance, without being disturbed by those who are playing other lessons. . . . For beginners he employs his *Chiroplast*, by which the children, even in their earliest lessons, acquire a proper position of the hand and arm. It cannot be denied, that this machine is admirably contrived for the object it is intended to fulfil; and it of course affords vast assistance to Mr. Logier in superintending a number of pupils at once. It might also be advantageously employed for learners in general; for though at the period of giving a lesson, the master has the opportunity of pointing out and correcting bad habits.[43]

Logier's method was generally used to teach a certain sort of student, the sort who would not necessarily go on to become professional performers. It was intended primarily for young girls and women who were playing at home. The British operatic composer and piano instructor William Shield wrote on 14 September 1814 that "many young ladies, who are formed by nature to express brilliancy and taste on the Piano-forte, have been impeded in their progress by practising the elements of Music with bad positions of the wrists and fingers."[44] The chiroplast could remedy these "bad positions."

From 1817 onward, however, Logier's system came under harsh attacks from various individuals and institutions. The London Philharmonic Society, which had been established four years earlier, claimed that his teaching method did not at all improve students' abilities to play. The conservative gentlemen were not impressed by the idea of using a machine to teach the art of piano playing. A series of missives by his enemies were answered by increasingly vitriolic defenses by Logier and his supporters. One major critique relevant to this chapter involved the means by which the facility of playing was improved, or the training of manual skills by an expert. An anonymous British critic defended the organic (as opposed to the mechanical) origins of manual dexterity, which "in the use of a musical instrument is the result of an infinite number of successive attempts. The success or the failure of such attempts depends upon the fitness or unfitness of the bodily organs applied to the purposes in view;—upon the talent and industry of the pupil;—and upon the mode of instruction employed."[45] The critic doubted that a machine was able to teach musical skills. Indeed, he was convinced that only a master could convey the requisite knowledge: "A well-formed boy or girl, whose joints are well-knit and flexible, requires no machine to direct his or her motions on a musical instrument; and those who do not possess such natural advantages, will derive very little benefit from musical gyres of any description. . . . The proper mode of fingering particular passages cannot be taught by any machine:—it requires individual attention and instruction from an able master."[46] Similarly, the German music critic G. L. P. Sievers found the chiroplast simply absurd:

Herr Logier's chiroplast can now be seen and purchased on the boulevards [of Paris]. The boulevards are, as everyone knows, the greatest fashion junk rooms [*Mode-Rumpelkammern*], where everything new can be viewed, and in which everything becomes old after twenty-four hours. The chiroplast belongs among such objects; the invention that proves that music, as well as everything else in politics, morality and art in Europe, stands at its point of culmination, in which sooner or later a complete crash will occur. If mechanical inventions, which are indeed more or less useful (in so far as they more or less make mechanical forces bearable), render the mechanism of handicraft easier, it seems to me an absolute stupidity to want to assist a spiritual [*geistigen*] or much more so—an ingenious [*genialische*] quality . . . through the alleviation of the movements of the hand. The movements of the hands while playing the piano are secondary, a product of an inner spiritual talent to this art. It therefore can only be led, modified and formed [*gebildet*] by this inner talent. Even if Herr Logier, or another one of a thousand such artists, had invented a *psychoplast,* in which the soul of the pianist, rather than the fist, were cut up and castrated as in a musical stable, his would be an idle work of music. Nature is the only sculptor, which forms the soul of the artist.[47]

He continued by wondering why anyone would think that such a mechanical contraption could somehow be a boon to music, belonging like the metronome "to the category of those mistakes of the human spirit," "believing one has made an improvement [to the art of music] through mechanical inventions."[48]

Supporters of Logier countered with the argument that if a mechanical technique (*Mechanismus*) could alleviate the problem, then there was no harm in employing it. Unfortunately for him and his defenders, history was, after all, on the anonymous critic's side, who noted that the virtuosi of the period had all acquired their skill from masters, not machines. The cultivation of taste and judgment was key in producing a pianist, and this "process require[s] the greatest skill and caution. Nothing inferior to a *first-rate* master can communicate just and valuable information in this branch of musical instruction."[49] Unlike Logier, the critic did not separate touch from taste. Although the critic admitted that Logier's chiroplast did position the hand and fingers well, thereby improving a player's touch, it could do nothing else: "That the chiroplast cannot teach the particular modes of fingering adapted respectively to different passages, must be quite obvious. It fixes the hand in one position upon five notes on any part of the key-board, and prevents the fingers from moving either to the right or left. Consequently, the performer, in this predicament, must always strike the same note with the same finger, and cannot learn the art of fingering passages of a greater compass than five notes, or of moving the fingers along the key-board, which constitutes so difficult and so important a branch of the mechanical art of piano-forte playing."[50] A. F. Kollmann, piano instructor and organist of the Royal Chapel of St. James in London, concurred. "First he [Logier] allows his students only one type of finger movement, namely always over the same key. And when he weans them through thirty pages of his work from drawing their hands together and stretching them apart, from setting them one beneath the other or crossing them, from leaping, all of which a child learns by means of a good technique, he removes the chiroplast and now expects them to be able to apply all these methods, particularly the stretching of the fingers, which can be quite impossible for the small hands of a young child."[51] Also, the aforementioned anonymous critic pointed out that by having many students play simultaneously, the effective teaching of the development of a single student's ear or his or her appreciation of sophisticated taste was seriously hampered. Finally, the master needed to divide his attention among many students. A letter to the editor of the *Edinburgh Weekly Chronicle*, written by one H. M. B. and published on 7 May 1817, echoed this sentiment: "Pupils [who have used Logier's methods and chiroplast] do not finger better than those taught by good teachers. They do not have a finer touch, nor more precision and judgment in performance than any master could give them."[52] Like the anonymous critic, H. M. B. too wondered how students could develop a good ear with up to thirty pupils playing together on pianos, which are not always in perfect tune with each other.

C. F. Müller, piano teacher, composer, and singing instructor of the Friedrich-Wilhelms-Gymnasium in Berlin, argued that "music cannot be executed in a factory-like fashion [*Fabrikenmässige*]."[53] Kollmann agreed, arguing that a major problem with Logier's system was that it "was the same for all students," sacrificing individuality of performance.[54] For Müller, *Fingerfertigkeit* (or manual skill) could only be taught by a good, attentive teacher.[55] And, he claimed that the chiroplast was not helpful for the correct position and velocity of the hand, the equivalence of touch, or that tender feeling on the fingertips.[56] Crucially, the chiroplast also hampered the pianist's development of touch by not allowing the pupil to lift his or her hands slowly from the keys, thereby causing the tone to fade away. Touch, for Müller, *contra* Logier, resided in more than just the fingers, but the hands and arms as well.[57] Once again, the charge had been levied against the chiroplast that a good teacher would render it superfluous, if not perhaps a hindrance. And Müller asserted that the ear could not be directed and taught by mechanical means.[58] One often reads how Logier's critics labeled his chiroplast and teaching method as "unnatural."[59] Kollmann argued that the finger-guide's only possible purpose was to ensure that the fingers of the beginner do not miss hitting the correct keys. He added that violins and other bowed instruments, where the tones are not generated by the keys and "where the pure grip requires the greatest accuracy and long hours of practice needed no such mechanical guide. . . ."[60] The chiroplast did not, in Kollmann's and Müller's views, enhance the teaching of mechanical skill. Finally, Kollmann ridiculed Logier's crass commercialism, which he labeled *Marktschreierei* (literally, "the screeches of the market") and which included the entrepreneur's emphasis on his secret teaching method (which could be purchased for hundreds of guineas) and his "chiroplast club" with special costumes and emblematic buttons.[61]

Prussia

Largely owing to these objections, Logier faced problems institutionalizing his pedagogical method throughout Europe. He initially enjoyed success in Britain, but his chiroplast and teaching method never achieved the status he hoped they would. And save the support of Kalkbrenner, the French were recalcitrant in accepting his technique.[62] In 1821, however, Logier's dreams of real success finally came to fruition when the Prussian government invited him to set up his institutes throughout the kingdom. Up to that point, the chiroplast had generally been ignored throughout the German territories.[63] Carl Friedrich Zelter had already embarked upon his unprecedented music reform throughout Prussia. As seen in chapter 3, his goal was to use music as a cultural resource to unite members of Prussia's *Bildungsbürgertum*. Such an ambitious project could only be realized by initiating major changes in schools and churches throughout the territories. Musical education was key. Although Zelter's reforms were far-reaching as he was professor of the Royal Academy of Art and the University of Berlin—including the improvement of church organs and choirs—the aspect most relevant to this chapter was Zelter's plan to create seminaries (*Seminarium*) in order to train future school teachers, choirmasters, organists, and music directors.[64]

The first seminary was established at the University of Berlin on 19 June 1815.[65] Less than four years later the East Prussian city of Königsberg (Kaliningrad) followed suit by founding its own seminary called the Institut für Kirchenmusik (Institute for Church Music), which provided musical education to theology students who wished to be trained as choirmasters and organists.[66] Zelter recommended in June of 1821 that the Prussian government send the music professor Franz Stöpel (also spelled Stoepel) to London for two months in order to observe Logier's technique taught at his London Institute. On 27 June 1821 Stöpel witnessed the adjudication of fourteen female pupils, who had been instructed using Logier's chiroplast and teaching technique.[67] Those present included representatives of Britain's Royal Family, other highly respected families and skilled musicians, and the Viennese composer and piano instructor Ignaz Moscheles. Stöpel labeled Logier's system "an important advance in the methodology of musical instruction," emphasizing how the chiroplast ensured "with precision a natural and methodical position of the body and hands."[68] He went as far as to claim that, *contra* the British critics, even "rare talent that develops into extraordinary virtuosity would or could be tested." Female pupils playing passages for their examination "demonstrated an admirable elegance of taste and feelings."[69] Indeed, the pupils combined both "taste and the greatest precision" when performing the pieces.[70]

At the completion of his two-month stay, Stöpel returned to Berlin recommending that Prussia now adopt Logier's method.[71] On 1 October 1821 Baron Karl Freiherr vom Stein zum Altenstein, Minister of State for Culture and Education, personally invited Logier to Berlin in order to open a Berlin academy, which would implement his teaching method, including his chiroplast. Logier opened the academy on 7 October, and then returned to London after staying only two weeks.[72] Stöpel's and Logier's accounts differ considerably at this point. According to Logier, Stöpel was unsuccessful in inculcating the necessary skills. Von Altenstein then invited Logier for a prolonged stay to assist with teaching at the academy. In August 1822 Logier departed once again for Berlin, and on 15 September he reopened the academy "in an elegant suite of apartments appointed for that purpose by the government. I commenced with sixteen pupils only, all females; some of them children of talent."[73] Logier himself wrote, "In Germany the boys are instructed in music, as well as the girls; it is considered a necessary part of their education. My professional friends may perhaps smile, when I tell them that I have as many classes for boys as for girls."[74] As Arthur Loesser has demonstrated, music making was central to the social development and discipline of the children of all ranks of bourgeois households.[75] In Britain in particular, piano playing was seen as a social obligation critical to the upbringing of daughters, whose ability to play music would impress potential suitors. In the German territories, music (in particular, playing the piano) was critical to the *Bildung* (both in the sense of education and cultivation) of both boys and girls.

Stöpel claimed that he continued to run the Institute after Logier's initial return to London and that on 22 March 1822 his students "passed their first public examinations to the satisfaction of the large numbers present."[76] Indeed, the audience was so pleased that Stöpel was encouraged to open another institute in the city, which he did indeed do a month later. In August of that year Logier returned to Berlin and visited Stöpel, noted the progress of the students, and decided to open yet another institute that he alone would run. He then proceeded to publish articles in German newspapers asserting that Stöpel did not teach the proper system as he had invented it. He also claimed that Stöpel's influence had ceased once he returned to Berlin, a suggestion that Stöpel vehemently denied. Stöpel continued to run his institute until April of 1823 and founded another in Potsdam.[77]

The dispute between the two notwithstanding, Logier offered private tuition at his academy for the children of the *Bildungsbürgertum*. His successful instruction "soon produced a revolution in the minds of the public in favour of my plan. The number of students increased. . . . The object which I had in view was, that as my system had once got into Germany, it should be permanently and firmly established there."[78] According to Logier, Zelter recommended to von Altenstein that Logier's pupils be evaluated. If the pupils improved, then the system should be introduced into the seminaries.[79] These institutions were state run, and both vocal and instrumental music formed a prominent part of the education. On 14 and 15 February 1823 the official adjudication of Logier's system took place. The examination lasted four hours, two hours each day. The judges included Zelter, the composer Bernhard Klein, the music director and organist A. W. Bach, and Schneider, concert master and director of the Berlin opera when Spontini was away.[80] They were all rather impressed with what they heard. Von Altenstein wrote to Logier that the "uncommon correctness . . . with which the pupils performed the theoretical and practical exercises had astonished him in no small degree."[81] Logier's method was now to be implemented at the seminaries. Von Altenstein chose twenty seminary professors throughout Prussia to travel to Berlin to be instructed by Logier himself. By 1824 Logier's technique and chiroplast were introduced in seminaries in Danzig (Gdansk), Stettin (Szczecin), Neu-Zelle, Potsdam, Weissenfels (near Erfurt), Karline (near Ilstenburg), Soast (near Munster), School Pforta (near Naumberg), Berlin, Breslau (Wroclaw), and Jenkau (Jankowa, near Danzig).[82] Dresden, Leipzig, Frankfurt am Main, Frankfurt an der Oder, Stuttgart, and Hamburg soon joined the list of cities with music institutes throughout the German territories employing Logier's chiroplast and technique.[83] His textbook,[84] in conjunction with his device, was to exert a powerful influence upon Prussian piano instruction for over half a century. (Johann) Karl (Gottfried) Loewe, music director at the Gymnasium, seminary and Church of Jacob of Stettin, found Logier's teaching method and chiroplast very useful for instruction in his seminary.[85] Berlin critic and editor of the *Berliner allgemeine musikalische Zeitung* Adolf Bernhard Marx praised the Prussian government for being the first

State to institutionalize Logier's teaching methods: "So in Germany it [Logier's method] received its scientific acceptance and basis, its general adoption in the science of sound [*Tonwissenschaft*]. Without a doubt Germany is the appropriate and sacred midpoint for scientific achievement generally speaking, especially for the science of music, since music has won its most active and warmest life only in Germany and has remained as an external, foreign entity for the British."[86]

A decade after Zelter's decision to implement Logier's chiroplast throughout Prussian schools, reviews of the device still appeared in major musical periodicals. An anonymous critic, labeling himself "the old piano school master," summed up his thirty years of experience in defending the hand-director: "I am actually no friend of machines in art, but this one here, please do not be frightened, ladies—I must place at the very top of my instruction, *conditio sine qua non*."[87] "Correct and methodical [*schulgerichte*] mechanics," was, according to the old piano master, fundamental to piano playing.[88] Such "mechanical skill" was predicated on "a correct position of the hand and correct movement of the finger joints, where movements are totally independent of the upper and lower arms."[89] Logier's chiroplast and Kalkbrenner's subsequently improved design of the hand-guide aided the student in attaining such skill.[90]

In 1841 the organist and "instructor of daughters" (*Töchter-Lehrer*) of Sanghausen Carl Breitung recommended the chiroplast for beginning students, although some of his poorer students could not afford the luxury. He argued that it was not sufficient for the instructor merely to possess a chiroplast: the students needed one at home for their daily practice.[91] Breitung incorporated the chiroplast in different stages of educating his pupils. He recommended Kalkbrenner's hand-guide, as it was less expensive, and it could also be used by "masters of greatest achievement."[92] These devices assisted the pupils in the correct positioning of the body, particularly the arms, hands, and fingers, "everyone of which must learn to hit [the keys] with the same strength."[93] During the 1820s the piano teacher Friedrich Wieck used Kalkbrenner's hand-guide to teach his daughter, Clara, future wife of Robert Schumann.[94] This emphasis of each finger possessing the same strength was underscored by one of Paris' leading piano instructors of the first half of the nineteenth century and one of Liszt's instructors, Carl Czerny. "You know, madam, that the five fingers are not by a long shot equally strong. For example, the thumb is much stronger than the others. The ring finger is, in comparison, for almost everyone the weakest of the digits. The pianist must therefore know how to use these different strengths so that for scales all fingers hit the keys with the same strength."[95]

Some twenty-one years after Logier's death, his chiroplast was still being used by beginning pianists. The piano teacher Franz Brendel's *Geist und Technik im Clavier-Unterricht* (Spirit and Technique of Piano Teaching) of 1867 included a section on the "methodical education of the hands and fingers," which recommended the chiroplast for the development of finger strength.[96] Also mentioned was the dactylion of Henri Herz, the Austrian pianist and composer who taught piano in Paris. It was an apparatus made up of ten rings through which the pupil placed his or her fingers. These rings were hung by threads, the ends of which were connected to springs, thereby increasing the pressure on the fingers adding to their strength after much practice.[97] Although the devices were helpful for children with particularly weak fingers, once the requisite strength was achieved, Brendel argued that the both the dactylion and hand-guide were rendered superfluous, and could even become harmful.[98] By the 1870s, Logier's chiroplast and Kalkbrenner's hand-guide were finally falling out of favor. They had been an example, however, of mass education via machines par excellence. They could teach piano students the skills relevant to touch. The universal applicability of mechanical principles enabled a larger cross-section of bourgeois culture to obtain fundamental musical skills.

Teaching the Mechanics of Virtuosity

During the early nineteenth century, several musicians and music scholars turned to physicists in order to explain virtuosity by drawing upon the universal principles of acoustical mechanics. Precisely because virtuosity was being increasingly defined as someone who could play rapid and difficult passages (or what was referred to as the mechanical aspects of performance) and decreasingly considered to be the performer who played slow, melodic passages, physicists seemed to be able to offer quantifiable answers to a seemingly nonquantifiable aesthetic phenomenon. Several music scholars thought that with the assistance of physicists, the analysis of the mechanical aspects of musical phenomena might enhance the education of music students with a view to improving their performance.

During the first half of the eighteenth century, virtuosity had been predominantly affiliated with music theoreticians, scholars noted more for their tomes on composition and temperaments than any ability to perform.[99] This early eighteenth-century definition of virtuoso owed much to Sébastien de Brossard's earlier definition of 1705, which had equated the attribute of being a virtuoso with a "*Superiorité de genie, d'adresse* ou *d'habileté,* qui nois fait exceller soit dans *la théorie,* soit dans la *Prattique* des *beaux Arts.*"[100] It had been a quality of "certain people, who excel in a particular art or science such as music and painting."[101] By the second half of the century, the term "virtuoso" had increasingly begun to denote the performer rather than the theorist.[102] The music theoretician Johann Adolf Scheibe, for example, defined virtuosity as a characteristic "only ascribed to those who have acquired a high degree of excellence," arguing that the term was more applicable to the "practical music than the theoretical."[103]

In 1821 E. F. F. Chladni complained how he was living in an age when speed and skill were the hallmarks of musical ability, rather than the playing of expressive, harmonic passages. The aesthetic purpose of Chladni's euphone and clavicylinder (discussed in chapter 2) was twofold: first, they attempted to stem the rise in interest of rapid passages in music played by virtuosi, and second, they reminded the audience how beautiful the slower, more melodic passages were. "The existence and possible proliferation of such instruments [the euphone and clavicylinder] are without a doubt true needs of our times, especially in order to put on hold those many individuals who are exclusively interested in the prevailing taste of speed and technique, whereby music is lowered to a form of handicraft. Rather these instruments are to achieve the opposite, to base the noble and true purpose of the art of measured taste on expressive and harmonic passages."[104] He adds a footnote to this passage: "One hears, more so in recent times than ever before, an awful surge of rapidly running and springing tones, whereby no other sensations are evoked other than astonishment. And this instead of what the purpose of music should be, to express sensitivities and to excite."[105] Aesthetic beauty was succumbing to digital gymnastics.

Chladni expressed his rather perceptive fear that the late eighteenth and early nineteenth century, when he had invented his two instruments, had witnessed a change in musical taste. Rather than the slow passages prominent in the musical performance in the late years of the century, rapid movements requiring a skilled hand were now, as had been the case in the first half of the century, the sign of both a talented performer and a trend-setting composer. As early as the 1790s, music critics had complained how glass-harmonica virtuosi were capitulating to the crass whims of contemporary audiences by playing rapid pieces, which were not well suited to such an instrument, using only the upper tones. The audience, the critics claimed, was only interested in "admiration, emotion, and terribly exaggerated sensitivity of the heart" evoked by lightening passages and exaggerated ornamentation.[106] This exaggerated sensitivity, as opposed to properly cultivated sensitivity, connoted vulgar superficiality. In a similar vain, the retired Royal Prussian Concert Master Carl Benda expressed his concern that an overemphasis on "mechanical skill" had taken away from the "tender feelings" needed to play properly adagio movements on the piano.[107]

This concern that technical activity was trumping other aspects of the performance seems well founded. A. B. Marx did his utmost to thwart the tendency of the period to succumb to the talents of the virtuosi, which at this time meant both a skilled performer and, rather pejoratively, a tasteless showman. Marx stressed the importance of the symphony (particularly Beethoven's) to the education of Berlin's *Bildungsbürgertum*. He, like Zelter, was interested in the role music could play in establishing a national identity.[108] In 1824 he reviewed a performance of Beethoven's Second Symphony, chastising the caprices of various virtuosi: "Not merely the actions of the principles, but also the extras, the dancers, the decorations, and the finery of the singers, capture [the audience's] attention and interest, while they imagine themselves to be occupied with the music. The content of the composition does not enthrall the greatest part of the listeners, rather this or that run or trill, the visual aspect—or more preciously, the singer's aigrette and cloak trimmings. . . . In the performance of a symphony nothing external should take part, not even an alluring personality or a conspicuous virtuosity."[109]

In 1830 Bremen's music instructor, Wilhelm Christian Müller, concurred. To him "the vulgar [*gemeine*] path of the technique, or the learning and mechanical practice of the art according to the best grammatical rules and theoretical principles" was anathema to the true purpose of music, which was "to excite, to penetrate into the heart of the sacred, in order to succeed in approaching the view of the ideal goddess of beauty. . . ."[110] As a result,

Müller bemoaned, "we find great virtuosi in singing and playing musical instruments, who view and handle music like a wooden jointed doll, which gives music all kinds of artificial and plastic positions, and which drape and decorate music with an endless diversity of multicolored styles of clothing so that the vulgar spectator finds great joy. But all of this is wrong, superficial."[111] Virtuosity, he continued, "strives for a tickling of the senses, styled fantasy, sentimental softness, and it subtracts from its higher purpose, of duty and prosperity. It is, hence, a product of corrupted taste."[112] Like all the sciences (*Wissenschaften*) and arts, the true purpose of music was "self improvement" (*Verschönerung*) and "civilized behavior" (*Versittlichung*).[113]

In addition to this gradual ambiguous transformation of the meaning of the virtuoso, there was a more subtle, and—apparently for several scholars of the late eighteenth and early nineteenth centuries—a more disturbing metamorphosis. As early as the 1730s, musical theorists and some composers had been criticizing the tendency of certain composers to test the skilled fingers of performers, particularly pianists.[114] This critique was heard with a much greater frequency and intensity throughout the early nineteenth century. The emphasis on "mechanical skill" as seen as contrary to the true idea of art, which is "the production of the real through the ideal."[115] Indeed, "mechanical acquisition in music never leads to an inner sacredness, where the true feeling of art and artistic originality bestow a higher consecration."[116] An article in the *Allgemeine musikalische Zeitung* from 1807 defined a "blind musician" as one who busied himself with "the mechanical aspects and technical aspects of the art" "without cultivating genius."[117] In 1830 W. C. Müller also placed part of the blame on contemporary composers, particularly Beethoven's instrumental works and Rossini's vocal scores.[118] He compared their compositions, which were filled with "lightning trills and runs," "wild leaps played in such dashing speed that neither performer nor listener has time to breathe," to the hideous fashion tastes, such as the "current kaleidoscopic color games of women's clothes."[119] Yearning for the eighteenth-century days of the simpler compositions, Müller lamented how the requisite level of technical skill to perform newly written sonatas and symphonies was so high that amateurs could no longer perform them. But he also admitted that traveling virtuosi were also responsible for the predicament, as they strove to shock and excite their audience.[120]

By the 1830s, the association of the term virtuoso with the music theoretician was totally shed. The virtuoso was a skilled performer, "the mighty, the superior on his instrument."[121] And virtuosity implied "the power, the perfection of the performance,"[122] defined as "an artiste's merit, an artiste's distinction."[123] Virtuosity was now associated only with the physical activity of playing an instrument or singing. The virtuoso was much less the thinker, and much more the doer. In his definition of virtuoso of 1838, Gustav Schilling actually contrasted two categories of artiste in music: those who composed pieces, and those who played them. Only the latter group of performers could be considered virtuosi, while the "performing composer" could be considered a "genius."[124] But repeating the concerns of a century earlier, Schilling was quick to point out that "when virtuosity searches for its task solely in the surmounting of technical difficulties, it sinks under all other forms of artistry, and becomes a handicraft. To make skills alone—the brilliance of technique alone, the highest goal of virtuosity, is to misunderstand it [the true meaning of virtuosity]."[125] Schilling continued by reiterating Chladni's earlier complaint: "We well know that in recent times this [playing difficult pieces] especially happens, and the brilliance of our instrumental music chiefly is the only part of virtuosity in which an enormous sum of mechanically acquired skills manifests itself. The guilt itself may lie partly with the composer."[126] The "trained mechanical skills" of the performer were seen as distinct from his or her necessary ability to comprehend the "latent spirit of the musical piece" (*schlummernde Geist des Tonstückes*).[127] This "spirit" was organic. Schilling noted that the mechanical skill of the virtuoso had risen tremendously when compared to the talents of virtuosi of the previous century, such that pieces, which eighteenth-century virtuosi had claimed to be unplayable, were now being performed with relative ease by beginners.[128] He lamented that few contemporary virtuosi were capable of playing well Beethoven's works, as they "depend on something more than mere technical skill," and hence were being played to show off one's skills, rather than for the beauty of the piece itself.[129] Rehashing the criticism levied by W. C. Müller on many contemporary performers, Schilling feared that instrumentalists were playing Beethoven merely to exhibit their technical brilliance rather than to demonstrate a deeper understanding of the piece. Unlike Müller, however, he did not place the blame on the great composer. He feared that the works of Muzio Clementi and Giovanni Battista Viotti, which did not require such a skilled hand, would be lost, as performers were not interested in playing them. Finally, he concluded his definition of virtuoso with the fear of the repercussions that this recent development would have on present and future compositions. According to Schilling, many contemporary composers were robbing audiences of the pure beauty of music by composing pieces that required mere technical mastery.[130] They, unlike Beethoven, had lost sight of the true purpose of composing music.

Paganini: The Devil's Violinist

The complaint against the exaggerated importance of *Fingerfertigkeit* to virtuosity was a response to the virtuoso's penchant for showmanship. This was perhaps best exemplified by Niccolò Paganini of the 1820s and '30s. Debates concerning Paganini's controversial virtuosity raged throughout European bourgeois and aristocratic circles. He himself reportedly started the legend that he had obtained his unparalleled skill from the Devil, continuing a centuries-old trope of violinists' deals with Satan. His fourth string, which was rumored to be composed of the intestine of his mistress whom he purportedly murdered, elicited wondrous and melodic tones. The rumors continued. He supposedly spent twenty years in prison for his murderous deed, accompanied only by his violin. During this time in solitary confinement, he was able to ferret out the secrets of his instrument, inventing a new fingering technique.[131] As fantastic as these stories were, they seem to pale in insignificance to his very real performances. Whenever he broke a string from his passionate and forceful playing, he compensated without missing a beat, by continuing the piece with only three strings. Should another break, he could play with two. Indeed, his coup de grace was his uncanny ability to play an entire piece on only one string. In order to safeguard his secret technique, Paganini rarely practiced in public, and he never permitted an accompanying orchestra to hear him play the cadenza until the actual performance. [132]

The German musician F. G. J. Schütz attempted to defend Paganini's virtuosity by claiming, contrary to what some of the Italian's critics claimed, that the violinist was not merely a technical wizard (*Fingerhexerei Paganinis* was often used to describe his talent), but also possessed the other necessary qualities of what he thought defined a virtuoso, such as "brilliance, agility, various nuances of playing, power and tenderness, clearness and precision, combined with the highest velocity."[133] Schütz added that Paganini not only possessed a "perfection in the entire technical component of his art," but his "intonation is always as pure as silver and extremely firm. Even in the most difficult leaps, he makes no mistakes. And his precision of the passages is just as unparalleled. . . ."[134]

One can begin to understand Schütz's spirited defense of Paganini by inspecting contemporary responses to his European tour of 1828–1831. The Belgian musicologist, composer, and music critic François Joseph Fétis first labeled Paganini a charlatan, who composed music that no one could play, but later changed his mind and called him the greatest living violinist.[135] Spohr claimed that Paganini's compositions were "a strange mixture of consummate genius, childishness, and lack of taste," and was both "charmed and repelled" by his playing.[136]

Perhaps the most revealing response to Paganini's playing was recorded by Schütz himself. When Paganini was touring Germany, Schütz decided to invite several friends to a performance which he himself was unable to attend. After the concert, he spoke to his friends individually, asking them what they had thought. Half of those who listened to Paganini considered the violinist to be "a divine, one-of-a-kind man, who could achieve unprecedented things, and who could thrill everyone."[137] The other half assured Schütz that he had not missed much, accusing Paganini of being a "charlatan."[138] How could it be that half a group experienced Paganini as the greatest living violinist, while the other half claimed he was a fraud? The split opinion revolved around the definition of virtuoso. Both groups agreed that Paganini possessed superb fingering technique. Indeed, Paganini's playing became synonymous with skill. But by the late eighteenth century, terms such as "dilettante" and "amateur" became inextricably and pejoratively linked to "virtuoso."[139] The ability to satisfy the curious, to quench the passion for the marvelous and to titillate the audience was often viewed with disdain by certain groups, particularly those belonging to the liberal professions. In a sense, Paganini was both charlatan and virtuoso. Indeed, the one needed the other. The charlatanic attributes, such as pandering to the crowds and not delivering what some musicians and critics saw as the true purpose of music, could not exist without true virtuosity. And true virtuosity could not thrive without charlatan-like attributes. The following anecdote illustrates this point. Henri Maréchal once told the elderly Franz Liszt that he was appalled by a pianist playing a difficult arrangement for the left hand of the *Rigoletto* paraphrase. In what Maréchal deemed as tasteless panache, the performer took out a handkerchief with his right hand and blew his nose, while still performing with his left. Liszt retorted, "My dear child, for a virtuoso that is necessary! It is absolutely indispensable."[140]

Paganini, musicologists inform us, revolutionized the violin's repertoire. In 1820, a year before Chladni condemned the vogue for playing rapid passages, Paganini published his twenty-four *Caprices*, op. 1, to which the twentieth-century violinist Yehudi Menuhin referred to as the "New Testament . . . of the violinist's manual" some 150 years later.[141] These pieces vastly extended the limits of the violin, encouraged new pedagogical techniques, and were seen at the time as studies of advanced technique signifying the highest level of skill.[142] More important to my account, of course, was the contemporaneous response to his *Caprices*. Despite the impressive number of copies sold throughout Europe, violinists of the day, including the very best ones, could not play Paganini's pieces.[143] Many argued that he had simply composed unplayable music. The renowned French violinist Baillot cried out in failure, "*Omnes vulnerant, ultimus necat!*"[144] The Italian, however, astonished audiences throughout Europe with his performances of the *Caprices*, which truly were a tour de force.

Standardizing the Virtuoso's Technique: German Musical Pedagogy

Throughout the early nineteenth century, then, many scholars were redefining the notion of the virtuoso in order to reflect technique and skill, distinguishing it from those who evoked musical spirit by playing harmonious pieces. For example, music critic Gottfried Weber and the Frankfurt *Kapellmeister* Carl Guhr tried to explain why Paganini's playing of the violin was far superior to anyone else's of the period.[145] Guhr accompanied Paganini on portions of his tour throughout the German territories, meticulously observing Paganini's recitals and thoroughly discussing with him his technique, in order to reveal "the secrets of Paganini's violin playing."[146] Of Paganini's ability, Weber wrote: "On this we all agree: his technical skill, the so-called mechanics of his playing is exceptional, as yet unprecedented and if not altogether incomprehensible, it is still to this point not comprehended."[147] Paganini's unparalleled technique separated him from the rest of the violinists.[148] Weber was puzzled that no one had previously provided a detailed account of the virtuoso's technique. He thought that by uncovering Paganini's clandestine technique, a generation of violinists could be educated to learn the master's skills.[149]

In 1829 Guhr published a textbook to assist young violinists to increase their level of technique, or as he himself hoped, "I would like to succeed in teaching something of the mechanical part, which Paganini has furthered so tremendously, and in so doing attain a higher level in general."[150] Although he knew quite well that it would be futile to hope that his students would play as well as the great Paganini, he did think that such a text, which included musical exercises for the students to play in order to master the various techniques, would bring about an overall improvement in play. It is critical to note that Guhr was interested in discussing that portion of Paganini's virtuosity that could be taught, namely the "mechanical part" (*mechanischer Theil*).[151] Guhr's text, which was subsequently translated into French, Italian, and English, is a prime example of the genre of textbooks for various instruments that first emerged throughout Europe during the second half of the eighteenth century and that saturated the market by the 1820s and '30s, dealing with pedagogical aspects of violin playing in the form of studies and exercises for students.[152] Although these texts could not teach "that part of music that results from the power of formation [*Einbildungskraft*] and feeling, the product of genius," they were nevertheless welcome additions for teaching artistic skill.[153] Several musicians, however, complained that these texts purported to teach taste.[154] By the 1820s and '30s these instructional aides for violins elucidated typical problems of technique, most notably posture, bow grip, left- and right-hand technique, and special effects as well as artistic interpretation, specifically ornamentation and tonal color.[155] Piano-teaching texts also enjoyed immense popularity during this period, including Johann Nepomuk Hummel's *Ausführliche theoretisch-practische Anweisung zum Pianoforte-Spiels* (A Detailed Theoretical-Practical Instruction on Piano Playing) of 1829, Czerny's *Systematische Anleitung zum Fantasieren auf dem Pianoforte* (Systematic Instruction to Fantasize on the Piano), also of 1829, and F. Kalkbrenner's *Méthode pour apprendre de pianoforte* (Methods for Teaching the Piano) of 1837.[156] All of these textbooks, aimed at amateurs and lesser professionals, were intended to complement individual private tuition, enhancing, but never totally replacing, the highly coveted master–teacher relationship.[157] They were used privately and in the newly created conservatories that were cropping up throughout Europe.[158]

Guhr argued that there were six ways in which Paganini differed from other leading violinists of the period. First, the tone that he summoned forth from his violin was superior as he tuned his strings sharper by as much as a full semitone. Second, he possessed a unique bowing technique that enabled him to spring effortlessly from one string to another. Third, he was able to mix bowing passages with pizzicato, using the second, third, and fourth fingers of his left hand, rather than the right. Fourth, as a result of his superior fingering technique, he often incorporated single and double harmonic tones in his performances. Fifth, he would often play pieces on only one string, usually the G string. And finally, his uncanny ability to play the violin as if it were a harp, mandolin, or similar instrument deceived listeners into thinking that they were hearing two performers playing simultaneously.[159] Using these differences as starting points, Guhr offered more elaborate explanations of Paganini's technique. According to Guhr, Paganini preferred to play the notes at the upper end of the violin's range, which very few violinists dared to attempt, and his ability to play harmonics on the thinner (i.e., higher) strings was unrivaled.[160] Guhr also commented on the performer's unique playing stance, noting that rather than holding his left elbow vertically underneath the midpoint of the violin, as is normally the case, Paganini pressed the point of his elbow quite tightly against his body, with his forearm turned outward (see figure 8.2). His right arm lay perfectly still and rigid against his body; therefore, his posture was apparently not as "noble" as other violinists, such as Baillot, Rode, or Spohr.[161] Paganini's bowing grip was also unique. The joint of the right hand did all the work, while the right arm remained still. The thumb and index finger held the bow very lightly, with only the small finger acting as support.[162] The violinist's center of gravity also attracted the German's attention. It lay on his left side, yet he was able to move his left shoulder more than any other violinist of note.[163] His right wrist could move with greatest ease, very rapidly directing the elastic movements of the bow across the strings. By disciplining the body to resemble Paganini's unique stance, students might improve upon their playing.

The attentive, skilled musician could emulate many of these characteristics. In a sense Guhr offered a blueprint (albeit a vague one) for how

Figure 8.2
Niccolò Paganini (1782–1840). Painting by Richard James Lane. Source: The National Portrait Gallery, London.

students might improve their playing. For him, emulation of Paganini was a worthy, yet unattainable, goal. As discussed below, some were convinced that such attempts were futile, owing to the uniqueness of Paganini's anatomy.

Explaining Anatomically the Talent of the Virtuoso

The Paganini phenomenon even ignited the minds and imaginations of European experimental natural philosophers and physicians. Francesco Bennati, the physician of the Italian Opera of Paris and personal physician of Paganini who accompanied the virtuoso during numerous performances, read to the French Academy a paper that attempted to explain Paganini's extraordinary virtuosity by referring to his medical history and anatomical build. Bennati, himself an amateur musician, did not want to analyze Paganini's facial features à la Johann Kaspar Lavater or Franz Joseph Gall, but rather he offered an anatomical composite of the virtuoso's entire physique. The physician argued that Paganini's virtuosity was "a result of the unique state of his body, rather than prolonged periods of practice."[164] Although his brain enabled him to compose brilliant pieces, "without his astonishing rhythmic feeling and the structure of his body, his shoulders, arms, and hands, he would have never become this incomparable virtuoso."[165] Bennati argued that Paganini's secret could be revealed by first appreciating that the "development of his general sensibility" was critical to his virtuosity.[166] The physician therefore gave a detailed account of the virtuoso's medical history, with a particular emphasis on the diseases affecting the irritability of the artiste's nervous system.[167] Second, Bennati argued that in order to understand Paganini's secret, one needed to study his physique, which he claimed, was unique. Indeed, unlike G. Weber and Guhr, Bennati implied that attempting to recreate Paganini's talent was ill advised, since it was completely predicated on his unique medical history and build.

> One studies the way he holds the violin, the position that he sometimes gives to his arm. And tell me, who could mimic this artiste? Who, for example, could cross elbows over the chest in order to produce a certain effect? . . . In whom else, other than Paganini, do we find a natural talent, which enables his playing, namely, that the left shoulder is an inch higher than the right? And with whom else when standing upright and letting his arms hang, does the right side appear to be longer than

the left? The flexibility of the capsular ligaments of both shoulders, the relaxation of the ligaments that attach wrist to the forearm and the wrist to metacarpus and which interconnect the various joints: who else possesses that and therefore the corresponding ability of Paganini's? The hand is not larger than normal, but by means of its flexibility, which is the same for all of its parts, his span is twice as large as normal. Therefore, without changing the position of his hand, he can uniquely bend the upper parts of the fingers of his left hand, which touch the strings so that without changing the position of his hand, the fingertips can bend laterally with the natural movement of the joint with great ease, reliability, and speed.[168]

Clearly for Bennati, Paganini's anatomy was sine qua non. He continued, "One might object by arguing that these physical talents are only developed by long hours of practice. Perhaps, but one still must admit that nature has given him an amazing form, so that he can come to this result. . . . All the advantages, which nature has bestowed upon him, were for him what the larynx is for the singer."[169] Although Bennati underscored the biological uniqueness of Paganini, his account must also be considered a mechanical analysis. In theory at least, Paganini's anatomical characteristics could be precisely measured and therefore quantified.

Bennati then proceeded to analyze the organs crucial to the perception of music, namely the brain and ear. Without offering any specific medical information on Paganini's inner ear, Bennati argued that "obviously, his ear is particularly well suited to receive exactly the tones, and there is a relationship between the external characteristics and the internal structure of the ear."[170] Continuing with Paganini's external characteristics, Bennati noted how the violinist had an enormous cerebellum, and "the convolutions of his occiput," which he linked to Paganini's sense of hearing, "are so prominent."[171] Apparently, the Italian could hear the slightest whisper from large distances, and speaking loudly in his vicinity caused him enormous amounts of pain, indicating that he possessed an extremely sensitive eardrum.[172] "His ears are constructed for the reception of sound waves. The auricle is wide and deep, the projections are sharply defined; all the lines are indented. It is impossible to find another ear whose parts are in a better proportion. . . ."[173]

Bennati's review caught the eye of Johann Wolfgang von Goethe, that titanic defender of organicism against the reduction of Cartesian mechanism. Bennati, according to Goethe,

sets out in an intelligent manner how the musical talent of this unique man, determined by the uniformity of his body and by the proportion of his limbs, is favorable to, even compelled to, produce the unbelievable, indeed impossible. This traces us back to the conviction that the organism in its determinations produces the wondrous manifestations of the living being.

Since there is still some room left, I only want to write down here one of the most important words, which our ancestors have left behind for us: "The animals are trained via their organs." Now one thinks how much of the animal has remained over in the human, and that the animal has the tendency to instruct its organs, so that one will always gladly return to these views.[174]

On the other hand, W. C. Müller thought that the relationship between the body and virtuosity was reversed: virtuosity was not predicated on unique anatomical forms, but rather it disfigured healthy human bodies. "[The] enormous rapidity of the fingers and arms, so that more than sixteen tones per second can be clearly heard, lies beyond the realm of human joints. One already has examples where youth have suffered paralyses of arms and fingers through excessive practicing on string instruments. Paganini's ghostly form bears witness, too, to his excessive practicing. With the current wriggling skill, this is a natural punishment of our luxurious times, which drives everything to the extreme."[175] Müller, too, ridiculed Guhr's attempt to capture the secret of Paganini's techniques in the form of a textbook with a view to establish a violin school, not because such an enterprise was doomed to fail from the outset, but because Paganini represented all that was wrong with music during the age of the virtuoso.[176] In his eyes, the true purposes of music, ethical improvement of the self and the worship of God, were being forgotten. Müller commented upon Paganini's appearance and performance posture, his "changing and unusual position of the violin," and "the entirely tragic-comical form of this emaciated man with long black locks of hair, pale face, fiery eyes and satirical-smiling mouth."[177]

The Revolt of the Pianists

The fascination with the Paganini phenomenon, however, was not limited to physicians and poets fascinated with nature. Pianists, perhaps more so than any other instrumentalists, were dazzled by the Italian diavolo. Musical pedagogy during the Classical period was deeply influenced by what Jean-Jacques Eigeldinger has referred to as the "mechanical conception of instrumental playing."[178] The acquisition of virtuosity was predicated on the successful playing of collected "recipes," referred to as *methods*, used to discipline the performer's body, particularly the fingers, hands, wrists, and forearms.[179] These were passages to be memorized by the student. Hours of repetitively practicing "digital gymnastics" and stubborn repetition with mechanical devices, as epitomized by *Études de mécanisme* of Czerny, Kalkbrenner, and Herz, were required.[180] One scholar labeled these Parisian performers "steel-fingered, chromium-laden virtuosos."[181] Kalkbrenner, in particular, was singled out as a "machine-tool pianist who functioned with perfect precision down to his last finger-joint."[182] One of Liszt's teachers, Ferdinando Paer complained that Kalkbrenner's fingers were "a well drilled company of soldiers," and that he sat motionless at the piano, much like a general surveying the battle from behind the front line.[183] During the early 1830s Paris was overrun by the mechanical virtuosi, or what the German writer, poet, and social critic Heinrich Heine called "a plague of locusts swarming to pick Paris clean."[184] The virtuoso's mechanical body, disciplined by the chiroplast or hand-guide, trumped his or her organic spirit. Leading pianists of the period, tired of practicing the *Études de mécanisme,* quickly adapted Paganini's *Caprices* for piano, marking a new era of skilled exercises for young pianists.[185] Robert Schumann was the first to draw upon the work of the Italian virtuoso, with his *Studien nach Capricen von Paganini* op. 3 of 1832, followed by his *Sechs Concert-Etüden nach Capricen von Paganini* op. 10 of 1833. Franz Liszt, whose epiphany at first hearing Paganini in 1832 is well known, composed his own variation in 1838, *Études d'exécution transcendante d'après Paganini,* which was subsequently revised in 1851 under the title *Grandes Études de Paganini.*

As the musicologist Plantinga has argued, "This [Liszt's piano playing] is a decisive move from the characteristic posture of the earlier nineteenth-century virtuoso, for whom the music performed has little purpose but the exhibition of his own lightning speed and thundering octaves."[186] Nevertheless, Liszt's early career was marred by accusations of being a showman. Nineteenth-century virtuosi (and their audiences) were just as interested in the visual aspects of performance as the auditory. The Hungarian pianist was renowned for his fiery gesticulations involving his torso, hands, arms, fingers, and even hair. His concerts were spectacular; they dazzled and entertained. He invented the modern piano recital, as he was the first to perform an entire program from memory and was the only performer of his day who could perform the piano's entire repertory, from J. S. Bach to Chopin. And perhaps most important, he was the first pianist consistently to turn his instrument perpendicular to his audience, allowing his spectators to observe his fingers, hands, torso, and facial expressions.[187] Clearly, his performances required more bodily involvement than the chiroplast or hand-guide would permit. Although the young Liszt advocated the use of the chiroplast to his pupils, he later abandoned it, arguing that it caused more harm than good.[188]

With Liszt the piano moved out of the aristocratic salon into the bourgeois public concert halls.[189] As a result of what some contemporaries saw as flashy pandering, mere panache, Liszt, too, was often lambasted as a charlatan by his critics for being too showy. Repeating Schilling's and Chladni's concerns earlier in the decade, the editor of *La revue musicale* Fétis wrote in 1828: "What a pity that natural gifts such as those possessed by M. Liszt are employed solely to convert music into the subject for a tumble-rigger and conjuror. This is not for what this enchanting art is destined. It should touch us, move us, not astonish us. . . . Profit from time where your still-virgin faculties permit your talent to change direction; take a step back and be the first among the young pianists, and have the courage to renounce brilliant frivolities for advances that are most substantial. You will reap the rewards."[190]

It seems that Liszt was well aware of this fine line demarcating true virtuosity from charlatanry. A decade later Liszt himself accused the Chassidic musician Michael Joseph Gusikow, who constructed an instrument out of wood and straw resembling a xylophone, of the same offense, describing him as "the musical juggler who plays an infinitely large number of notes in an infinitely short period of time, and draws the most possible source out of two of the least sonorous materials. This is the prodigious overcoming of difficulty that all of Paris is now applauding."[191]

In his necrology of Paganini published in *Gazette musicale* on 23 August 1840, Liszt lauded the violinst's unsurpassed virtuosity. Yet, his praise was now not without reservation. Paganini's art was flawed by his own arrogance: "His god was never any other than his own gloomy, sad, 'I.'"[192] As Walker correctly points out, Liszt was criticizing Paganini's self-aggrandizing virtuosity. For Liszt, art was "a sacred power," which should be used to serve all of humanity. Reminiscent of Sulzer, Zelter, and W. C. Müller, this moral view of music was, to use Liszt's own term, a *génie oblige*. "May the artist of the future gladly and readily decline to play the conceited and egotistical role which we hope has had in Paganini its last brilliant representative. May he set his goal within, and not outside himself, and be the means of virtuosity, and not its end. May he constantly keep in mind that, though the saying is *Noblesse oblige*!, in a far higher degree than nobility—*Génie oblige*!"[193] Clearly Liszt was agonizing over the similarities between himself and the Italian. Three years earlier, in 1837, he reflected on his youth as a *wunderkind*, or the period when he was anything but a *génie oblige*: "[K]nowing . . . nothing of the powers with which I felt endowed, there came over me a bitter disgust against art, such as it appeared to me: vilified and degraded to the level of a more or less profitable handicap, branded as a source of amusement for distinguished society. I would sooner be anything in the world than a musician in the pay of the exalted, patronized and salaried by them like a conjuror or the learned dog Munito. Peace to his memory."[194]

Liszt's distinction between virtuosity—the awe-inspiring skilled manipulations and performance for its own sake—and the *génie oblige*—the application of virtuosity to a greater moral and nobler good—is informative, but certainly not unique. The music critic Schilling, too, differentiated between "genius" and "mere virtuosity." The virtuoso was simply "raw mechanism," whereas the genius "produced original works through his own productivity."[195] For Schilling the composer was the genius, the inventor, "the creator of the true art form," unlike the virtuoso performer, who simply reproduced what had already been created.[196]

In his revealing article of 1810, "Ueber die ästhetische Bildung des componirenden Tonkünstlers" (On the Aesthetic Education/Cultivation of the Composer), Steuber had argued that the genius of the composer comprised an "element of the perception of an aesthetic being, which represents itself as an inner amalgamation of certain proportions of psychic, moral and intellectual sentiments."[197] The aesthetically educated composer must have taste, meaning that "he has learned to apply the sentiments, which emerge directly from the organism, through intellectual mastery as a guiding approach to the consummate representation of the art."[198] Tastelessness, on the other hand, resulted when the "fantasy of the artist loses itself in the unnatural."[199] The genius was organic, the antithesis of the mechanical, and hence could not be calibrated. This view was perhaps best expressed by E. T. A. Hoffmann's renowned review of Beethoven's Fifth

Symphony, which he claimed was a "product of genius," "romantic taste . . . and talent," and "the infinite" (or *Unermessliche*, literally meaning "the immeasurable").[200]

That other famed pianist of the nineteenth century, Chopin, offered an artistic interpretation of technical work as an alternative to mechanical exercises. He stressed the refinement of the ear and muscular control and relaxation.[201] Rather than equalizing each finger's touch by executing tedious exercises, Chopin cultivated each finger's unique characteristic.[202] He famously exploited the uniqueness of the fourth and fifth fingers for chromatic effects in his Étude in Thirds, op. 25, no. 6.[203] Touch was paramount to Chopin. Renowned for his lightness of touch, he revolutionized piano instruction—in stark contrast to Liszt's technique, in which "the fingers should be strengthened by working on an instrument with a heavy, resistant touch, continually repeating the required exercises until one is completely exhausted and incapable of going on."[204] Chopin, his students inform us, wanted absolutely nothing to do with such a gymnastic treatment of the piano. Liszt's marathon practicing sessions were diametrically opposed to Chopin's, who at times would instruct his pupils not to practice for more than three hours a day and advise them to break up their practicing routine "by reading a good book, observing masterpieces of art, or by taking an invigorating walk."[205] More important, Chopin criticized pedagogical techniques that overemphasized mechanical, technical exercises. As Karol Mikuli, one of his students, recalled: "He never tired of inculcating that the appropriate exercises are not merely mechanical but claim the intelligence and entire will of the pupil, so that a twenty-fold or forty-fold repetition (even nowadays the worshipped arcanum of so many schools) does no good at all; while one hardly dare mention the kind of practice advocated by Kalkbrenner, during which one may occupy oneself reading."[206] Initially fascinated with Kalkbrenner, Chopin quickly came to think of his teaching and hand-guide as cold and unnatural. He proclaimed, "I am firmly convinced that I shall not be an imitation of Kalkbrenner: *he* has not the power to extinguish my perhaps too audacious but noble wish and intention to create for myself a new world."[207]

Chopin's organic creativity was thus deliberately contrasted with Kalkbrenner's mechanical imitation. *Contra* Kalkbrenner and Logier, Chopin first stressed that the pianist's touch was not restricted to his or her fingers. Indeed, the entire body played a role in touch. Second, he insisted on the suppleness and *independence* of the fingers, and was often cited as saying, "Have the body supple right to the tips of the toes."[208] He instructed his students that there were "[a]s many different sounds as there are fingers—everything is a matter of knowing good fingering. . . . Just as we need to use the conformation of the fingers, we need no less to use the rest *of the hand,* the wrist, the forearm and the upper arm. One cannot try to play everything from the wrist, as Kalkbrenner claims."[209] Like C. F. Müller, another German critic of Logier, Kollmann had attacked this restricted view of piano playing back in 1824: "Mr. Logier claims and indeed insists that all movement and force of the finger's touch while playing the piano-forte must originate with the fingers only, which need to be seen as hammers of the instruments. The participation of the weight and muscles of the hands and arms in the movement and strength of playing is detrimental. This is, however, undoubtedly the same if he were to claim that the only necessary way to speak properly would be to clench the teeth together so that the jaws could not move and only the lips could."[210] Specifically, one thinks here of Chopin's Étude in A Minor, op. 25, no. 11. The climax of that piece requires a simultaneous twist of both hands: the extreme muscular exertion causing great physical pain for most pianists.[211] According to Rosen, practicing an étude of Chopin "is an athletic exercise: it stretches the hand, develops the muscles, increases suppleness, [and] enlarges physical capacity."[212] Indeed, his Étude in A Minor required such muscular control that several dedicated pianists trying to master it suffered from tendinitis and needed to halt their playing temporarily. In some cases, it actually destroyed their careers.[213]

Physics in the Service of Musical Pedagogy

Acoustical theory actually played a role within this debate as to whether the virtuoso was something organic and spiritual (either divine or demonic), or whether one needed to understand the delicate, intense, unparalleled features of the skilled hand and body of the performer. And acoustical principles shaped the discussions on the related issue of whether or not physics could assist musicians in the teaching of their discipline. Being committed to the belief that music both strengthened an individual's *Bildung* and bolstered German nationalism, Gottfried Weber was particularly interested in using scientific principles to unravel and demystify the property of virtuosity with a view to pass on the virtuoso's skill to a younger generation of musicians.

Weber wanted to ascertain why different virtuosi generated a different tonal color on the same instrument, such as a violin, "for example, to research why one and the same violin, which is bowed with the same bow and under the exact same conditions, elicits sounds of completely different impressions in different hands, and which physical causes give rise to this difference of timbre."[214] Weber continued by asserting that he was not interested in "the superficial response" of "the dilettante," namely that the difference in timbre was merely a result of the individuality of the various performers and their method of playing. Such responses might satisfy the "pure aestheticists [*reinen Ästhetiker*]" and the "practical technicians [*praktischen Techniker*]," but not the "acousticians [*Akustiker*]," "who will not be satisfied with the words and catchphrases of individuality, but rather will want to know in particular how and in what part player A has a different physical effect [*physischen Einwirkung*] on the instrument than player B, and how the difference of the physical effect exerts itself. . . ."[215] He conjectured that the difference in tonal color might be due to the different ways in which the virtuosi bow the string. The violinist can bow closer to the bridge, alter the angle of the bow to the string, or change the pressure and speed of bowing. In each case, the violinist produces a different sound. He admitted that there were innumerable combinations of these variations corresponding to a myriad of ways in which a string can be set into motion.[216] But, he continued, it was the goal of the physicist to establish empirically the link between the various ways a string vibrates and summons forth the corresponding unique timbre. That explanation was Chladni's task, whose response was published alongside of Weber's essay. Weber postulated that the different ways of bowing a string resulted in differently shaped curves. The resulting sound waves thereby assume unique forms, which then affect our eardrums in different ways.[217]

Weber then turned his attention to the piano. He argued that, as opposed to the manner in which sound was produced by a violin, the sound of a piano was not, according to the laws of physics, determined by the way in which the pianist touches the keys. The pianist cannot alter the position of where the hammer strikes the strings. That mechanical feature is determined by the piano builder. The only variable that the virtuoso could introduce via his or her touch was the strength with which the hammer strikes the string by altering the pressure of his or her fingers on the keys.[218] "The physical, evident truth is clearly opposed to common belief that a pleasant tone is more or less—or indeed very much dependent upon the differing manner in which a key of a piano is depressed. One speaks of the hard, wooden or elastic melodious touch both for this and other types of *Je ne sais quoi*:—regardless of how common such a belief is, one can easily see that there is no reason for such a belief through the contrast with the distinct tone produced by violin playing."[219] Logier's mechanical method of piano playing rendered such a reductionist view plausible. If mechanical principles inherent in the chiroplast were applicable to piano playing as Logier suggested, then perhaps the mechanical principles of physics could also explain the virtuoso-pianist's touch.

Weber's message was directed to "the spokespersons of art" and "the art critics" whose "specious drivel on completely unique, pleasant tones . . . of a particular pianist through his . . . unique touch could be heard daily."[220] Such judgments, Weber asserted, were without any physical basis. The pianist can only vary his or her touch by altering the speed and pressure of his or her fingers.[221] This force then directs the hammer to strike the string with the corresponding velocity. Any other attempts to explain the variety of touch "are nothing more than empty dreams, which surely no physicist nor anyone else familiar with the fundamental laws of dynamics will deny."[222] In short, Weber argued that the physical laws of dynamics and mechanics were sufficient to explain the pianist's touch.[223] He wished to dismiss any explanation based on "magic,"[224] "secrets"[225] or that "*je ne sais quoi*"[226] in order to understand the tone that a pianist generates.[227] Because a pianist could not alter the way in which a hammer hits a string (as can a violinist vary the way the bow strikes a string), Weber argued that the mechanical principles known to the physicist (such as the mass and velocity of the hammer and mass of the string) could contribute to an account of the pianist's technique. His desire to explain the virtuoso's technique in terms of mechanical principles must be put in a context of pedagogy; he wished to glean critical information to pass on to future generation of musicians. Weber saw the universal principles of acoustics, with which he was very familiar, as providing the pedagogical tools he and his music colleagues sought in order to demystify virtuosity.

Chladni then responded to G. Weber's short essay.[228] Chladni's contribution is just as mechanical as Weber's and Bennati's, arguing that, for the violin, the player sets a string in motion by means of the bow. Every difference in the physical shape which the vibrating string experiences results in a different timbre.[229] He then briefly discussed the physics of curves as detailed by early eighteenth-century savants. In 1715 Brook Taylor had published the theory of transverse vibrations of a string and provided the mathematical relationship between the number of a string's vibrations per second and its length, weight, the weights attached to it, and the length of a seconds pendulum. Daniel Bernoulli had determined the shape that a perfectly flexible string assumes when acted upon by two forces, one of which is perpendicular to the curve's direction and the other parallel. And in the 1760s Count Giordano Riccati had provided mathematical calculations of vibrating strings and cylinders. All of these savants had argued that the curve, which the shape of a vibrating string assumes, must be considered mathematically as a cycloid. Or, mathematically expressed, if L is the length of a string, the largest displacement occurring at the midpoint of the vibrating portion of that string with the first type of vibration is expressed by A, the second type of vibration is expressed by B, the third by C, and so forth, while x is the arbitrary abscissa and y the corresponding ordinate:

$y = A \sin(\pi x/L)$ for the first type of vibration;

$y = B \sin(2\pi x/L)$ for the second type of vibration;

$y = C \sin(3\pi x/L)$ for the third type of vibration.[230]

If the initial curve of a string can be expressed by this equation, then it is in the instant when a vibration of the entire string terminates: $y = -A \sin(\pi x/L) + B \sin(2\pi x/L) - C \sin(3\pi x/L) \ldots$[231] Leonhard Euler had originally proposed in the 1730s (and Joseph Louis Lagrange subsequently confirmed in his *Mélanges de Turin* of 1759) that other (noncycloidal) curves also occur, which are dependent upon the initial curvature resulting immediately after the string has been struck. According to Euler, the curved shape a vibrating string assumes can be arbitrary, and depends only on the first curve that is given to it. Not even the different parts of this curve follow any type of law of continuity.[232] Such curves can arise that cannot be expressed by any mathematical equation. Although Jean de Rond d'Alembert agreed with Euler and Lagrange that in addition to Taylor's cycloids other curved lines occur, he argued that parts of a string assuming such curves must indeed conform to the law of continuity.[233]

The debate between Euler and d'Alembert was published in numerous articles in the proceedings of the St. Petersburg and Berlin Academies of Science. It was finally resolved by considering the initial shape of a string to be more angular than curved after immediately being struck, which then rounded out over time and formed more or less a cycloid. According to Chladni, these great geometers could not agree because one initially needed to determine the differential equations of three different dimensions through the investigation of a vibrating string. The integration of these differentials yielded arbitrary and variable functions, and one still could not determine whether these functions were totally arbitrary and whether all possible curves (including those that are discontinuous) could be understood as such, or if only those that could be expressed by an algebraic or transcendental equation could be understood as such.[234] As Ernst Heinrich and Wilhelm Eduard Weber demonstrated in *Wellenlehre* of 1825, when the first impulse is delivered to a string, that impulse does not immediately result in the string assuming a perfectly regular form. Rather, the string initially behaves as an advancing wave that slowly becomes a standing vibration. Hence, it partially follows Euler's theoretical predictions, but it also behaves as a long rope or string. The resulting curves of the sound waves produced depend upon the spot on the string where the bow strikes, the string's elasticity, and the angle of the bow to the string. All these factors affect the violin's timbre.

Chladni continued, however, by underscoring the importance of the peculiarities of the particular performer. Other critical factors, he argued, included the performer's "character, partly in the anatomical differences, for example in the various degrees of flexibility of the hand which holds the bow, the various strength, softness, elasticity and power of the finger over the string, and also partly the way he was taught. . . ." [235] All of these attributes, save the last, could in principle be measured. But just as the mechanical should not, in Chladni's view, be equated with virtuosity, the essence of what makes a unique and characteristic sound of a violinist could not, at least up to that point, be reduced to mere acoustical principles.

Chladni then discussed piano playing. *Pace* Weber, Chladni argued that where the hammer strikes the string greatly influences the various shapes of the string's vibrations. But, unlike the case for violin virtuosi, Chladni claimed that "in no way is it to be believed that a player could transfer something of his individual character in a manner inexplicable to us directly to the instrument, rather it occurs only indirectly, through the mechanics of the treatment."[236] Since the hitting of the hammer always occurs in the same direction and in the same point on the string, only the velocity or strength with which the hammer hits the string determines the timbre. Here, according to Chladni, physics could determine the aesthetic quality of touch.

Chladni and Weber's article sparked a response from Schwinning, a pianist and piano instructor from Frieland in Mecklenburg-Strelitz, in July of 1835.[237] Schwinning commenced his essay by denying Chladni and Weber's distinction between the excitation of the string, which occurs first, and the subsequent shape of the string's vibration, which gives rise to the tone. Schwinning asserted that these "moments of touch" cannot be separated, but occur simultaneously in order to produce a "pleasant tone."[238] Schwinning continued by attacking Chladni and Weber's assertion that the uniqueness of the individual's touch, upon which a pleasant tone more or less depends, is unfounded for pianists. The hammer constantly hitting the string in a particular spot in the same direction and the pianist's inability to affect the tone after striking the keys were, Schwinning concurred, characteristics of the mechanical fittings of the instrument. These factors, however, do not determine the characteristic touch of the pianist; they merely limit it.[239]

This point concerning touch formed the crux of Schwinning's protest. He conceded that the range of influence of a pianist's unique touch is more limited than a violinist's, harpist's, or guitarist's.[240] However, by reducing touch to the degrees of force in which the fingers hit the keys, Schwinning argued that both scholars had ignored a large range of tone modification, which depended upon the feelings on the pianists' fingertips. The degree of force that the performer employed is only one aspect of a more encompassing "gradation of feeling [*Gefühlsgradation*]" or "varying modification of feeling [*verschiedenene Gefühlsmodification*]," which is transferred from the finger to the string via the key.[241] The pianist's touch "is the manner in which the force of the finger is transferred to the key. It is not the sum of force, which the finger carries to the key, but the manner in which the sum of the touch's force is developed at the moment of the finger's activity."[242] In short, the tone's force was the effect of touch, not the touch itself. The force's strength and weakness were but material concepts, which could be used to describe touch, but were not touch in themselves.[243]

Schwinning quickly realized the pedagogical ramifications of Chladni and Weber's misconstrual: "the teacher of touch must not be content with the force's strength or weakness as actually being touch, rather he must develop touch from the available elements."[244] The instructor needed to consider touch as the activity of all the working organs and the strength of their combined force, which results in a pleasant one. This is the "moment of developing [*Entwicklungsmoment*]," which musicians call the "unique type of touch [*eigenthümliche Anschlagsart*]."[245] The force must be understood as an "act of formation [*Bildungsart*]." Unlike Chladni's or Weber's, Schwinning's depiction of the development of touch's strength was not mechanical, predicated on the instrument's mechanism, but rather was organic. And a particular form of the hands and fingers as well as the fingers' activities were all necessary to be able to transfer the developing force in a purposeful manner.[246] The teacher needed to convey that organic process, which evolved over time, to his pupils. The fingers and hands needed to be disciplined in order to achieve the desired effect, the sought-after touch. Schwinning argued that the "secret," which Chladni and Weber were attempting to elucidate as the uniqueness of the pianist's touch, was based on the disciplined activity of the fingers.[247] "This conscious development of the touch's strength can be taught: the available concepts for such a development sharpen the feeling of the fingertips carrying out the touch, whereby the realm of experiences and with it the entrance to every secret of a good performer is opened."[248]

In short both Weber and Chladni approached the problem as one of measurability. Since they were convinced that touch was simply the contact of the pianists' fingers on the keys, quantifiable mechanical principles seemed to be the only issue. Schwinning countered by arguing that the organic feeling on the fingertips is truly touch: a quality that could not be reduced to physical principles.

In 1863 Hermann von Helmholtz was the next scientist to address the physics of piano and violin strings in his seminal *Die Lehre von den Tonempfindungen als physiologische Grundlage für die Theorie der Musik*[249] (On the Sensation of Tones as a Physiological Basis for the Theory of Music). Like Chladni, Helmholtz was interested in breaking sound down into its component parts with the view, in part, of contributing to musical pedagogy. He argued that the musical tones of piano strings were influenced by the upper partial tones elicited by a struck string, which in turn were based on three factors: the nature of the stroke, the place struck, and the density, rigidity, and elasticity of the string.[250] In a piano, the hammer strikes the strings, and the strength and number of the upper partials of the tone increase with the number and abruptness of the discontinuities of the string's vibration. The string rebounds instantly once the hammer strikes it; therefore, only one single point of the string is set directly into motion, not moving until a wave of deflections results, traveling backward and forward over the string. The string generates the prime tone and a number of upper partials. If the hammer is soft and elastic, which was much truer of pianos built in the 1850s and '60s than those built a half-century earlier, the motion has time to spread along the string before the hammer rebounds. The point on the string is not set in motion with a sudden jolt, but increases gradually and continuously in velocity during the contact. The motion is less discontinuous, and as a result, the force of the upper partial tones is decreased considerably.[251] *Pace* G. Weber and Chladni, Helmholtz argued for the importance of piano builders' skills to the quality of the sound produced: "Clearly the makers of these instruments have here been led by practice to discover certain relations of the elasticity of the hammer to the best tones of the string. The make of the hammer has an immense influence on the quality of the tone. Theory shews that those upper partial tones are especially favoured whose periodic time is nearly equal to twice the time during which the hammer lies on the string, and that, on the other hand, those disappear where periodic time is 6, 10, 14, &c., times as great."[252]

The second factor influencing the role of the upper partials in determining the tone of a string, the point where the string is struck, was also determined by artisanal knowledge: "The selection is not due to theory. They [the points on the string that are struck] are results from attempts to meet the requirements of artistically trained ears, from the technical expertise of two centuries."[253] The third factor, the material and thickness of the string, was also clearly a result of the work of the manufacturer, far removed from the influence of the particular performer. The thinner, finer wires (particularly those made of gold lace or catgut) produced the higher upper partials.[254] The weight and elasticity of the hammers and the nature of the felt were also crucial to a piano's tone. Instrument makers typically chose heavier and softer hammers for the lower octaves, and lighter and harder hammers for the upper octaves. Helmholtz claimed that "in no other instrument is there so wide a field for alteration of a quality of tone; in no other, then, was a musical ear so unfettered in the choice of a tone that would meet its wishes."[255] It is interesting to note that Helmholtz, who was an accomplished amateur pianist, never discussed the role of touch to his theory of musical tones. His account was just as mechanical as that of the piano makers' skills. When the pianist and piano instructor Frederic Clark-Steiniger, fearing a total loss of the pianist's control over the quality of the sound, queried Helmholtz on this issue, Helmholtz responded, "As far as I know, on the newer mechanisms [pianos] the rate of speed with which the hammer flies against the string, i.e. the force of the blow from the key is the only way to modify tone."[256]

Such a view stood in stark contrast to his theory of musical tones of bowed instruments. Here, a mechanical explanation failed, since "no complete mechanical theory can yet be given for the motion of string excited by the violin-bow, because the mode in which the bow affects the motion of the string is unknown."[257] And Helmholtz argued that "notes of the same pitch can sometimes be sounded on the same instrument with several qualitative varieties. In this respect the 'bowed' instruments (i.e. those of the violin kind) are distinguished above all other."[258] In order to analyze the vibrations of a violin string, he deployed a method of observation invented by Jules Antoine Lissajous, discussed in the previous chapter. Using a version of a Lissajous vibration microscope, described below, Helmholtz researched vibrating violin strings.

Helmholtz wished to calculate the entire motion of the string and the intensity of its corresponding upper partials by concentrating on the vibrational form of a particular position on the string. He fastened a lens L, which was a combination of two achromatic lenses similar to the object lenses of microscopes, to one end of a prong of a tuning fork, G (see figure 8.3). He peered through a magnifying glass made of a strong convex lens at a grain of starch reflecting a flame, appearing as a bright point of light. If one moves the lens while the point remains stationary, the point itself will actually appear to move.[259] After arranging the instrument, which he called a vibration microscope, such that a fixed luminous point could be clearly seen, he attached a tuning fork to a metal plate, setting it into motion, causing the doublet L to move up and down. The observer would see the point of light vibrate. Since the vibrations occur so rapidly, the path of light appears to the human eye as a fixed line, increasing in length with

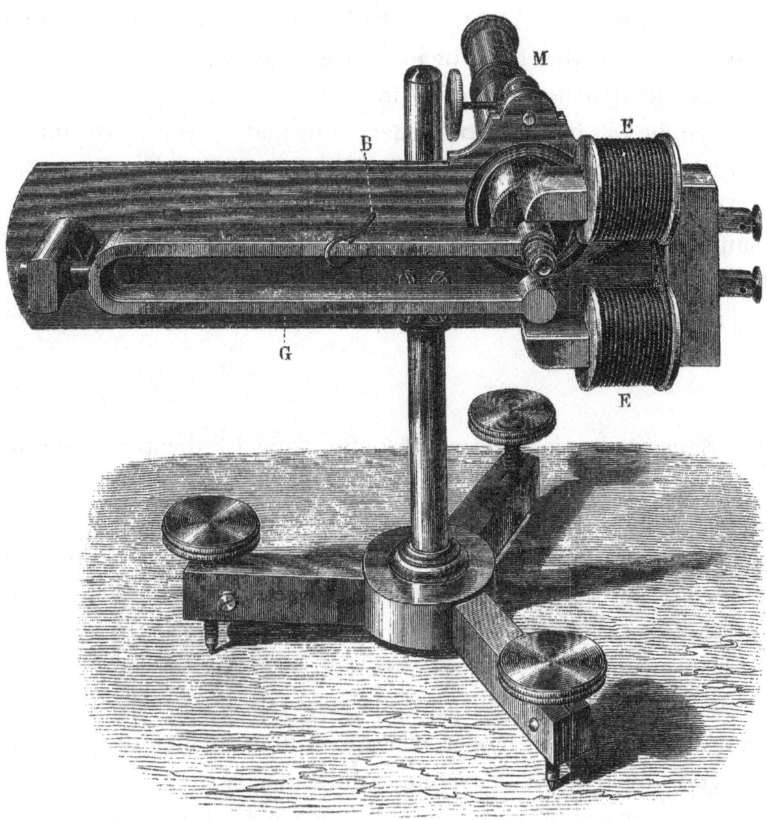

Figure 8.3
Helmholtz's vibration microscope. Source: Helmholtz 1863. Reprinted with the permission of The Burndy Library, Cambridge, Mass.

the movement of the tuning fork.[260] The grain of starch is fastened to the vibrating string such that the grain moves back and forth horizontally, while the doublet moves up and down vertically. This combined motion permits the observer to see the resulting curvilinear motion. The string must be attached to the luminous point by first darkening a spot on the string with ink and then rubbing the dried dark spot with wax. Powder is then sprinkled over the starch so that a few grains stick to the wax. The violin is set up such that its strings run in a vertical direction with respect to the microscope. The bow is drawn across the string, parallel to the fork's prongs. Helmholtz then observed the resulting curves of vibrations through the microscope.[261] He tuned the a′ string of the violin to b♭′, precisely two octaves higher than the microscope's tuning fork.

Since the vibrational forms of the individual points of the string were now known, he was in a position to determine mathematically the intensity of each upper partial generated by the violin string by using a Fourier series. The intensity of the upper partials diminishes as the pitch increases. The amplitude of the second partial is one-fourth the prime tone, that of the third partial is one-ninth of the prime tone, the fourth is one-sixteenth of the prime tone, and so on.[262] The parabolic arc of the string is equal to

$$\frac{8V}{L^2} \quad [(L-x)],$$

where V is the amplitude in the middle of the string, L is the string's length, and x is the distance of a given point on the string from the string's endpoint.[263]

By analyzing the string's curves, Helmholtz noted that any scratching of the bow along the violin string resulted in sudden jumps, or discontinuous displacements of the resulting curves. He observed that "inferior bowed instruments seem to be distinguished from good ones by the frequency of such greater or smaller irregularities of vibration."[264] In addition to contrasting bowing techniques, he was able to maintain a steady vibrational figure for a well-manufactured contemporary violin, most likely made in Mittenwald. But an old violin made by the Italian master Guadanini generated an even steadier vibrational figure from which Helmholtz could barely count any irregularities. He concluded, "this great uniformity of vibration is evidently the reason of the purer tone on these old instruments, since every little irregularity is immediately felt by the ear as a roughness or scratchiness in the quality of tone."[265] Hence, in a sense, Helmholtz had developed an assay to test the quality of a violin. The structure of the instrument and the quality of the wood (one possessing the most perfect elasticity possible) affect the instrument's tones. But, while the pianist's touch was conspicuously absent from his analysis of the vibration of the piano string, he conceded that "the art of bowing is evidently the most important condition of all. How delicately this must be cultivated to obtain certainty in producing a very perfect quality of tone and its different varieties, cannot be more clearly demonstrated than by the observation of vibrational figures."[266] In the case of bowed instruments, physics illustrated how the art of the bowing technique could not (at least to this point) be reduced to mechanics. Indeed, by labeling it an art (*Kunst*), and arguing that it must be cultivated, it seems that Helmholtz was not holding out much hope for a future mechanical explanation of this craftlike activity.

The debate on the ability of physicists to explain the pianist's touch and virtuosity started with G. Weber and Chladni, continued with Helmholtz, and lasted throughout the century. By the 1850s, the concept of touch began to encompass several techniques. While in the 1820s and '30s "*Anschlag*" had referred exclusively to the manner in which the fingers struck the keys, resulting in the "connections and separations between the tones and their gradations," discussions about piano pedagogy of pedaling arose much later in the century.[267] Hugo Riemann, Heinrich Gerner, Ludwig Riesmann, and Hans Schmidt—all pianists and piano instructors of the second half of the century—argued that the piano's sounds and its various possibilities of articulation were organic, rather than mechanical.[268] The renowned London piano builder of the Broadway Piano Company, Alfred Hipkins, allegedly proclaimed circa 1896, "The sounds different pianists produce are as distinct as their thumbprints."[269]

Such a statement was more than mere hyperbole. In the 1890s the pianist and piano instructor Marie Jaëll and the Parisian physiologist and chief physician of the Kreqlin-Bicetre, Charles Féré, attempted to measure various pianists' technique, including touch, while playing certain pieces.[270] In 1896 Jaëll took fingerprints of various students' left and right hands while playing a Beethoven sonata. And Féré measured Jaëll's speed of attack to the nearest one hundredth of a second, using a chronometer made by the artisan Arsonval. Jaëll recalled:

> [F]rom 1897 to 1907 [the year of Féré's death] Dr. Féré communicated all his research to me, as I communicated mine to him. . . . We have made repeated [finger]prints of the same chord. He compared the touches made by each finger and informed me, "They are similar; therefore, it is conclusive." I continued the work with a passionate interest. . . .
>
> I researched the correct movements, and by means of these movements, I found the harmony of touch, the musical memory, the improvement of the ear, all the facilities, which seem to be dormant in each of us.
>
> These movements were relentlessly controlled in the laboratory experiments of Dr. Féré in Bicetre. Continued for ten years, the chronometric experiments established the unexpected fact that the movements of the automated fingers undergo

delays, which result in some pianists appearing to be backward or deficient. While on the contrary, the movement of thought, the correct movement, accelerates—as it improves—the response times allowing my pupils to respond to the signal of the chronometer with a precision only ascertained with great intelligence.[271]

The sciences of physiology and physics had joined forces to explain the art of touch.

In conclusion, during the early nineteenth century, machines were used to discipline the human body. The beauty of specialized handicraft was giving way to the aesthetic of mass production. Individuality of expression was being sacrificed for standardization in many forms of culture during that period. In music, chiroplasts and hand-guides were performing these functions. Pupils using these mechanical devices were all trained in the same fashion; everyone using these machines could supposedly achieve similar levels of skill. The universal principles of mechanics were seen as leveling the proverbial playing field. As we have seen, debates arose as to whether machines could more efficiently convey skilled techniques than a master. Much of the argument centered around the theme of the organic individual versus the factory-like machine and its methods. And in part as a result of these mechanical devices, critical questions concerning the true purpose of music dominated discussions in musical circles. Was music about the ability of the individual to free him- or herself from the influence of the State in order to cultivate his or her own taste? Or was the purpose of music to assist civil servants in fostering camaraderie and nationalism? Or should music simply reflect the extravagant lifestyles of the upper classes? And could mechanical contraptions assist in these endeavors?

The virtuoso, of course, was someone special. His or her skills were not mass-produced; they were simply desired by the masses. Interestingly, musicians attempted to uncover the "mechanical parts" of the virtuoso's body, specifically his or her fingering techniques, by applying the laws of mechanics and dynamics to his or her performance. The key aspect of skill in all themes of culture—science included—is that it lends itself to quantification. Once a skill has been measured, inevitably the next step is to produce a machine that will mimic it. Some musicians turned to physicists to see if the mechanical principles of acoustics could communicate certain skills of the virtuoso to students. Interestingly, the virtuosi themselves were seen as machine-like, as they dazzled their audiences with machine-like rapidity and precision.

This story has offered a history of one aspect of the relationship between music and physics. Far removed from cosmic orbitals and Pythagorean ratios, this account is firmly anchored in the material culture of the nineteenth-century *Bildungsbürgertum*. One might be tempted to wax poetic about how science and music were inextricably linked in nineteenth-century Germany; surely, that is true. But more to the point, this chapter has proffered a glimpse into a period when debates throughout the musical community emerged on what constituted virtuosity, how taste could be developed, and the true role of *Geist*. And physicists, with their penchant for quantification, wondered just what could be measured and what could not. The freedom of the individual to cultivate his or her own character and taste, the role of the State in defining those attributes, and the relationship between the organic and the mechanical were all at stake.

Appendix

An explanation of pitch notation is necessary. I am using late-eighteenth- and early-nineteenth-century notation for pitch throughout this work. Here is a guide to comparing pitches.

Historical pitch notation	Modern scientific pitch notation	Frequency in equal temperament	Example of the pitch
C	C2	~65.4 Hz	8-foot organ pipe
c	C3	~135.8 Hz	4-foot organ pipe
a	A3	220 Hz	Cello's open A string
c′	C4	~261.6 Hz	Middle C on the piano
a′	A4	440 Hz	Violin's open second string and concert pitch for many orchestras
c^2 (or c″)	C5	~523.3 Hz	The space above the middle B line of the treble clef
c^3 (or c‴)	C6	~1046.6 Hz	1/2-foot organ pipe

Note that in just intonation, rather than equal temperament, C is 64 Hz, c is 128 Hz, c′ is 256 Hz, c2 is 512 Hz and c3 is 1024 Hz. Also, two types of measurements for pitch were used during this period: simple vibrations, or the vibrations of a fork from one extreme position to the original (simple vibrations per second, or svps), and complete vibrations, or the vibrations of a tuning fork from one extreme position to the other and then back to the original position (or vibrations per second, vps, or Hz).

Notes

Chapter 1

1. Palisca 1961, pp. 91–92.

2. For the craft aspect of science, see Ravetz 1972; Polanyi 1994; Collins 1985; Hacking 1983; Jackson 2000.

3. Palisca 1961, pp. 91–137, and Walker 1967, pp. 228–250. For more recent works, see Cohen 1984 and Stephenson 1994.

4. Gouk 1999.

5. Note that I am not arguing that only German scientists engaged in aesthetic issues; rather, I am interested in offering an explanation of why they in particular were doing so by drawing upon their historical circumstances, many of which were indeed unique to Germany.

6. Lenoir 1997, pp. 131–178.

7. Ibid., p. 136.

8. Dettelbach 1996, p. 266.

9. Humboldt 1989, p. 48, as translated in Dettelbach 1996, p. 268.

10. Dettelbach 1996, p. 271.

11. Schiller 1967, letter 2, paragraph 5. See Dettelbach 1996, p. 275.

12. Bourdieu argues for the links among politics, aesthetics, and epistemologies and their embeddedness in daily practices. Bourdieu 1977, pp. 164–167, 188–190.

13. Dettelbach 1996, p. 276.

14. Wise 1999, pp. 107–140.

15. Moltke 1892, p. 3. (As translated in Wise 1999, p. 109.)

16. Wise 1999, pp. 110–112.

17. Ibid., p. 116.

18. Ibid., p. 113.

19. Ibid., p. 114.

20. Ibid.

21. Ibid., p. 130.

22. Lenoir 1997, pp. 174–175.

23. Cahan 1993, pp. 559–601, and Jurkowitz 2002, pp. 291–317.

24. Jurkowitz 2002, p. 304.

25. Lenoir 1997, p. 175.

26. Hatfield 1993, pp. 523, 525.

27. Ibid., p. 533.

28. Ibid., p. 535.

29. Lenoir 1997, p. 177.

30. Berg 1980, pp. 2, 83, 153, 211, 325.

31. Marx 1857–1858 (1973), pp. 692, 704.

32. Schaffer 1994, pp. 203–277.

33. Ashworth 1996, pp. 629–653.

34. Shapin 1994; Schaffer 1997, pp. 456–483; and Morus 1998.

35. For an account of the status of skill relevant to natural philosophy in the German territories during the early nineteenth century, see Jackson 2000, esp. pp. 99–142.

36. This part of my story is reminiscent of Michel Foucault's work on regimes of institutionalized training during the early nineteenth century. See Foucault 1977.

Chapter 2

1. See, e.g., Ernst Heinrich Weber and Wilhelm Eduard Weber 1825, dedication page, and Bernhardt 1856, p. 96. For a most informative recent biography of Chladni, see Ullmann 1996.

2. Chladni 1817, p. xi.

3. Chladni 1802, p. xviii.

4. Chladni 1817, p. xi.

5. See, e.g., Chladni 1821–1826: 1821, cols. 393–398; 1824, cols. 809–814; 1825, cols. 725–730; 1826, cols. 40–41, 693–696; and Chladni 1821b: cols. 369–374.

6. Chladni 1802, pp. xv–xvi, and Melde 1866, pp. 7–8.

7. Chladni 1802, pp. xv–xvi.

8. Ibid., p. xvii; Melde 1866, p. 4; and Bernhardt 1856, p. 14.

9. Chladni 1787, p. 2.

10. Ibid., p. 6n.

11. Ibid., pp. 6–12. For example, he bowed rods with one completely fixed end and one free end, one end that was only partially fixed, both ends completely free, both ends only partially fixed, both ends completely fixed, and finally one end totally fixed and the other only partially.

12. I would like to thank Elizabeth Cavicchi for pointing this out to me.

13. Ibid., pp. 24–50.

14. The most common method of determining an unknown pitch during the first three decades of the nineteenth century was to match it with the number of vibrations of a string stretched vertically across a sounding box, known as a monochord. If the number of vibrations of the two pitches are identical, no beats can be detected. Using Brook Taylor's formula, the frequency of a string, n, is $\sqrt{\frac{2gP}{LG}}$, where g is the gravitational constant, P is the string's elasticity, L the strength's length, and G its mass per unit length. See Fischer 1825, col. 503. Although Chladni argued that monochords remained useful in certain circumstances for determining the pitch of a given tone (particularly a string's), he also suggested a new method for objects that were not amenable to his nodal methods. One takes a long strip of metal (he suggested iron or tin), and ensures that it has the same thickness throughout. One then places the end of the strip of metal in a vise so that the protruding end is long enough to produce "slow, transverse vibrations," too slow to be audible, but which the eye can count with the beats of a seconds pendulum. One notes the length of a strip, which is required to give the observed number of vibrations, with a line drawn by a pencil. If one wants to know how many vibrations a given pitch makes in a second, one simply places the strip in a vise so that the protruding end is short enough to produce the same number of vibrations per second as the pitch to be determined. By comparing the length of the short end, which produced the pitch, with the length of the long end whose vibrations could be counted, one can simply calculate the pitch, $1/l^2$, where l is the length of the metal strip. For example, Chladni recommended that one use a long strip of metal approximately 2 feet long, 1/2 inch wide, and 1 Parisian line thick that would produce 4 simple vibrations per

second (or 2 vibrations per second, cycles per second, or Hz), since the human eye easily and accurately detects this number of vibrations. If one shortens the length of the vibrating part of the metal by one half with a new clamp, the resulting string will produce 16 svps, which one deduces from mathematics, since the eye cannot count that many vibrations per second, nor can the ear hear so few. If one repeats the procedure of shortening the strip's length by one half with another clamp, the resulting strip will produce 64 svps, which corresponds to the pitch generated by a 16-foot organ pipe (32 Hz). Halving this length once again will yield 256 svps (128 Hz), or c. Chladni called this strip of metal a "tonometer." Chladni 1800a, pp. 4–6. See also Chladni 1802, pp. 35–36. In short, Chladni wished to offer a method for determining the pitch of all sounding bodies, not just vibrating strings. Hence, he determined pitches not merely by comparing them with the length of a string (as was the case with the monochord), but also by counting the number of vibrations the actual body produced. Chladni 1802, p. 8.

15. In 1677 the mathematician and natural philosopher Georg Philipp Harsdörfer had spoken of *lustige Wein-Musik* ("merry wine music"), in which he took eight glasses and filled each with differing amounts of wine and then stroked the rims of the glasses with moistened fingers. Unfortunately, he claimed that the tones generated could be quite painful to the ear. See Quandt 1800, cols. 151–152.

16. Mendel 1870–1877, here 4 (1874), pp. 533–534. To listen to passages played by a glass harmonica, download the tracts at http://www.amazon.com/gp/product/B00005QISL. I recommend tracks 7 and 10.

17. Sachs 1940, pp. 404–405.

18. Mendel 1870–1877, here 4 (1874), pp. 534–535.

19. Kunzen and Reinhardt 1793, part I, p. 25.

20. Ibid., p. 148.

21. Sachs 1920, p. 384.

22. It was generally argued that women, who—it was claimed—possessed a soft, subtle touch, were better suited to play the harmonica than the piano.

23. Koch 1802, p. 740.

24. Müller 1788.

25. Ibid., *Vorbericht* (no page numbers given).

26. Ibid.

27. Mendel 1870–1877, "Harmonika," 4 (1874), p. 535.

28. Koch 1802, p. 741, and Mendel 1870–1877, "Harmonika," 4 (1874), p. 536.

29. Mendel 1870–1877, "Harmonika," 4 (1874), p. 536.

30. Müller attempted to alter the design of his harmonicas in order to reduce costs. Müller 1788, *Vorbericht*.

31. Mendel 1870–1877, "Harmonika," 4 (1874), p. 536. Her name was also spelled Kirchgassern and Kirchgässern.

32. Mendel 1870–1877, "Harmonika," 4 (1874), p. 536.

33. Ibid., p. 537.

34. Ibid.

35. Röllig 1787. See also Koch 1802, p. 741.

36. Kunzen and Reinhardt 1793, part II, pp. 148–149.

37. Mendel 1870–1877, "Harmonika," 4 (1874), p. 537; and Koch 1802, p. 741.

38. Koch 1802, pp. 741–742.

39. Ibid., p. 828.

40. Klein 1799, cols. 675–679, and Koch 1802, pp. 742–743.

41. Sachs 1940, p. 405.

42. Chladni 1821a, pp. 5–7. He was later to become acquainted with the numerous articles in musical journals, such as the *Allgemeine musikalische Zeitung* and *Caecilia*, describing the acoustical properties of those innovative instruments as well as the types of music best suited for each.

43. Chladni 1787, pp. 26–28.

44. Ibid., pp. 29–51.

45. Ibid., pp. 51–65.

46. Ibid., p. 65.

47. Chladni 1790a, pp. 539–543, and Chladni 1790b, pp. 201–202.

48. Chladni 1802, pp. xviii–xix.

49. Ibid., p. xx.

50. For Chladni's own description of his euphone, see Chladni 1821a, pp. 135–176, and tables IV and V.

51. Chladni 1790a, p. 540, and Chladni 1802, p. xix.

52. Chladni 1793b, part II, p. 77, and Chladni 1795, pp. 309–313.

53. Kunzen and Reinhardt 1793, part II, p. 118.

54. He finally did make the design to his instrument known in 1821, when he was sixty-four years old and could no longer undertake such long journeys. He published the details of the euphone and clavicylinder in Chladni 1821a. It should be noted that Chladni did have a priority dispute with the physician Dr. Christian Friedrich Quandt of Jena. Chladni 1793b, part II, p. 78; and Chladni 1791, pp. xxiii–xxiv.

55. Kunzen and Reinhardt 1793, part II, p. 11.

56. Chladni 1790a, pp. 541–543; Chladni 1793b, p. 77; and Chladni 1790b, pp. 201–202.

57. Chladni 1800b, cols. 305–313, and Chladni 1801a, pp. 310–313.

58. Chladni 1821a, tables II, III and IV; Melde 1866, p. 11; and Bernhardt 1856, pp. 84–87. On Chladni's description of building the clavicylinder, see Chladni 1821a, pp. 33–121.

59. Chladni 1800b, col. 306. By 1801, he increased the range from D to f^3, or four octaves and a minor third, and by 1803, he increased it again, this time by a mere whole tone, from C to f^3. See Chladni 1803, p. 138, and Chladni 1801a, pp. 310–313.

60. Kunzen and Reinhardt 1793, part II, pp. 11, 122. See also Chladni 1801b, cols. 386–387, and Chladni 1800b, cols. 307–308.

61. Chladni's own description is published in Chladni 1803, pp. 136–139.

62. Chladni 1800b, col. 306.

63. Chladni 1803, p. 138.

64. Kunzen and Reinhardt 1793, part II, p. 12.

65. For example, the andante from the symphony to the "Opera of the Alchemist" by Schuster, and the choral, "Nun danket alle Gott," were performed on the euphone by Chladni. Popular pieces for both the harmonica and euphone included: Karl L. Röllig's "Un poco adagio" and "Siciliano" (Röllig c. 1786); *Harmonikasonate* by Naumann; and Adagio for Glass Harmonica in C major, KV 617a, by Mozart.

66. Chladni 1821a, p. 13. For the clavicylinder, several piano and organ pieces were recommended, such as the Court Organist of Darmstadt C. H. Rink's *Praktische Orgelschule,* the twelve adagios, various chorales, and the preludes, and Gerber's *Sammlung von Orgelstücken,* particularly "Befiehl Du Deine Wege." Ibid., pp. 125–126.

67. "Vollständige und gebundene Sätze," in Chladni 1803, p. 139.

68. His performances on the clavicylinder in Weimar in February of 1803 were very well received. See Chladni 1803, pp. 136–139.

69. Anon. 1807, p. 395.

70. Ibid.

71. Chladni 1817, p. vii.

72. Chladni 1826b, pp. 137–144. See also Chladni 1817, pp. viii–ix.

73. Chladni 1826b, p. 137.

74. Ibid., p. 138. Laplace was the patron of both Poisson and Biot. Cuvier's employment at the Collège de France was secured by Lucien Bonaparte. For Cuvier's ability to socialize with important contacts, see Outram 1984, pp. 55–68.

75. Chladni 1826, p. 139.

76. See, e.g., Chladni 1800b, cols. 306–307.

77. Chladni 1817, p. 140.

78. Ibid., p. 141.

79. Ibid., pp. 143–144.

80. As quoted and translated in Bucciarelli and Dworksy 1980, p. 34.

81. Appendix to Chladni 1809. "Rapport adopté par la Classe des Sciences Mathématiques et Physiques et par celle des Beaux-Arts, dans les séances du 13 février et de mars 1809, sur l'ouvrage de M. Chladni relatif à le theorie du son," as translated in Bucciarelli and Dworksy 1980, p. 5.

82. As translated in Bucciarelli and Dworksy 1980, pp. 353–357. The authors speculate that Laplace desired the new competition to further the career of his protégé, Simeon-Dénis Poisson, who at the time was not yet a member of the First Class (ibid., pp. 6, 37). Poisson, however, never completed an essay worthy of submission. As stated later in the text, I argue that Laplace was interested in Chladni's use of longitudinal vibrations in calculating the speed of sound through various substances.

83. For the most comprehensive account of Sophie Germain's entry, see Bucciarelli and Dworksy 1980, pp. 40–84.

84. Ibid., pp. 40, 54.

85. Chladni 1815, col. 14. Chladni seemed delighted that the likes of Lagrange, Poisson, and Plana from Turin had dedicated themselves to solving the problem. In this particular article, Chladni does not mention Sophie Germain's name, but he does so in 1821. He thought that the only person in the German territories who might be able to offer a solution to the posed problem was Carl Friedrich Gauss, who apparently did not seem all that interested. Ibid., col. 14.

86. Lindsay 1966, p. 633. Reprinted in Lindsay 1973, pp. 5–20, here p. 9. Gustav R. Kirchhoff provided the correct theory in 1850.

87. Wilhelm Weber 1830a, p. 187. Weber actually hoped to use longitudinal vibrations to test the purity of metals.

88. Chladni 1787, p. 76.

89. Chladni 1793a, part I, pp. 3–35.

90. Ibid., p. 34.

91. Busse 1793, part II, pp. 177–181, 185–187.

92. Ibid., p. 177.

93. Ibid.

94. Chladni 1793c, p. 33.

95. Chladni 1796.

96. Chladni 1802, pp. 65–66.

97. It turns out that Chladni later realized that the thickness of the material does indeed play a role in determining the pitch of longitudinal vibrations. In 1802 he claimed that longitudinal vibrations might be inversely proportional to the square of the string's weight. See Chladni 1802, p. 109. In 1827, he argued that perhaps that the longitudinal vibrations were proportional to the square root of the string's longitudinal elasticity and inversely proportional to its weight. Chladni 1827a, p. 31. See also Ullmann 1996, p. 61.

98. Chladni 1796, pp. 4–5.

99. Chladni 1802, p. 104.

100. Chladni 1796, pp. 7–10. For the different physical properties between longitudinal and transverse vibrations, see Chladni 1802, p. 109.

101. Chladni 1798a, p. 66.

102. Ibid.

103. Chladni 1796; Chladni 1798c, pp. 7–17, and Chladni 1802, pp. 265–266. See also Melde 1866, p. 9, and Ullmann 1996, p. 70.

104. Melde 1866, p. 9, and Ullmann 1996, p. 70. Today, physicists calculate the speed of sound through tin at 2730 m/s, silver 2640 m/s, copper 3580 m/s, for glass anywhere between 3490–5300 m/s depending on the type of glass, and 5170 m/s for iron. Ullmann 1996, p. 70. See also Ernst Heinrich Weber and Wilhelm Eduard Weber 1825, p. 467.

105. Chladni 1798b, p. 276.

106. Ibid., p. 277.

107. Chladni 1798a, pp. 65–79, translated in Chladni 1798b, pp. 275–282. See also Chladni 1802, pp. 226–230.

108. Chladni 1802, pp. 226–227, and Chladni 1798b, p. 276.

109. Chladni 1802, p. 227. The pitches could be raised and lowered by altering the length of the strings suspended between two pegs of the monochord. Using Taylor's formula, the pitch of the string could be determined from its thickness, weight, length, and elasticity.

110. Ibid., pp. 227–229. Hydrogen gas formed by combining iron with sulfuric acid produced a tone one octave higher than air, while hydrogen gas generated by mixing zinc with muriatic acid (now called hydrochloric acid) produced a tone one octave and a minor third higher (ibid., p. 229). These claims were a culmination of the work of Newton, Euler, D. Bernoulli, Lambert, and Riccati. Chladni 1798b, p. 276.

111. Chladni 1802, p. 228.

112. Ibid., p. 230. Physicists today provide the following measurements: the speed of sound through oxygen is 324 m/s, nitrogen 343 m/s, carbon dioxide 265 m/s, nitrous oxide 333 m/s, and hydrogen 1300 m/s. See Ullmann 1996, p. 76.

113. Chladni 1802, p. 229.

114. Biot 1810, pp. 407–432.

115. Ibid., pp. 409–410.

116. Ibid., p. 411.

117. Ibid., p. 410.

118. Ibid., pp. 411–412.

119. Ibid., p. 413.

120. Ibid.

121. Ibid., p. 414.

122. The sound traveled through the air of the cylinders in 2.76 seconds, as compared to the value obtained of 2.79 seconds. Ibid., pp. 418–419.

123. Laplace 1817, pp. 225–243. See also Chladni 1821c, cols. 594–595.

124. Laplace 1817, pp. 233–234, and Chladni 1821c, cols. 594–595.

125. Laplace 1817, p. 236.

126. Ibid.

127. This is not to say that no musicians constructed instruments during the mid-nineteenth century. One thinks of Theodor Boehm of Munich, who was a composer, virtuoso, and renowned flute maker. But he was rather the exception.

128. See Chladni 1821a, pp. 34–35.

Chapter 3

1. See, e.g., Sudhoff 1922; Fischer 1952, pp. 349–352; Degen 1956, pp. 333–340; Pfannenstiel 1958; Schipperges 1972; Siefert 1969; Schipperges 1968; and Flehr 1983.

2. Kümmel offers a brief account of songs sung at the annual meetings, but he limits his discussions to late-nineteenth-century meetings. Kümmel 1987, pp. 470–472.

3. For an excellent discussion of the use of music in creating a German national identity, see Applegate and Potter 2002.

4. Te Heese 2001, pp. 62–84.

5. Ibid., p. 71.

6. Platinga 1984, p. 10.

7. Rosen 1995, p. 363.

8. Ibid., p. 410.

9. Oken 1823, col. 553.

10. Ibid., p. 554. It should be noted that Oken's Versammlung deutscher Naturforscher und Aerzte was not the first attempt to unite German investigators of nature and physicians. From 1801 to 1805 the Vaterländische Gesellschaft der Aerzte und Naturforscher Schwabens was an ephemeral attempt to achieve for Schwabia what Oken wished for all German territories. And, J. S. C. Schweigger called for the aged Leopoldinisch-Carolinische Akademie der Naturforscher to become a German academy. See *Journal für Chemie und Physik* 23 (1818), p. 350. Oken himself took the Gesellschaft Helvetischer Naturforscher in Geneva as a model for the Association. See Fischer 1952, p. 349. See also *Isis* 1 (1821), *Litterarische Anzeiger*, cols. 196–198.

11. Oken 1823, col. 555. See also Schwägrichten und Kunze 1823, cols. 1–3.

12. Oken 1823, cols. 555–556.

13. Schweigger 1823, p. 458. Reprinted in Pfannenstiel 1958, p. 35.

14. Ibid.

15. *Goethes Gespräche mit Eckermann*, 3 May 1827, reprinted in Schipperges 1972, p. 4.

16. *Goethes Gespräche mit Eckermann*, 27 January 1830, reprinted in Schipperges 1972, p. 4.

17. Oken 1823, col. 557.

18. Ecker 1883, p. 68.

19. Ibid., p. 43.

20. As quoted in Sudhoff 1922, p. 32.

21. C. E. H. 1820, col. 37.

22. Ibid., col. 38.

23. Riemer 1833–1834, part I (1833), pp. 352–353.

24. Ibid., p. 353.

25. As reprinted in Schünemann 1932, p. 35.

26. Ibid.

27. *Zweite Denkschrift*, December 1803, as reprinted in Schröder 1959, p. 82, and as translated in Applegate 1998, p. 294.

28. Schröder 1959, pp. 129, 132, 135, and Applegate 1998, p. 295.

29. Schröder 1959, p. 111, and Applegate 1998, p. 295.

30. Pfeiffer and Nägeli 1810. For the importance of the singing of hymns and *Lieder* to the moral *Bildung* of the *Volk*, see Horstig 1806, cols. 129–132. The importance of music as a source of *Bildung* (particularly for children) was the subject of numerous articles in *Allgemeine musikalische Zeitung* and other musical journals during the second decade of the nineteenth century, owing in large part to the publication of Pfeiffer and Nägeli 1810. See, e.g.: Dreist 1811, cols. 833–842, 858–879, 876–878; Lindner 1811, cols. 3–8, 17–23, 33–43, 49–59; Wendt 1811, cols. 281–287, 297–303, 316–324, 333–341; anon. 1818, cols. 533–539; Benelli 1818, cols. 697–702; J. C. H. 1819, cols. 565–576; Marx 1827, pp. 395–397; and Marx 1828b, pp. 211–214.

31. Blumner 1891, p. 9.

32. Ibid., p. 12.

33. Staudinger 1913, p. 57.

34. Wilhelm Christian Müller 1830, part II, p. 365.

35. Staudinger 1913, pp. 64–65.

36. Ibid., p. 65, and Nipperdey 1972, pp. 5, 18.

37. Staudinger 1913, p. 70.

38. Plantinga 1984, p. 109. See also Henkel 1829, pp. 413–414.

39. Koch 1802, p. 901, as translated in Plantinga 1984, p. 109.

40. Mosse 1996.

41. Staudinger 1913, p. 69.

42. Oken 1827, col. 356.

43. Ibid., col. 357.

44. Ibid., cols. 358–359.

45. Oken 1828, cols. 443–444.

46. Ibid., cols. 444–445.

47. The opening day's festivities began with a choir of eight hundred singing a hymn composed by Mozart and Handel's Hallelujah chorus. See Strombeck and Mansfeld 1842, p. 22.

48. Anon 1842. Reprinted in Pfannenstiel 1958, p. 146.

49. Hoffmann-Donner 1867, p. 7.

50. Ibid., p. 14.

51. Ibid., p. 32.

52. Ibid., pp. 29–30.

53. "Das Lied von den Speculanten," written by Stiebel, argued that "[i]n the mystical regions, where the Ascarids live in hyperpolygamy, where in the murky creases of the mucous membrane bloody hemorrhoids reign, the Speculator is always present." Ibid., p. 51.

54. Heinrich Hoffmann-Donner was also the editor of the *Liederbuch* distributed at the Frankfurt meeting in 1867. Although it is unfortunately beyond the scope of this chapter to provide an in-depth analysis of all of the authors of these choral-society songs, it should be noted that they included scientists and physicians as well as renowned scholars not affiliated with the natural or medical sciences, such as the poet and theologian Gustav Schwab, and the author, painter, and statesman Victor von Scheffel.

55. Ibid., p. 12.

56. Ibid., p. 13.

57. "Die Scoperei" claimed rather sarcastically that physicians now had an instrument for every human orifice. Ibid., pp. 55–56. Similarly, "Sonst und Jetzt" talked about how stethoscopes, thermometers, and specula were dominating medical diagnosis. Ibid., pp. 57–58.

58. "Lied der Hydrophisen," ibid., pp. 49–50.

59. Ibid., pp. 64–66.

60. Ibid., p. 73.

61. Ibid., p. 33.

62. Ibid., pp. 115–116.

63. Johnston 1831, p. 226.

64. Ibid.

65. Ibid., p. 222.

66. Ibid., pp. 237–238. For an account of the music played at the Hamburg meeting, including the *Liedertafel* under the direction of Methfessel, see Bartels and Fricke 1831, pp. 33–35.

67. Kielmeyer and Jäger 1835, pp. 45–47.

68. Ibid., p. 49.

69. Ibid., p. 50.

70. Ibid., p. 51.

71. Morrell and Thackray 1984, p. 23.

72. See, e.g., Brewster's letter to Babbage, 21 February 1831, ibid., p. 33; and Brewster to Phillips, Secretary of the Yorkshire Philosophical Society, 23 Februrary 1831, ibid., pp. 33–34.

73. Ibid., p. 38.

74. Johnston 1831, p. 237.

75. Phillips to W. V. Harcourt, 10 August 1838, in Morrell and Thackray 1984, p. 273.

76. Ibid., p. 194.

77. Of these 466 participants, 192 were from Berlin, while 130 were from outside Prussia. Oken 1829, col. 242. Thirty-eight members were at the Halle Meeting in 1823, 37 in Würzburg in 1824, 88 in Frankfurt am Main in 1825, 115 in Dresden in 1826, and 156 in Munich. Sudhoff 1922, p. 19, and Fischer 1952, p. 351.

78. Pfannenstiel 1958, pp. 41, 44, 45, 48.

79. Degen 1956, pp. 333–340, here pp. 334–335.

80. Sudhoff 1922, p. 23, Degen 1956, p. 338.

81. As quoted in Degen 1956, p. 339.

82. For a detailed account of the Berlin meeting, see anon 1828.

83. Oken 1829, col. 271.

84. Ibid., and Rave 1962, pp. 363–364.

85. Hobsbawm and Ranger 1983.

86. Goethe was the only living savant whose name appeared on the curtain.

87. Humboldt and Lichtenstein 1829, p. 13. Handel's *Alexanderfest* had been a part of the Singakademie's repertoire for over twenty years. See Blumner 1891, p. 38.

88. Dryden 1979, p. 1.

89. Ibid.

90. Ibid., p. 2.

91. Ibid.

92. Ibid., p. 3.

93. Ibid., p. 4. See also p. iii.

94. Riemer 1833–1834, part V (1834), p. 119.

95. Humboldt and Lichtenstein 1829, p. 13.

96. Ibid., pp. 13–14.

97. Ibid., p. 14.

98. Ibid.

99. Ibid., p. 15.

100. Ibid.

101. Ibid.

102. See the two letters from Alexander von Humboldt to Felix Mendelssohn-Bartholdy, in Mendelssohn, "Green Books" I, 29 and 30, the Bodleian Library, The University of Oxford.

103. Sweet 1980, vol. 2, pp. 107, 193, and Applegate 1998, p. 283n.

104. Applegate and Potter 2002, pp. 9–10.

105. Ibid., p. 9.

106. Felix Mendelssohn to Karl Friedrich Zelter, 15 February 1832, in Elvers 1986, p. 172, as quoted in Applegate and Potter 2002, p. 9. His sense of a national identity, forged by previous composers, however, should not be confused with political unification, which he did not desire.

107. Blumner 1891, p. 243.

108. Rudorff 1900.

109. Fanny Mendelssohn to Carl Klingemann, 12 September 1828, in Hensel 1995, pp. 226–227.

110. Ibid., p. 227.

111. A number of amateur musical societies of the early nineteenth century, much like the Versammlung, did not permit women performers, unlike the popular-music scene, where they thrived. See William Weber 1975, pp. 35–36, 55.

112. Humboldt and Lichtenstein 1829, pp. 20–21. See also Oken 1829, cols. 272–273. This cantata was performed in Berlin on 26 June 2003 by the Berlin Cappella Male Choir and members of Ensemble Cosmpolitano Berlin. See the Alexander-von-Humboldt Foundation 2003, CD (see mitpress.mit.edu/catalog/item/default.asp?ttype=2&tid=11023 ["musical samples"]). The following is an English translation of the text:

Welcome! We gladly call out to you,
It is the greeting of friendship ringing out to you.
A blessing rules over this festival,
Ordained with the power to reach out to us.
With pride and emotion we are moved
By the salvation brought to us by the ruler's benevolence.
So then let the thankful greeting ring out
To the protector of all that is good and beautiful.
Out of the dreaded night of chaos
The power wrests itself from the elements;
The earth stood firm. Rigid and defiant
She presented the storm with proud peaks
And thrusted the shore's cliff-breast towards the ocean.
Fire, air and waves fought a raging battle.

The storm breaks mighty blocks loose,
And they thunder down into the valley below.
The waves foam full of wrath at the mountain's barrier
And carve themselves out deep chasms.
And dreadful is the wild force of the fire as its
Destruction penetrates to the womb of the deep.

Loud is the fury of the savage battle.
Discord causes destruction.
Flames, storm and flood
Are threatening grim devastation.
What God created in the power of his wisdom,
Is sinking back into dark night!

Cease!—the sound of a wonderful voice rings out
And suddenly the elements' anger is fettered.
Storm and waves hold still;
The sea of flames sinks down to peaceful embers.

Then the wonderful clarity of light, full of blessings,
Breaks through the ethereal dreams.
And shining bright the truth is clear to all,
The chorus of the elements is now reconciled.

Together the powers' keen striving has effect,
As only unity can bring forth true salvation.

Now the related powers take effect and work to form
And build the most marvelous world.
The earth is resplendent, the fire gleams,
And sweet breezes move the flood.
The ether arches up high and twinkling stars gently
Draw golden circles in their shining path.

And as the mighty building of the world is formed,
So, too, it is formed in the human breast!
The savage power of the earth lives
In the soul with ruinous effect,
Unless a shining, lofty goal overcomes with unity
The quarrel of our powers.

For although the impulses swell on all sides:
Every branch belongs to the trunk
And the tree blessed with knowledge
Will be resplendent in full bloom
And Heaven will protect it with its blessing.

Oh Lord, bless where we are going,
What the united power strives for,
So that in the fleeting stream of time,
The work may stand solid as rock.

And as it rises up and towers above in honor,
Power and glory,
So will it praise only you, for it is consecrated to your greatness.
Welcome!

113. Oken 1829, col. 272.

114. Ibid., cols. 274–247.

115. Pfannenstiel 1958, pp. 50–51, and Kruta 1973, p. 163.

116. Riemer 1833–1834, part V (1834), p. 119.

117. Humboldt and Lichtenstein 1829, p. 27.

118. Oken 1829, col. 352.

119. Kruta 1973, p. 155.

120. Ibid., p. 161.

121. Universitätsarchiv Göttingen Kur. 4Vb 95a, folia 36–46.

122. Ibid., folio 20.

123. Ibid., folio 21.

124. Ibid.

125. Ibid., folio 23.

126. Ibid., folio 24.

127. Ibid., folia 16–17, here folio 17.

128. Ibid., folia 26–27.

129. See also Applegate 1998, p. 284.

130. Frevert 1988, 1991.

Chapter 4

1. Kaestner 1830, pp. 297–300.

2. Ibid., p. 298.

3. Ibid. During the Romantic period, the definition of genius metamorphosed from a peculiar ability possessed by a creative individual into a power that possessed and engulfed him. See Schaffer 1990, p. 83. See also Kivy 2001.

4. Jackson 1994, pp. 416–417.

5. Goethe 1887–1919, here II, 6, 8, as translated in Mueller 1952, p. 23.

6. Ibid., II, 6, p. 295, as translated in Mueller 1952, p. 89.

7. Ibid., II, 28, pp. 69–70.

8. Ibid., pp. 71–77.

9. "Über das Marionettentheater," in Sembdner 1970, 2, pp. 338–345, here p. 340.

10. Ibid., pp. 341–342.

11. Ibid., p. 342.

12. Bleiler 1967, p. xiii.

13. Ibid., pp. 183–214, here pp. 211, 203.

14. Ibid., pp. 71–103, here p. 89.

15. Ibid., p. 81.

16. Ibid., p. 94.

17. Ibid., p. 95.

18. Ibid., pp. 95–96.

19. Ibid., p. 96.

20. Ibid., pp. 96–97.

21. Hegel 1974, p. 956. See also Kramer 2002, p. 318, n. 19.

22. Letter of 20 March 1843, published originally in the *Augsburger Gazette* and shortly thereafter in *Lutèce: Lettres sur la vie politique, artistique et sociale de la France* of *Heinrich Heines Werke* 1970ff (vol. 19, 1977—originally published in 1855), p. 176, and Metzner 1998, p. 143.

23. Brewster 1819, p. 116.

24. Sprengel 1771 and Poppe 1810.

25. Anon. 1998, p. 27.

26. Nöding 1823.

27. Ibid., pp. 6–7.

28. Ibid., pp. 7–10.

29. Ibid., p. 14.

30. Ibid., p. 23.

31. "Apollonion" in Gustav Schilling 1835–1838: 1 (1835), pp. 247–248, here p. 247.

32. Nöding 1823, pp. 16–20.

33. Ibid., pp. 32–33.

34. There are two types of reeds used in musical instruments, the free-vibrating reed and the beating (or pulsating) reed. The free reed vibrates inside an opening that is slightly larger than the reed itself so that the opening does not interfere with the vibrations of the reed. The beating reed vibrates against an aperture, that is slightly smaller than the reed; hence, the reed continually beats against the edges surrounding the opening. See Ord-Hume 1986, p. 14.

35. Ahrens 1996, pp. 20–38, here p. 20; Wilke 1823, col. 118; Gottfried Weber 1829a, p. 184. The first free-reed instrument was the Jew's harp, dating back to the mid-sixteenth century, which was in use until the late eighteenth century. Ord-Hume 1986, p. 16.

36. Roeber 1839, p. 80.

37. Kempelen 1791.

38. Schmitz 1983, p. 11.

39. Ahrens 1996, p. 20. See also Wilke and Kaufmann 1823, col. 150, and Owen 2001, p. 64.

40. Ord-Hume 1986, p. 21, and Wilke and Kaufmann 1823, col. 150. For Vogler's technique of organ building, see Wilke 1824, cols. 673–677, 689–694.

41. Ord-Hume 1986, p. 21; Vogler 1801, col. 523, n.; and Wilke and Kaufmann 1823, cols. 150–154. For Vogler's mechanism of crescendo in his orchestrion, see Sauer 1824, cols. 370–375. See also Gottfried Weber 1829a, p. 184.

42. As cited in Bowers 1972, p. 345.

43. Wilke and Kaufmann 1823, pp. 152–153.

44. Gottfried Weber 1829a, p. 184.

45. Ibid., and Sauer 1813, cols. 117–120.

46. Gottfried Weber 1829a, p. 184; Sauer 1813, col. 118; Uthe 1811, col. 429.

47. Vogler 1801, col. 517.

48. Ibid., col. 521. For a summary of Vogler's organ building, see Wilke 1824, cols. 673–677, 689–694.

49. Strohmann 1811, col. 153. Strohmann wished to establish the priority of his invention over the organ builder Uthe. See Strohmann 1810, col. 968; Uthe 1811, cols. 429–431; and Wilke and Kaufmann 1823, col. 149. See also Gottfried Weber 1829a, p. 185.

50. Strohmann 1811, col. 154.

51. Ibid.

52. Ibid., col. 155.

53. Uthe 1811, col. 430. Uthe mentioned that the organ builder Scheidler in Bennungen was working on similar pipes years before. Ibid., col. 431.

54. "Johann Nepomuk Mälzel," in Schilling 1835–1838: 4 (1837), pp. 509–511, here p. 509.

55. Buchner 1992, p. 95.

56. Ibid., p. 96.

57. Ibid., p. 97.

58. Entry for "Kaufmann," in Brockhaus 1824–1837; 6, p. 128, and Schmitz 1983, p. 14.

59. Schilling 1835–1838: 4 (1837), entry for "Kaufmann," p. 57; Carl Maria von Weber 1812, cols. 633–636; entry for "Kaufmann" in Brockhaus 1824–1837: 6, p. 128; and Schmitz 1983, p. 14.

60. Mechanicians who built clocks and musical automata included the Kaufmanns, Leonhard and Johann Nepomuk Mälzel, Anton Wist, Heinrich Steiner, and Christian Seyffert. Weißflog 1999, p. 1, and Wonneberger 1989, pp. 9–10.

61. Most historical accounts list 1805 as the year of the invention, while others cite 1812 as the date. Most likely, the 1812 was an improved version over the first, which had been completed in 1805. Schilling 1835–1838: 4 (1837), p. 57, and 1 (1835), p. 544. See also Schmitz 1983, p. 15.

62. "Bellonion," in Schilling 1835–1838: 1 (1835), pp. 544–545.

63. Fleck 2001, p. 8. See also anon 1887, pp. 121–123, 133–137.

64. Carl Maria von Weber 1812, col. 636.

65. Ibid.

66. Schilling 1835–1838: 2 (1835), pp. 223–224. See also Bowers 1972, p. 480.

67. Kaufmann 1823, cols. 119–120. Kaufmann cited Wilke's article in *Allgemeine musikalische Zeitung* for "a good description of such a process," col. 119.

68. Audsley 1965: 2, pp. 37–38.

69. The interval was approximately two-fifths of a diatonic semitone.

70. Kaufmann 1823, col. 120.

71. Ibid.

72. See "Crescendo—und Decrescendo—Zug," in Schilling 1835–1838: 2 (1835), pp. 328–330, here p. 330.

73. Kaufmann 1823, col. 120.

74. Tiggelen 1987, p. 169.

75. Schilling 1835–1838: 4 (1837), p. 58.

76. Carl Maria von Weber 1812, cols. 633–636.

77. As quoted in anon. 1887, p. 133.

78. *Allgemeine musikalische Zeitung* 43 (1841), col. 1075.

79. This periodical is also referred to as the *Deutsche Illustrirte Zeitung*. Anon 1851 (1976), p. 380, reprinted in anon 1976, p. 8, and anon 1887, p. 133.

80. The Blessing family in the Black Forest built orchestrions, which were musical automata resembling Mälzel's panharmonicon, from 1820 onward. See Blessing 1991, pp. 5–26.

81. Bowes 1972, p. 480. Original: *London Illustrated News*, 5 July 1851.

82. Ibid.

83. Ibid.

84. Ibid.

85. Ibid.

86. Ibid.

87. Anon 1887, p. 122.

88. Anon. 1847 (1976) "Kunstnachrichten: Mechanische Musik," as quoted in *Das mechanische Musikinstrument* 2 (1976): pp. 4–6, here p. 4.

89. Ibid., p. 6.

90. Paddison 2002, p. 321.

91. Koch 1782–1793: 1 (1782), p. 16, as translated in Baker and Christiansen 1995, p. 144.

92. Sulzer 1771–1774: 2, "Empfindung," (1771), p. 54, as translated in Baker and Christiansen 1995, p. 28.

93. Sulzer 1771–1774: 2 (1771), pp. 58–59, 54, as translated in Baker and Christiansen 1995, p. 31.

94. Paddison 2002, p. 322.

95. Friedrich Schlegel, on the other hand, held another view, equally hostile to eighteenth-century affect theories as the first, which was taken up by Eduard Hanslich in the mid-nineteenth century. Paddison 2002, p. 322. See also Bowie 2002, pp. 33–36.

96. Buschmann 1988, p. 8. Original: *Zeitschrift für Instrumentenbau* (1910), p. 4. To listen to a harmonium, go to www.harmoniumservice.de. Click on "downloads," then "Klangbeispiele."

97. Sachs 1912–1913, p. 1324.

98. Heyde 1980, p. 447. See also Sachs 1912–1913, p. 1324, and Gleichmann 1820, cols. 507–508.

99. Gleichmann 1820, col. 506.

100. Buschmann 1988, p. 8.

101. Sachs 1913, p. 14.

102. Gleichmann 1820, cols. 507–508, and Sachs 1913, p. 14.

103. Heyde 1994, p. 149, and Elste 1987, p. 92.

104. Das Geheime Staatsarchiv Preußischer Kulturbesitz, Rep. 120A, Abt. XIV, Fach 2, Nr. 32, Bd. 1, folia 14–26, here folio 14; and Chladni 1822, col. 178.

105. Ord-Hume 1986, p. 23.

106. Ibid., p. 24, and *Allgemeine musikalische Zeitung* 12 (1809), col. 469.

107. Buschmann 1988, p. 8.

108. Ibid.

109. Haeckl was the son of a renowned Viennese physician, Johann Christoph Haeckel, who also possessed an impressive range of musical knowledge as an accomplished singer and instrumentalist. Although Anton began his medical studies at university, he needed to discontinue his formal education in order to raise his five siblings after the death of their parents. He refocused his attention on music, with which he had been raised as a child, composing numerous pieces for worship. Schilling 1835–1838: 3 (1836), pp. 414–415.

110. Ord-Hume 1986, p. 23, and Sachs 1913, p. 297.

111. Sachs 1913, p. 297, and Chladni 1821–1826: 23 (1821), col. 395.

112. Ahrens 1996, p. 23. See also Österreiches Staatsarchiv; Haus-, Hof-, und Staatsarchiv, Hoftiteln der Musikinstrumentenbauer, Obersthofmeisteramt 1845, Rubric 12/33.

113. Ahrens 1996, p. 24.

114. *Allgemeine musikalische Zeitung* 41 (1839), col. 668.

115. Das Geheime Staatsarchiv Preußischer Kulturbesitz, Rep. 120A, Abt. XIV, Fach 2, Nr. 32, Bd. 1, folia 49–63.

116. Ibid., folio 54.

117. Ibid.

118. Owen 2001, p. 66.

119. Das Geheime Staatsarchiv Preußischer Kulturbesitz Rep., 76, Ve 9, Abt. XVc, Nr. 8; and Heyde 1994, p. 149.

120. Heyde 1994, p. 149.

121. Ibid., p. 150.

122. Ibid. Original: *Allgemeiner Anzeiger der Deutschen*, 9 April 1829.

123. Fischer-Dieskau 1997, p. 90.

124. Das Geheime Staatsarchiv Preußischer Kulturbesitz, Rep. 76, Ve 9, Abt. XVc, Nr. 8, f. 25; and Heyde 1994, p. 150.

125. Das Geheime Staatsarchiv Preußischer Kulturbesitz, Rep. 76, Ve 9, Abt. XVc, Nr. 8, f. 27; and Heyde 1994, p. 150.

126. Friedrich Kaufmann's note to Wilke 1823, pp. 118–119. It should be noted that Wilke disagreed. See Wilke 1825, col. 542.

127. *Wiener Zeitung,* Amtsblatt 4 July 1826, as quoted in Ottner 1977, p. 123.

128. Stadtsarchiv Wien, Hauptregistrar 1826, Departmentbücher M, 16152; *Wiener Zeitung,* 12 September 1826; Ottner 1977, p. 123. See also Haupt 1960, pp. 168–169.

129. As early as 1820 Gleichmann suggested that the techniques relevant to the construction of an aeolodicon could be applied to the organ. Gleichmann 1820, col. 505.

130. Schulze 1829, cols. 190–191.

131. Schünemann 1932, pp. 6–10. From 1804 to 1809, Zelter wrote six proposals for musical reform in Prussia. See Fischer-Dieshau 1997, p. 90. The quotes are taken from his sixth proposal of 1809, as quoted in Schünemann 1932, p. 10.

132. Forkel 1788–1801: 2 (1801), pp. 22–23, as translated in Butt 2001, p. 213.

133. This document was first published in Schünemann 1932, pp. 10–18. It is the second draft. The origins of this essay date back to September 1803, Schünemann 1932, p. 8. See also Schünemann 1937, pp. 30–31.

134. As quoted in Schünemann 1932, p. 11.

135. As quoted in Schröder 1959, p. 79; translated in Applegate 1998, p. 292.

136. As quoted in Schünemann 1932, p. 14.

137. Applegate 1998, p. 292.

138. Schünemann 1932, pp. 19, 28.

139. Appelgate 1998, p. 282.

140. Ibid.; Engel 1960, pp. 90–92; Edler 1983, p. 89.

141. Applegate 1998, p. 288.

142. Letter of 14 May 1809, as quoted in Schünemann 1932, pp. 30–31.

143. As quoted in Schünemann 1932, p. 32.

144. *Allgemeine musikalische Zeitung* 15 (1813), cols. 717–724.

145. Anon 1814, cols. 657–662, and anon 1819, cols. 517–523.

146. Wilke 1821, cols. 625–630, 641–646.

147. Hermann 1822, cols. 678–682.

148. H[äser] 1827, cols. 723–724.

149. Billroth 1830, col. 409.

150. See, e.g., Seibt 1998 and Schmidt 2002.

151. Fröhlich 1820, col. 389.

152. Ibid., col. 391.

153. Ibid., col. 392.

154. Ibid.

155. Ibid., col. 405. See also col. 421.

156. Wilhelm Christian Müller 1830, part II, p. 365.

157. Pellisov 1834, cols. 723–724.

158. Ibid., col. 735.

159. See, e.g., Wolfram 1815; Schneider 1834, 1835; Werner 1823; Reichmeister 1828; Zang 1804; Lehmann 1830; Reichmeister 1832; Kützing 1836.

160. Dearling 1996, p. 131.

161. Wilke 1823, pp. 113–119, 149–155. Indeed, Christian Friedrich Gottlob Wilke was to comment in 1823 that "an organ, which would be built of such pipes of different character and sizes, receiving different degrees of wind from the compression bellows, would produce a shrill, but impressive effect" (p. 117).

162. Wilke 1823, p. 114, and Audsley 1965: 2, p. 36.

163. Audsley 1965: 2, pp. 37–38.

164. Wilke 1823, pp. 114–115, and Audsley 1965: 2, p. 39.

165. Audley 1965: 2, p. 39.

166. Schneider 1834, p. 55 and note.

167. Wilke 1823, p. 119.

168. Chladni called for further developments along these lines. He wanted the loudness or softness of each organ's tone to be controlled by the key, rather than the bellows, as Friedrich Kaufmann was able to accomplish on the harmonichord. Chladni 1826d, cols. 40–41.

169. Wilke and Kaufmann 1823, cols. 117–118, and Perne 1821, cols. 133–139, 149–156.

170. Perne 1821, col. 138.

171. Ibid., col. 136. The mechanical design of his *orgue expressif* was kept secret until he published its description of it in a series of articles in *Journal des débats* in 1829. See Buschmann 1988, p. 8.

172. Perne 1821, p. 136.

173. Ibid.

174. Ibid., p. 138.

175. Ibid., p. 150. See also Biot 1817, part I, pp. 386–389.

176. It is interesting to note that the Bernhard brothers, organ builders from Romrod in the Catholic Grand Duchy of Hessen, worked on perfecting a mechanism that would purposely sharpen or flatten a pitch by as much as a full tone. Apparently, such a mechanism was "particularly of worth to Catholic services, when often the tone of the organ either cannot be given before the hymn, or cannot always be reached immediately while singing." Chladni 1821–1826, 26 (1824), col. 813. See also *Allgemeiner Anzeiger der Deutschen*, no. 247 (1824), p. 2784.

Chapter 5

1. Werber 1976, p. 8.

2. Wiederkehr 1960, p. 4.

3. Weber and Weber 1825, dedication page.

4. Wilhelm Weber 1892–1894: 3, p. 355. As translated in Olesko 1996, p. 121.

5. Weber and Weber 1825, pp. 511–530. Although I shall refer to both brothers when citing this text, Wilhelm was the one responsible for this particular portion of the work.

6. The Weber brothers defined "*eine stehende Schwingung*" as "a modified movement of a wave that resonates perfectly." Ibid., p. 511. See Weber and Weber 1826, cols. 186–199, 206–213, 222–235. We would refer to this phenomenon today as a standing wave.

7. Weber and Weber 1825, p. 513.

8. Since he was dealing with a density wave, Weber discussed the particles' velocities and densities at different points in time. It should be noted, however, that he did not take into account the entire trigonometric analysis of Augustin Jean Fresnel on the subject of reflection, which had been carried out just a few years earlier.

9. Ibid., pp. 521–525.

10. Gottfried Weber 1824: 1, p. 4. The Weber brothers argued that this was the first work that dealt with the study of music with scientific rigor. See Weber and Weber 1825, p. 522.

11. Weber and Weber 1825, pp. 523–524.

12. Ibid., pp. 524–525.

13. Wilhelm Weber 1828a, pp. 397–407.

14. Weber's dissertation was entitled "Theoria efficaciae laminarum maxime mobilium arcteque tubose aerem sonantem continentes claudentium" (Theory of the efficiency of highly mobile reeds, which tubes with resonating air tightly enclose). A year later he completed his *Habilitationsschrift*, "Leges oscillationis oriundae, si duo corpora diversa celeritate oscillantia, ita conjunguntur ut oscillare non possint nisi simul et synchronice" (Laws of vibrations, which arise when two bodies with differing number of vibrations are connected in such a manner that they can only vibrate simultaneously and synchronistically). Wilhelm Weber 1827.

15. Wilhelm Weber 1828a, p. 401.

16. Ibid. Weber's interest in reed instruments can also be seen in his translation of Charles Wheatstone 1828, pp. 175–183. See Wilhelm Weber 1828b, pp. 327–333. Of particular interest was the reed of the Jew's harp.

17. Wilhelm Weber 1828a, p. 401.

18. See also Gottfried Weber 1829a, pp. 184–186. Note that tuning forks of the 1820s were far from being perfect harmonic oscillators.

19. Weber argued that transversally vibrating bodies possess a low pitch during their intensification due to refraction (*Beugung*), while longitudinally vibrating bodies have a higher pitch with an increased intensity due to compression and attenuation. Wilhelm Weber 1828a, p. 402. See also Wilhelm Weber 1829a, p. 194. The acoustics scholar Bindseil argued that tuning forks, and indeed all transverse vibrating rods (particularly rod-shaped reeds) vibrate more slowly when the paths of vibration are large, i.e., when the rods are struck hard. As the initial force that set the rod or reed in motion ceases, the paths of vibration decrease, resulting in a slight increase in pitch. See Bindseil 1839, pp. 450, 481–482. As the air pressure in the reed pipe decreases, the layer of air vibrating with the reed approaches the maximum vibration of the air column. The pitch is relatively high, as the reed pipe's pitch approximates an open labial pipe. As the pressure increases, the layer of air vibrating with the reed approaches a surface of nodal points, the tone is relatively lower, and the pitch of the reed pipe approximates a covered labial pipe. See Harleß 1853: 4, pp. 503–706, here p. 659. The French organ builder Aristide Cavaillé-Coll demonstrated experimentally that an increase of pressure from 2 1/4 to 4 inches of pres-

sure results in an increase in pitch by 1 in 180. For A = 440.0 vps, the increase in volume raises the pitch to 442.4 vps. See Ellis 1880, p. 296. For an excellent account of Cavaillé-Coll's work, see Schuster 1991. See also Cécile and Emmanud Cavaillé Coll 1982.

20. Wilhelm Weber 1828a, p. 401.

21. Ibid., p. 406.

22. These conditions included the length of the air column.

23. Ibid., p. 403. See also Wilhelm Weber 1827. Helmholtz commented on how a brass reed tuned to b♭ and connected to a resonator flattened in pitch when the pressure in the bellows increased. See Helmholtz 1954, p. 102. See also Bindseil 1839, pp. 481, 659.

24. Wilhelm Weber 1828a, p. 404.

25. Ibid., pp. 407–408. Weber provided measurements for pipes corresponding to a♭, a, b♭, b, and c.

26. Wilhelm Weber 1829b, pp. 193–206.

27. Covered labial pipes are also known as stopped or closed labial pipes, as they have a cap on the top. As a result of that cap, a covered labial pipe sounds at an octave lower than an open labial pipe of the same length.

28. Ibid., pp. 199–203. See also Wilhelm Weber 1830b, pp. 177–213.

29. Wilhelm Weber 1829b, p. 203.

30. Ibid., pp. 203–206.

31. Ibid., p. 204.

32. Ibid., p. 206.

33. Weber's condition of accepting Göttingen's offer of professorship in 1831 was the hiring of a *Mechanikus*. Clearly, this was critical to his research. Universitätsarchiv Göttingen Kur. 4Vb, 95a, folio 49.

34. Wilhelm Weber 1828a, p. 408; Wilhelm Weber 1829a, p. 202; Wiederkehr 1967, p. 25.

35. *Leipziger Adreßkalender* 1824, p. 210.

36. *Leipziger Adreßkalender* 1826, p. 80.

37. Stadtarchiv Leipzig, II, Sekt. H 1518, folia 1–2.

38. Stadtsarchiv Leipzig, II, Sekt. H 1821, folio 1.

39. Entries under Oertling in the Adreßbücher, Berlin, from 1830 to 1865. See, for example, *Adress-Buch für Berlin* 1835 and *Adreß-Kalender* 1830–1837, 1840–1845, 1847–1866.

40. Oertling 1843, pp. 60–72.

41. Wilhelm Weber 1829c, pp. 415–438.

42. Wilhelm Weber 1829a, p. 196.

43. Wilhelm Weber 1829c, p. 417. "*Mensur*" is the ratio of the pipe's width to length for each tone. It varies for each type of organ pipe, the environment where the organ is played, etc., and is even today a tightly guarded secret for each organ builder.

44. Ibid., pp. 417, 420.

45. Wilhelm Weber 1829c, pp. 192–246.

46. Ibid., pp. 201–223. An ideal reed pipe differs from a real one in two significant ways. First, the vibrating reed of an ideal pipe is perpendicular to the length of the air column. The reed of a real reed pipe, however, is parallel to the length of the pipe, forming a part of its wall. Second, all parts of an ideal reed pipe undergo the same motion, whereas the reed of a real pipe oscillates the greatest at its free end, and its fixed end does not move at all.

47. Ibid., p. 196.

48. Ibid.

49. Ibid., pp. 216–241, esp. pp. 230–231.

50. Ibid., pp. 227–228.

51. Recall from chapter 2, Taylor's formula for the frequency of a vibrating string, $\sqrt{\frac{2gP}{LG}}$, where g is the gravitational constant, P is the string's elasticity, L the strength's length, and G its mass per unit length.

52. Ibid., p. 235. In a footnote, Weber explained the discrepancy between the observed and predicted values of third reed pipe as arising from an error in his notebook, where he thought he might have switched a + sign with a – sign. Once corrected, the observed value was 439.5.

53. Ibid., pp. 235–238.

54. Ibid., pp. 238–243.

55. Ibid., p. 236.

56. Ibid., pp. 237, 224.

57. Ibid., pp. 238, 241.

58. Ibid., pp. 238–240.

59. Ibid., p. 239, and Wilhelm Weber 1829b, pp. 199–201n.

60. Wilhelm Weber 1829d, pp. 240–241.

61. See Lagrange 1761–1773: 2 (1760–1761), pp. 152–154. See also Fox 1971, p. 83, and Heilbron 1993, p. 170.

62. Wilhelm Weber 1829d, p. 241.

63. Ibid.

64. Fox 1997, p. 199. The Newtonian equation for the speed of sound through air is $c = \sqrt{(P/\varrho)}$ where P is the pressure of the gas and ϱ is its density. Lindsay 1966, p. 637.

65. Fox 1997, p. 199; Finn 1964, p. 12; Biot 1802, pp. 173–182.

66. Fox 1971, p. 83.

67. Ibid., pp. 84–85.

68. Poisson 1808, pp. 319–392.

69. For the derivation of Poisson's argument, see Heilbron 1993, pp. 174–175.

70. Neither savant was aware of such experimental evidence provided by John Dalton in the early 1800s, who noticed that the maximum rate at which a thermometer rose when air filled the evacuated receiver of an air pump was 1°F in 3.5 seconds. A similar decrease in temperature took place whenever the temperature inside the receiver was 50°F above the surrounding air. In another set of experiments, he placed a capillary tube with one closed end, which contained a mercury thread inside a receiver. He compressed the air inside the receiver to two atmospheres, halving the length of the air column trapped by the mercury thread. The air was then permitted to escape rapidly until the thread assumed its original position, and the tap was closed. The length of the air column caught in the capillary tube decreased by 1/10 and was restored to its original length after the tap was reopened. Dalton concluded that the tap had been temporarily decreased by cold in the one case and increased by heat in the other. See Fox 1971, pp. 87–88.

71. *Mémoires de la classe* 1811, p. xcv, as translated in Fox 1971, p. 134.

72. Finn 1964, p. 11.

73. Fox 1971, pp. 136–137; Finn 1964, pp. 12–15; Heilbron 1993, pp. 107–114.

74. Heilbron 1993, p. 107.

75. My account owes much to Fox 1971, pp. 139–140, and Heilbron 1993, pp. 107–114.

76. Reproduced from Heilbron 1993, p. 110.

77. Fox 1971, p. 139.

78. Heilbron 1993, p. 112.

79. Ibid., p. 114.

80. Laplace 1816, pp. 238–241.

81. Finn 1964, p. 14, and Lindsay 1966, p. 637.

82. Laplace argued in this paper that the constant $(1 + k)$ was equivalent to the ratio of the specific heat at constant pressure (c_p) to the specific heat at constant volume (c_v), or γ in modern notation. See Fox 1997, p. 200.

83. Heilbron 1993, p. 176, and Fox 1971, p. 164.

84. Laplace 1816, p. 240.

85. The agreement between Laplace's value for c_p/c_v and the present-day accepted value of 1.40 is purely coincidental, owing to errors in his argument being compensated by the errors made by Delaroche and Bérard's data. Fox 1971, pp. 164–165, and Finn 1964, pp. 13–15.

86. Laplace 1816, p. 241.

87. Fox 1997, pp. 200–201; Crosland 1978, pp. 123–124; Gay-Lussac and Welter 1822, pp. 436–437.

88. As translated in Crosland 1978, p. 124. See also Laplace 1822, pp. 267–268, and Finn 1964, p. 17.

89. Crosland 1978, p. 124, and Fox 1971, p. 170. Note that Crosland claims Laplace's corrected value for the speed of sound in air to be 337.8 m/s. Laplace argued that the 1% discrepancy was due to water vapor in the atmosphere. Heilbron 1993, p. 178.

90. Carnot 1824.

91. Dulong 1829a, pp. 113–159. See also Fox 1971, p. 253. Dulong was a very talented amateur musician. See Girardin and Laurens 1854, p. 10.

92. Dulong 1829b, pp. 470 n., 471.

93. Ibid., p. 471.

94. Ibid. See also Fox 1971, p. 253. Wilhelm Weber was very much aware of the details of Dulong's experiments. See Wilhelm Weber 1829b, pp. 199–201n. Indeed, he himself most likely translated this essay into German. See Dulong 1829b, pp. 438–479.

95. Dulong 1829b, pp. 471, 473–475. See also Fox 1971, p. 253.

96. Dulong 1829b, pp. 476, Wilhelm Weber 1829b, p. 201 n.

97. Wilhelm Weber 1829d, p. 240 (italics in the original).

98. Ibid., p. 242.

99. What Weber failed to appreciate here was the tacit skill necessary to produce a properly functioning reed. Indeed, reed-instrument players inform me that it is impossible to make the same reed twice. And, they are convinced that no scientific, algorithmic description can accurately represent how to produce a reed. They often speak of a certain knack, or innate ability of gifted reed makers.

100. Wilhelm Weber 1829d, p. 243.

101. Wilhelm Weber 1830c, pp. 3–26.

102. Wilhelm Weber 1833, pp. 1–17. Weber did author an article entitled "Akustik" two years later, but it did not offer any original contribution to the field, only a summary of important developments up to that point. Schilling 1835–1838: 1 (1835), pp. 99–119.

103. Wiederkehr 1967, p. 33.

104. As quoted in Wiederkehr 1967, p. 57.

105. Wilhelm Weber 1833, p. 1.

106. Ibid., p. 3.

107. Ibid.

108. Ibid.

109. Ibid., p. 5.

110. Helmholtz was later to show that a string produces a "regime of tones," including the fundamental and its upper partials.

111. Wilhelm Weber 1833, p. 14.

112. Ibid., p. 16.

113. Chladni 1826c, pp. 189–212.

114. Chladni 1827b, cols. 281–284. Readers of the *Allgemeine musikalische Zeitung* were already familiar with Wilhelm Weber acoustical work, as Chladni had reviewed *Wellenlehre* a year earlier. See Chladni 1826a, cols. 17–23. And during this period, the *Allgemeine musikalische Zeitung* published articles on the scientific basis of music. See, e.g., Häser 1829, cols. 53–57, 69–71, 91–94, 110–114, 143–147.

115. Wilhelm Weber 1829a, pp. 186–204. The abridged version did not go into the mathematical or experimental details; it merely summarized the final results.

116. Wilke 1834, p. 64.

117. Ibid.

118. Schilling 1835–1838: 6 (1838), p. 821.

119. Kützing 1838, p. 49.

120. K. Stein, "Gottlob Töpfer," in Schilling 1835–1838: 6 (1838), pp. 670–672.

121. Töpfer 1833.

122. Weber and Weber 1826, cols. 185–200, 205–213, 221–236.

123. Töpfer 1834, p. 54.

124. Ibid.

125. Audsley 1965: 1, p. 403.

126. Töpfer 1834, pp. 56, 60–61, 66–69, 80–81.

127. Ibid., pp. 71–72.

128. Töpfer 1843, pp. 93–94.

129. Tacke 2002, pp. 53–54.

130. I would like to thank Schulze Organ Builders and Restorers of Potsdam, Germany, for their helpful discussions on the historical practices of organ building and restoration.

131. Audsley 1965: 1, p. 403. Throughout his tome, Audsley criticizes attempts by physicists, in particular Helmholtz and Tyndall, to offer practical laws for organ builders, dismissing them as "futile." He adds sarcastically, "Certain ingenious authorities, learned in mathematics, have essayed to formulate laws and lay down rules, only to find them upset and rendered utterly valueless in practice, proving in this, as in many other artistic and tonal matters, that an ounce of hard fact is worth a hundredweight of supposition or theory, however much either may be bolstered up by mathematical formulae and apparently conclusive demonstration, arrived at, not in the practical atmosphere of the pipe-making and voicing rooms of an organ factory, but in the manner breathing-space of the physical laboratory, furnished with the usual special and insufficient means of investigation." Audsley 1965: 2, p. 587. Although he is undoubtedly correct that these equations on their own can never be sufficient, and that one needs the tinkering of the artisans, Audsley was clearly unaware of Töpfer's praise of the 1830s, '40s, and '50s of Weber's equation for reed pipes. Indeed, Audsley only discusses the posthumously published work *Die Theorie und Praxis des Orgelbaues* of 1888, heavily annotated by Max Allihn. See Audsley 1965: 1, pp. 401–403, and 2, pp. 587–588.

132. Töpfer 1855, pp. 316–317.

133. Tacke 2002, pp. 133–166.

134. Wilhelm Weber 1830c, pp. 3–26.

135. Cécile and Emmanuel Cavaillé-Coll 1982, pp. 90–94; Owen 2001, p. 65.

136. Dearling 1996, p. 131.

137. Audsley 1965: 2, p. 41.

138. *Glasgow Herald*, 3 November 1877, as reprinted in Audsley 1965: 2, p. 42.

139. As quoted in Audsley 1965: 2, p. 38.

140. Ellis's note in Helmholtz 1954, p. 96.

141. Bindseil 1839, p. 499n., and Johannes Müller 1838–1840: 1 (1838), p. 178.

142. Bindseil 1839, p. 448.

143. Ibid., pp. 508–510. See also Harleß 1853, pp. 540, 601.

144. Harleß 1853, pp. 619–620, 649–660.

145. 4 (1852), pp. 124–125, and 5 (1853), pp. 93–98, reprinted in Helmholtz 1882–1895: 1 (1882), pp. 235–255. See also Darrigol 1998, pp. 5–6.

146. Darrigol 1998, p. 6; Helmholtz (1882): 1, p. 250; and Wertheim 1848, pp. 434–475.

147. Darrigol 1998, p. 8; Helmholtz 1859, pp. 1–72; and Helmholtz (1882): 1, pp. 303–382, here pp. 303–307.

148. Darrigol 1998, pp. 3, 8–9.

149. Ibid., p. 10.

150. Ibid., pp. 5–29.

151. Helmholtz 1861, pp. 321–327.

152. Helmholtz 1954, p. 391, and Helmholtz 1861, pp. 322–323.

153. Helmholtz 1954, p. 391.

Chapter 6

1. Schmidt 1909, p. 4.

2. Abrogast 1992, p. 157, and Klusen, Stoffels, and Zart 1979–1980: 1 (1979), p. 32. See also Haase 1957, pp. 57–59.

3. Schmidt 1909, p. 5.

4. Scheibler 1816, cols. 505–511.

5. Ritter 1920, pp. 48–50. Other Scheiblers also contributed to the technical development of the silk industry: Emil Scheibler (1820–1863) made several critical improvements to the silk-spinning frame, and his brother Guido (1826–1889) also invented a series of machines, which greatly increased the efficiency of silk manufacturing. See Schmidt 1909, pp. 3–4.

6. Brauns 1906, p. 3.

7. Chung 1974, p. 92.

8. Thun 1879, pp. 23–24, 28.

9. Ritter 1920, p. 50, and Rbt. 1931, p. 171.

10. Ritter 1920, pp. 112–113. See also Rösen 1965, pp. 38–39.

11. By 1846, 3,000 of the 8,000 Krefeld looms had been shut down. Thun 1879, p. 113.

12. Brauns 1906, p. 3, and Thun 1879, p. 114.

13. Noble 1984.

14. Brauns 1906, p. 30.

15. The musical inversion of a minor third is a major sixth (5:3), of a major third is a minor sixth (8:5), of a perfect fifth is a perfect fourth (3:4). The octave of the tonic is, of course, also consonant.

16. Lindley 2001a, p. 248.

17. Barbour 1951, pp. 3, 25.

18. Lindley 2001b, p. 643.

19. Lloyd and Boyle 1963, p. 52.

20. Lindley 2001a, p. 248.

21. Ibid. This difference is referred to as a lesser diesis.

22. Ibid. This interval is called a greater diesis.

23. Barbour 1951, p. ix.

24. Ibid., p. x.

25. Ibid., pp. 26–27. See also Jorgensen 1991, p. xvii.

26. Barbour 1951, pp. x, 27–44.

27. This temperament was recommended by Giseffo Zarlino in the sixteenth century. Each fifth remains diminished and imperfect by 2/7 of a comma. Ibid., p. 32. A comma is defined as a very small interval between two notes, which is introduced by computing their frequencies.

28. Christiaan Huygens recommended this comma, whereby fifths were flattened by 2/9 of a comma. Ibid., p. 37.

29. This temperament was developed by Pietro Aaron, also in the sixteenth century. The interval C–E is pure, C–G is a little flat, as is G–D, and A is tuned so that the intervals D–A and A–E are equal. Ibid, p. 26.

30. This temperament, recommended by Francisco Salinas, results in pure minor thirds, tritone, and major sixths. Fifths and major thirds are reduced. Ibid., pp. 33–34.

31. In the eighteenth century, other variations of meantone temperament were added, such as 3/10 comma, 5/18 comma, 1/6 comma, 1/7 comma, 1/8 comma, 1/9 comma and 1/10 comma. Ibid., pp. 40–44.

32. Lloyd and Boyle 1963, p. 146.

33. Lindley 2001c, pp. 290–291.

34. Marpurg 1756.

35. Marpurg 1776.

36. Chladni 1802, p. 48n., and Scheibler, "Ueber mathematische Stimmung," in Scheibler 1838, folio 9, and Scheibler 1837, foreword.

37. Kirnberger 1771–1779.

38. Rasch 2002, p. 220.

39. Lindley 2001a, p. 257.

40. Rasch 2002, p. 220. It should be noted that tuners were claiming to tune their instruments in equal temperament at this time. As Jorgensen has argued, a true equal temperament (as we now understand it) was not achieved until the late nineteenth century. He refers to these early methods, starting with Scheibler's, as quasi-equal temperament. Since I am more interested in the transition to what the tuners themselves called equal temperament (rather than comparing it with modern standards), I shall simply refer to it as equal temperament. For a somewhat different take, see Jorgensen 1991, pp. 1–7.

41. Lehmann 1827.

42. Hummel 1828.

43. Lindley 2001a, p. 259.

44. As quoted in Jorgensen 1991, p. 417; Jousse 1832, p. 28.

45. Montal 1836 (1976).

46. Ibid. Jorgensen 1991, p. 2. Although Jorgensen argues that the English were able to tune organs in equal temperament since 1810, Schwiening claimed in 1837 that

neither the English nor the French were interested in tuning organs in equal temperament. See Jorgensen 1991, p. 467, and Schwiening 1837, p. 220.

47. Woolhouse 1835, p. 39. As quoted in Jorgensen 1991, p. 425.

48. De Morgan 1864, p. 141. The paper was written in 1857, although it did not appear in print until seven years later.

49. Ibid.

50. Ibid., p. 142.

51. Ibid.

52. Ibid.

53. Scheibler 1834a, p. 1, and Roeber 1834, pp. 333–362, 492–520.

54. See the appendix.

55. Roeber 1834, pp. 337–338.

56. Wehran 1853, p. 33. Wehran translated portions of Loehr's original essay, as well as added his own material.

57. Scheibler's first tonometer was composed of fifty-two tuning forks spanning the octave of a, corresponding to 434 svps (217 Hz) to a', 868 svps (434 Hz). This tonometer, unfortunately, no longer exists. Each fork was approximately 8.3 svps higher than the preceeding fork, or 4.15 beats. Letter from Scheibler to Spohr, 3 February 1832, Handschriftenabteilung der Gesamthochschul-Bibliothek Kassel-Landesbibliothek und Murhardsche Bibliothek der Stadt Kassel, 4° Ms. Hass, as reprinted in Abrogast 1992, p. 163. See also Roeber 1834, p. 192. He later built a second tonometer ranging over the octave from a to a', which corresponded to 440 and 880 svps (220 and 440 Hz) respectively. Sadly, this tonometer also no longer exists. Alexander J. Ellis, translator of Helmholtz's *Die Lehre von den Tonempfindungen*, tested Scheibler's 56-fork tonometer, finding that 32 of the forks produced precisely 4 beats per second with the preceeding fork, the other 23 (one does not, of course, count the first fork) ranged from 3.8 to 4.2 beats per second. By using electrographic techniques of the 1870s and 1880s, McLeod and A. Mayer determined that Scheibler's 56-fork tonometer was accurate to at least within one beat every ten seconds (0.1 beats per second), and often within one beat every twenty seconds (or 0.05 beats per second). Ellis, "appendix" to Helmholtz 1954, p. 444.

58. Scheibler 1838, fol. 8–9. The name is sometimes spelled Kemmerling. See, e.g., Kraushaar 1838, p. 9.

59. Scheibler 1834a, p. 53.

60. Scheibler 1837b, foreword.

61. Scheibler 1837b, foreword and p. 9.

62. Wehran 1853, p. 30. Jorgensen claims that starting in 1810, English organ tuners could have drawn upon Robert Smith's technique of 1749 to tune in equal temperment. Jorgensen 1991, p. 468. Apparently, they chose not to, as British organ builders would first consider tuning in equal temperament some fifty years later.

63. Wehran 1853, p. 30.

64. Ibid., pp. 30–31, and Loehr 1836, pp. 17–18.

65. Wehran 1853, p. 33. For the inadequacies of the ear, see also Loehr 1836, p. 18.

66. Wehran 1853, p. 42.

67. Ibid., p. 43n.

68. Ibid., pp. 44–45. Scheibler had Kämmerling engrave Mälzel's graduated index scale on the pendulum itself, making measurements more precise. And, he had Kämmerling make ten subdivisions between each number of Mälzel's scale. Ibid., p. 43n.

69. Scheibler 1836, pp. 31–32.

70. To ensure that it is the case that a is 440 svps (220 Hz) and the a pulsating fork is 432 (216 Hz), rather than the pulsating fork 432 and a is 424 (212 Hz), the tuner tunes both fork and string to unison, and then tightens the string sharper until four beats occur during every oscillation.

71. Letter from Scheibler to Spohr, 3 February 1832, Handschriftenabteilung der Gesamthochschul-Bibliothek Kassel-Landesbibliothek und Murhardsche Bibliothek der Stadt Kassel, 4° Ms. Hass, as reprinted in Abrogast 1992, p. 164.

72. Wehran 1853, p. 45, italics in the original. Presumably the fork corresponding to the octave of the tonic was not used; hence only twelve forks were used.

73. Scheibler 1834a, p. 43, and Loehr 1836, pp. 30–31.

74. Scheibler 1834a, pp. 13–29; Scheibler 1838, fol. 3; and Loehr 1836, pp. 23–30.

75. The number of combination tones depends on the interval of the initial two pitches. Some intervals generate one combination tone (such as the octave), while others produce three or four. See Loehr 1836, p. 27.

76. Scheibler 1837b, pp. 10–11; Scheibler 1838, fol. 7–8; Loehr 1836, pp. 33–39.

77. Wehran 1853, p. 52.

78. An organ stop is the control on an organ console that selects a particular pipe to sound.

79. Scheibler 1837b, p. 21; Scheibler 1838, fol. 24; Loehr 1836, p. 41.

80. Scheibler 1838, fol. 1–9; Wehran 1853, pp. 53–54; Loehr 1836, pp. 32–33.

81. Scheibler 1837b, pp. 29–30, and Wehran 1853, p. 54. If the organ tuner possessed a modified metronome, he could tune the semitones as follows. For example, the note b♭′ was tuned as an ascending major third with *Hülfston* d′, such that the combination tones would produce two beats with the metronome marking at 77.4. The note b′ could be tuned as a fourth with *Hülfston* e to produce four beats between combination tones at pendulum marking 78.5. As an ascending fourth from g, c′ could be tuned by hearing one beat at pendulum marking 53.1, while c#′ can be tuned with mathematical c#′, with two beats at 65.4. The d′ pitch must be higher than *Hülfston* d′ by three beats at 66.6, while the pitch d#′ produces one beat with b′ at 63.1, and one beat with *Hülfston* a′ at 82.3. The f′ pitch generates one beat with c′ at 70.6, f#′ one beat with c′ at 75.0; g′ is the octave of g, while g#′ forms a pure fifth with mathematical c″ (512 vps); therefore, g# and c must produce four beats at 84.0 Finally, a″ is the octave above a′. After tuning the octave of the tuning stop, the other notes of the same stop, the remaining stops are tuned by unisons and octaves. Scheibler 1838, fol. 2–3. For an earlier version, see Scheibler 1834a, pp. 14–15.

82. Scheibler 1837b, pp. 30–31, and Wehran 1853, pp. 58–59.

83. Loehr 1836, p. 3.

84. Ibid., p. 4. Mathematical temperament takes the mathematical ratios of notes for the establishment of pitches. It is the same as Pythagorean temperament.

85. Ibid., p. 16.

86. Ibid., p. 45.

87. Töpfer 1842, p. iii.

88. Ibid., p. iv.

89. Ibid., pp. 25–27. Three major thirds are tuned in the mathematical ratio 4:5: a–c#, c#–e#, e#–g##; therefore, the interval a–g## is 64:125. The question that needed to be resolved was: what is the difference in vibrations between a and g##? Given the mathematical ratios of musical intervals: a = a, c# = 5/4a, e# = $5^2/4^2$a, g## = $5^3/4^3$a, and therefore, a′ = 2a. The difference (*D*) between a′ and g## is therefore 2a–$5^3/4^3$a, or a = 21.33D. What pendulum marking is necessary to count four beats between a′ and g##? The answer: 60:*N* = 8 vibrations:*D*, where *D* equals the number of vibrations in which g## is flatter than a′.

90. Ibid., pp. 22–24.

91. Ibid., pp. 41–45. The major third a to c# will produce four beats with a pendulum at 65.2, while c# and f will produce four beats at 82.8. The f can now be used to tune a (440 svps, 220 Hz), as this third generates four beats at 104.0. Töpfer continued: a and c produce four beats at 89.0; c and e four beats at 77.8, e and g# four

beats at 98.2; g# and B four beats at 84.4; B and d# four beats at 73.6; d# and g 4 beats at 93.0; g and B♭ four beats at 79.5, B♭ and d four beats at 69.5, and finally d and f# four beats at 87.5.

92. Kraushaar 1838.

93. Ibid., pp. 15–20. For example, using a c′ tuning fork, one would tune the octave higher, c″. One would then tune a fifth above c′, or g′, and c″ with f′. The pitch g′ would then be tuned with its lower octave g, while f′ would be tuned with b flat; g would be tuned with d′; and b flat with its octave b flat′. One now has tuned two chords: b flat, d′, f′ and g, b flat, d′. One would then proceed to tune d′ and a′ and b flat′ with e flat′. Hence, the following chords would be in tune: f′, a′, c″; d′, f′, a′; e flat′, g′, b flat′; and c′, e flat′, g′. The a′ pitch would be tuned with its lower octave a, e flat′ with a flat, a with e′ and a flat with a flat′. As a result, the resulting chords would be in tune: a, c′, e′; c′ e′ g′; a flat, c′, e flat′ and f′, a flat′, c″. Finally, the tuner would tune e′ with b′, a flat′ with d flat′, b′ with b, d flat′ with g flat and b with f sharp′. As a result e′, g′, b′; g, b, d′; b, d′, f sharp′; d′, f sharp′, a′; b flat, d flat′, f′; g flat, b flat, d flat′; and g flat, a, d flat′ would be in tune. Kraushaar 1838, pp. 16–17. If one were using an a′ tuning fork, rather than a c′ tuning fork, one would tune a with its lower octave a, a with e′, a′ with d′, e′ with b′, d′ with g, b′ with its lower octave b, and g with its upper octave g′. The two chords, g, b, d′ and e′, g′, b′, would now be in tune. One would then proceed to tune b with f sharp′ and g′ with e′. This would yield the following tuned chords d′, f sharp′, a′; b, d′, f sharp′; c′, e′, g′; and a, c′, e′. The f sharp′ pitch would be tuned with its lower octave, f sharp, c′ with c″, f sharp with c sharp′ and c″ with f′, resulting in the following tuned chords: f sharp, a, c sharp′; a, c sharp′, e′; f′, a′, c″; and d′, f′, a′. Finally, c sharp′ would be tuned with g sharp′, f′ with b flat, g sharp′ with g sharp, b flat with b flat′, g sharp with d sharp′, and b flat′ with e flat′. The last chords would then be c sharp′, e′, g sharp′; e′, g sharp′ b′; g sharp, b, d sharp′; b, d sharp′, f sharp′; g sharp, c′ d sharp′; f′, g sharp′, c″; b flat, c sharp′, f′; c sharp′, f′, g sharp′; b flat, d′ f′; g, b flat, d′; e flat′, g′, b flat′, and e flat′, f sharp′, b flat′. Kraushaar 1838, pp. 18–19.

94. Scheibler 1834a, pp. vi–vii.

95. Ibid., pp. vii–viii. See also *Allgemeine musikalische Zeitung* 39 (1837), cols. 240–242, here col. 241.

96. Scheibler 1837b, p. 10, and Schwiening 1837, pp. 237–238.

97. "Vorbemerkung der Redaction," 1835, cols. 485–488. See also Schwiening 1837, pp. 231–232.

98. *Allgemeine musikalische Zeitung* 39 (1837), cols. 240–242, reprinted in *Intelligenzblatt für Crefeld und die umliegende Gegend*, no. 112, 25 August 1837, p. 3.

99. G. W. F 1837, col. 241, reprinted in *Intelligenzblatt für Crefeld und die umliegende Gegend* no. 137, 22 May 1837, pp. 2–3.

100. G. W. F. 1837, col. 241, reprinted in *Intelligenzblatt für Crefeld und die umliegende Gegend*, no. 112, 25 April 1837, p. 3.

101. Schwiening 1837, p. 238. See also G. W. F. 1837, col. 242.

102. As quoted in Schwiening 1837, p. 13, and Wehran 1853, p. 29.

103. G. W. F. 1837, col. 240, reprinted in *Intelligenzblatt* no. 112, 25 August 1837, p. 3.

104. Schwiening 1837, p. 238.

105. We only have the letters written by Scheibler to Spohr, Landesbibliothek und Murhardsche Bibliothek der Stadt Kassel 4° Ms. Hass. 287, reprinted in Abrogast 1992, pp. 160–168.

106. Abrogast 1992, p. 162.

107. Ibid., p. 163.

108. Ibid., p. 164.

109. Ibid., p. 165. Such precision was grossly exaggerated.

110. Ibid., p. 166. The "physicists in Stuttgart" refers to the Versammlung deutscher Naturforscher und Aerzte meeting detailed below.

111. Ibid., p. 166.

112. Ibid. Orchestral musicians tune their instruments by ear, using fifths and octaves.

113. See Scheibler 1837a.

114. Haas 1886, p. 2.

115. Ibid., p. 3.

116. In 1885 the International Conference on Pitch in Vienna decided that a′ should be 435 Hz, or 870 svps, or the French standard. See the next chapter.

117. Ibid.

118. Ellis 1880, p. 299.

119. Ibid., p. 300.

120. Ibid., p. 304.

121. *Frankfurter Oberpostamtszeitung*, no. 29, September 1832, reported by Schnyder von Wertensee in Scheibler 1834, pp. iv–v. See also Haase 1957, p. 61.

122. Scheibler 1834, p. v.

123. Muncke 1833, pp. 392–403.

124. Ibid., p. 390.

125. Ibid., p. 399 n. Dulong seems to have been the French physicist responsible for drawing German physicists' attention to la Tour's siren. In his 1834 article, Roeber argued that Scheibler's method for determining the vibrations of a pitch was superior to the use of sirens (ibid., p. 344). Five years later, however, he actually argues that Cagniard de la Tour's "renowned" work on the siren served as proof of Scheibler's determination of the dependence of the pitch of a tone as well as the velocity of the resulting beats (Roeber 1839, pp. 90–91). Hence, German acousticians began to take la Tour's work seriously during the mid-1830s, despite the fact that the Frenchman's invention dates back to 1819. See Turner 1977, p. 4.

126. Roeber 1839, p. 2, and Maley, Jr., 1990, p. 36.

127. Maley, Jr., 1990, p. 38. Roeber states the year as 1754. Roeber 1839, p. 2.

128. Maley, Jr., 1990, p. 42.

129. Ibid., p. 44. In earlier works, Tartini claimed that the combination tone was equal to the 1/2 term. Roeber 1839, pp. 2–3.

130. Maley, Jr., 1990, p. 65. See Vieth 1805, pp. 265–314.

131. Young 1855: 1, pp. 64–98. See also Maley, Jr., 1990, p. 67.

132. Young 1855: 1, p. 84, and Maley, Jr., 1990, p. 68.

133. Maley, Jr., 1990, p. 73.

134. Young 1802, p. 146.

135. Blein 1827.

136. Maley, Jr., 1990, p. 91.

137. Ibid., p. 92.

138. Wilhelm Weber 1829e, pp. 216–222.

139. Ibid., pp. 218–219.

140. Ibid., pp. 219–220.

141. Ibid., p. 222.

142. Hällström 1832, pp. 438–466.

143. Maley, Jr., 1990, p. 97.

144. Ibid.

145. Hällström 1832, p. 444, and Roeber 1839, p. 8.

146. Turner 1977, p. 12.

147. Roeber 1834, pp. 333–362, 492–520, and Roeber 1839, pp. 1–141, esp. pp. 1–53.

148. Scheibler 1834, p. vi.

149. Roeber 1839, p. 22.

150. Ibid., pp. 33–34.

151. Ibid., p. 28.

152. Ibid., pp. 29–30.

153. Ibid., pp. 31–32.

154. Ibid., p. 39.

155. Ibid., pp. 30–31.

156. Poggendorff 1834, pp. 520–528. See Young, "Outlines of Experiments and Inquiries Respecting Sound and Light," in *Philosophical Transactions of the Royal Society of London* in 1800, reprinted in Young 1855: 1, pp. 64–98. See also Kipnis 1991, pp. 26–32.

157. Poggendorff 1834, p. 520. See also Young, "Inquiries Respecting Sound and Light," in *Philosophical Transactions of the Royal Society of London* in 1800, reprinted in Young, 1855: 1, pp. 64–98. See also Kipnis 1991, pp. 26–32.

158. Poggendorff 1834, p. 524.

159. Poggendorff 1834, p. 521, and Maley, Jr., 1990, pp. 102–103.

160. Poggendorff 1834, p. 525, and Maley, Jr., 1990, p. 103.

161. Poggendorff 1834, p. 525.

162. Turner 1977, p. 13.

163. Ohm 1839, pp. 463–466.

164. Turner argues that Ohm means that both pitches must be simple harmonic vibrations. See Turner 1977, p. 13.

165. Ibid., p. 14.

166. Ibid.

167. Ibid., p. 16.

168. Ibid.

169. Helmholtz 1856, pp. 497–540.

170. Helmholtz 1856, pp. 518–521, 532–536. See also Turner 1977, p. 17.

171. Turner 1977, p. 17.

172. Helmholtz 1856, pp. 536–549, 532–536. See also Turner 1977, p. 17.

173. See Pantalony 2005, pp. 57–82.

174. Lecomte 1857, pp. 47–136. The foreword was written in 1846.

175. Ibid., p. 47.

176. Ibid., pp. 48, 101 n. 3.

177. Ibid., p. 49.

178. Ibid., pp. 51, 101 ftn. 3, and 103 n. 1.

179. Ibid., p. 101 n. 3.

180. Ibid., pp. 59–112.

181. Ibid., p. 128.

182. Jorgensen 1991, p. 1.

183. As quoted in Lloyd and Boyle 1963, p. 66.

Chapter 7

1. Anon. 1836, p. 9.

2. Actually, I tuned the a string on my cello to 220 vps, tuning the lower octave of 440.

3. See, e.g., the collection of essays in Frängsmyr, Heilbron, and Rider 1990. Alder offers a very enlightening account of the quantification of the meter in late-eighteenth-century France, illustrating that the rhetoric of "natural" standards belied the social and economic interests of those seeking a national standard. See Alder 1995. Schaffer provides a compelling account of how standardization in Victorian Britain, in the form of metrology, linked the inner workings and relationships of British physics laboratories with technical and economic projects in society, such as cable telegraphy. See Schaffer 1992, pp. 23–56. See also Schaffer 1995, pp. 135–172. For an interesting account dealing with the ethics and trustworthiness of measurement, see Gooday 2004. And Galison has written on the politically charged negotiations involved in the formation of the international bureau of weights and measures in 1875 and the final sanctioning of the meter in 1889. He details the scientific, technological, social, economic, and political pressures (particularly from the European and American railway companies) to standardize time in the 1880s. See Galison 2003, pp. 84–155.

4. Olesko 1995, pp. 118–125.

5. Olesko 1995, pp. 118–121, and Dove 1828.

6. Note that standardization was critical not only to German scientists, but to British, French, and U.S. scientists as well. German territories had particular reasons to standardize weights and measures relevant to unification.

7. Olesko 1996, p. 131.

8. Ibid.

9. Ibid.

10. Similarly, William Thomson (Lord Kelvin) spoke of the aesthetics of precision measurement. See Smith and Wise 1989, pp. 127–128, 243–244, 282, 454–456, 685–686, 700.

11. Simpson 1665. See also Harding 1983, p. 1, and Wehmeyer 1993, pp. 15–16.

12. Harding 1983, p. 6.

13. Purcell 1696, unpaginated. Harding 1983, p. 7.

14. Loulie 1696, pp. 82–92. See also Wehmeyer 1993, p. 29.

15. Tiggelen 1987, pp. 140–141.

16. L'Affillard 1705, p. 55, and Harding 1983, p. 11. See also Sauveur 1701, and Wehmeyer 1993, p. 29.

17. Harding 1983, p. 12, and Pajot 1735, pp. 182–195. See also Wehmeyer 1993, p. 29.

18. Tans'ur 1756, pp. 41–46; Choquel 1762; Gabory 1770; Weiske 1790; Wright 1795; and Harding 1983, pp. 18–19.

19. Harrison 1775.

20. Fétis 1830, pp. 41–42. See also Tiggelen 1987, pp. 141–142, and anon. 1836, pp. 7–8.

21. *Allgemeine musikalische Zeitung* 5 (1803), col. 705. See also Bürja 1790.

22. Wenk 1798.

23. Stöckel 1800, col. 658.

24. Ibid.

25. Ibid., col. 660.

26. Ibid., col. 677.

27. Ibid., col. 665.

28. Ibid., col. 678, and Wehmeyer 1993, p. 41.

29. Stöckel 1800, col. 679.

30. Guthmann 1806, col. 118.

31. Ibid.

32. Ibid., col. 119. The French composer, dancer and professor of the Conservatoire de Paris Enfin Despréaux, however, thought that all of the aforementioned mechanical devices should be disposed of, and one simply use an ordinary weight suspended by a cord, which was attached at the top to a graduated scale and whose length could be varied. See Fétis 1830, p. 41.

33. For a biographical sketch of Gottfried Weber, see Schilling 1835–1838: 6 (1838), pp. 832–833.

34. Gottfried Weber 1813, cols. 441–447.

35. Ibid., col. 443. Italics in the original.

36. Ibid. Italics in the original.

37. Ibid., col. 444.

38. "Nachrichten" in *Allgemeine musikalische Zeitung* 15 (1813), cols. 784–788, here col. 784. Italics in the original. See also Tiggelen 1987, p. 148.

39. The standard metronome today beats from 40 to 208 times a minute.

40. For the controversy between Mälzel and Winkel, see Rochlitz 1818, cols. 468–473. See also Tiggelen 1987, pp. 52–64, 144–156.

41. Tiggelen 1987, p. 145.

42. Ibid., pp. 148–149.

43. Rochlitz 1818, col. 470.

44. Leonhardt 1990, pp. 133–134.

45. Harding 1983, pp. 27–28.

46. *Wiener allgemeine musikalische Zeitung*, no. 6, 6 February 1817.

47. *Wiener allgemeine musikalische Zeitung*, no. 5, 30 January 1817.

48. Gottfried Weber 1814, col. 445.

49. Ibid., col. 447.

50. Ibid., col. 448.

51. Ibid.

52. Ibid., col. 461.

53. Ibid., col. 462.

54. Ibid., col. 464.

55. Ibid., cols. 464–465. Italics in the original.

56. The letters from Mälzel to Breitkopf and Härtel of Leipzig were destroyed during World War II. They are partially reproduced in Haupt 1927, p. 126.

57. Ibid., p. 134.

58. "Mälzels Metronom," in *Allgemeine musikalische Zeitung* 19 (1817), cols. 417–422, here cols. 417–418.

59. Anon. 1821, p. 303. The range of the metronome was from 50 to 160 beats a minute.

60. W. H. Cutler and the Editor 1824, pp. 31–33.

61. Wehmeyer 1993, p. 52, and Stadlen 1979, pp. 12–33. See also Riehn 1979, pp. 70–84, and for the metronomic markings for his Ninth Symphony, see *Caecilia* 6 (1826), p. 158.

62. Wehmeyer 1993, p. 52.

63. Hamburger 1951, p. 161. See also Leonhardt 1990, pp. 145–146.

64. Haupt 1927, p. 129, and Schindler 1840, pp. 212–214.

65. Thayer et al. 1901–1911: 4 (1907), p. 65.

66. Sievers 1819, col. 599.

67. Ibid., col. 600.

68. Ibid.

69. Ibid.

70. Anon. 1836, p. 10.

71. Ibid., p. 5.

72. Ibid.

73. Reichardt 1776, p. 86; Mendel 1968, p. 215; and Haynes 2002, p. 316. Two measurements were used during this period: simple vibrations, or the vibrations of a fork from one extreme position to the original (simple vibrations per second, or svps), and complete vibrations, or the vibrations of a tuning fork from one extreme position to the other and then back to the original position (vibrations per second, or vps, or Hz).

74. Ellis argued that once corrected, Marpurg's pitch for the Prussian capital during that period was closer to 420. Ellis 1880, p. 318.

75. Ribock 1782, and Haynes 2002, p. 314.

76. Translated in Powell 1791, as cited in Haynes 2002, p. 313.

77. Haspels 1987, p. 122, and Haynes 2002, p. 338.

78. There were a number of ways in which tuning-fork pitches were measured. In 1823 Ernst Gottfried Fischer measured the pitches of several tuning forks by means of a monochord. See Fischer 1825, pp. 501–510. See also Delezenne 1854 and Haynes 2002, p. 330. For the values of historical pitches, see Ellis's "appendix" to Helmholtz 1954, pp. 495–500.

79. Ellis 1880, pp. 310, 318. A few remarks are in order to put these values in their proper perspective. First, I rounded off Ellis's values to the nearest vibration per second, as such precision seems meaningless in historical context. Second, and more to the point, such precision is not necessary. Such a minute difference is extremely difficult to detect. Wapnick and Friedman 1980, pp. 176–184. The ability to discern two pitches not sounding simultaneously depends on the skill of the musician, specifically on his or her musical memory. The longer the interval between the sounding of the pitches, the more difficult it is to discern the difference. And the human ear is more sensitive to certain frequencies than others. Third, temperature affects pitch. In older, unheated churches, a difference of as much as 18 vps of the same pipe has been detected between the hottest and coldest days of the year (Haynes 2002, p. xliii). And pitch is also dependent upon amplitude. A violin bowing an open string with a continued force for a forte volume can produce a pitch differing as much as 5 vps than when playing pianissimo, and an oboist's performing a staccato pitch is slightly sharper than the same pitch played legato (ibid., p. xliv; see also Mendel 1978, n. 103, p. 90). As the renowned instrument restorer Rainer Weber remarked in 1992, "If we approach [early instruments] with the numerical mindset of the technologist, looking for absolute answers accurate to the last decimal, we will deceive ourselves. . . . We would do well to look less precisely." Weber 1992, p. 298. See also Haynes 2002, pp. xlv–xlvi. It also important to ask: how do we know about historical pitches? Perhaps the best source of information comes from original instruments (particularly fixed-tone instruments, such as cornets, Renaissance flutes, transversos, and pitch pipes). The pitches of these instruments do not change all that much over time, unless their designs were altered throughout their histories (Haynes 2002, p. 4). Tuning forks, of course, keep their pitch relatively constant (for musical purposes there is an insignificant change due to temperature), but before the nineteenth century, it is very difficult to determine where and when they were manufactured (ibid., p. 31). By the nineteenth century, tuning forks provide the most accurate method of historically determining pitches, as the year and city of manufacture are generally well known.

80. Koch 1802, p. 822, and Haynes 2002, p. 312.

81. Ellis 1880, pp. 309–310.

82. Haynes 2002, p. 349.

83. Sievers 1817, col. 302, and Haynes 2002, p. 330.

84. Anon. 1835, col. 206, and Haynes 2002, p. 349.

85. Schindler 1855, pp. 66–68, and Näke 1862, p. 30.

86. Näke 1862, p. 28.

87. Ellis 1880, p. 310.

88. Näke 1862, p. 29.

89. Ibid., p. 13.

90. Ibid., p. 28.

91. Tuning forks from a theater in Berlin indicate this rise. In 1822, concert pitch was about 437 vps, increasing to 440 in 1830. By 1834 it has risen to approximately 442, and had reached nearly 451 by 1858. See Ellis 1880, pp. 310–311.

92. Ellis 1880, p. 319.

93. See, e.g., *Allgemeine musikalische Zeitung* 3 (1801), col. 76; Knecht 1803, cols. 529–535; Michaelis and Schicht 1814, cols. 772–776; anon. 1829, col. 292; and anon. 1835, cols. 205–207.

94. *Allgemeine musikalische Zeitung* 3 (1801), col. 76, and Haynes 2002, p. 333.

95. C. F. Michaelis was professor of the philosophy of aesthetics at Leipzig. See Michaelis 1795, 1800a,b.

96. Schicht 1812.

97. Michaelis and Schicht 1814, cols. 772–776.

98. Haynes 2002, p. 330.

99. A. Cavaillé-Coll 1859, p. 5, and Sauveur 1701.

100. Chladni 1800a, p. 8. See also A. Cavaillé-Coll 1859, p. 13.

101. Euler had measured two 8-foot C pipes of two organs and determined their pitches to be 118 and 125 svps (59 and 62.5 Hz). The music theoretician Marpurg proposed 124 svps (62 Hz) for this pitch, while the St. Petersburg *Kapellmeister* Sarti noted this pitch of his organ to be 131 svps (65.5 Hz), which corresponds to an a′ of 436 vps. Chladni 1800a, p. 8. See also A. Cavaillé-Coll 1859, p. 5. He has Marpurg measuring the 8-foot C pipe as 125 svps, rather than 124.

102. Chladni 1800a, p. 9.

103. Chladni criticized the method recommended by Sauveur for a standard pitch, namely the pitch generated by an organ pipe with the precise length of one Parisian

foot. Chladni correctly argued that the Frenchman had neglected the influence of the air, which in turn would affect the pipe's pitch. Ibid.

104. From 1841 to 1852 Weber famously conducted his electrical resistance measurements in Leipzig and Göttingen. See Wilhelm Weber 1892–1894: 3, pp. 25–214.

105. Olesko 1995, p. 118. Original: Das Geheime Staatsarchiv Preußischer Kulturbesitz, Rep. 120A, Abt. IX, Fach 1, Nr. 2, Bd. 1 (1814–1862), fol. 67–68.

106. Wilhelm Weber 1829b, pp. 193–206.

107. Ibid., p. 194.

108. Ibid., p. 195.

109. Ibid.

110. Bindseil 1839, pp. 627–628. The first serious description of la Tour's siren and its use for providing a constant pitch relevant to piano and organ building is Kützing 1838, pp. 3–7. He also mentions reed pipes for the same purposes (p. 7).

111. Wilhelm Weber 1829f, pp. 1–19.

112. Ibid., p. 19.

113. Ibid., pp. 1–2.

114. Ibid., p. 5.

115. Ibid., p. 7.

116. Weber sought the advice of the *Oberbergrath* Schaffrinsky of Berlin, who was well versed in precision mechanical work, for assistance in the design model of the clamp.

117. Weber 1829f, pp. 8–9.

118. Ibid., p. 13. Clearly Weber's monochord was much more successful at producing pleasant longitudinal tones than Chladni's first attempts some four decades earlier.

119. Ibid., pp. 13–14.

120. Ibid., p. 14. For other applications of Weber's monochord/tonometer, see pp. 14–15.

121. Ibid., p. 17.

122. Ibid., pp. 17–18.

123. Ibid., p. 18.

124. Ibid., p. 19.

125. Scheibler 1834, p. 53.

126. Ibid.

127. Scheibler, "Mittheilung des Wesentliche musikalischen und Physikalischen Tonmessers," in Scheibler 1838, §15. This value is slighly flatter than de Prony's average pitch of 442 measured in 1832. A. Cavaillé-Coll 1859, pp. 7, 13.

128. Kielmeyer and Jäger 1835, p. 77.

129. Scheibler 1838, §1–15 and Scheibler 1837a.

130. Bindseil 1839, pp. 604, 621, 625, 629, 630, 681 n.

131. Hdt. 1847, col. 803.

132. Haynes 2002, p. 350.

133. As quoted in anon. 1859, p. 492.

134. Ibid., p. 492.

135. Ibid., p. 493.

136. Ibid. Italics in the original.

137. Ibid. Italics in the original.

138. Ibid.

139. Ibid., p. 496.

140. Ibid.

141. Ibid.

142. Ibid.

143. Ibid.

144. Alder 1995, p. 51.

145. Anon. 1859, p. 494.

146. Ibid., pp. 494–495.

147. Ibid., p. 494.

148. Ibid.

149. Ibid., p. 495.

150. The Commission gathered tuning forks made in 1858 with the following pitches from European cities and concert halls: Brussels, 456 vps; London, 455; Théatre royal in Brussels, 453; Marseille and Lille, 452; St. Petersburg and Berlin,

453; Milan 450; Prague, 450; Leipzig, 449; Munich and Paris, 448; Pest and the Hague, 446; Turin, Weimar, and Wuerttemberg, 445; Brunswick, 444; and Toulouse, 442. They also received six older tuning forks. In 1854 tuning forks from Stuttgart, Gotha, and Vienna all sounded at 443 vps. A tuning fork in St. Petersburg from 1796 made 436 vps, while Mozart's tuning fork in Berlin sounded at 422, and the Grand Opera of Paris tuned to 404 vps in 1699. See Meerens 1864, p. 11. Again, I have rounded off to the nearest full vibration per second. A. Leipp and Castellango 1977, p. 11.

151. Ellis 1880, p. 298.

152. Lissajous 1885a, pp. 93–95; Lissajous 1885b, pp. 814–817; Lissajous 1857, pp. 147–231; and Lissajous 1873, pp. 878–890.

153. Haynes 2002, p. 348, and Näke 1862, p. 4.

154. Subsequent, more precise, measurements of the *diapason normal* revealed its actual pitch was 435.4 vps. Cavaillé-Coll measured the *diapason normal* by using his copies of Scheibler's forks. Also Hipkins of Britain determined the *diapason normal* to be 435.4, as it made 4.1 beats with a fork measured at 439.5 vps. Ellis 1880, p. 323.

155. Anon. 1859, pp. 496–497.

156. Ibid., p. 497.

157. Ibid.

158. Ibid.

159. Ibid.

160. Ellis 1880, p. 328. The mean pitch of these forks, excluding the Baden one, which was too flat, was 435.4 at 15°C.

161. Haynes 2002, p. 355.

162. Ellis 1880, p. 314.

163. Ibid.

164. Ibid.

165. Ibid.

166. Ibid., p. 312.

167. Näke 1862, p. 4.

168. Ibid., p. 3.

169. Ibid., p. 8.

170. Ibid., p. 9.

171. As quoted in Näke 1862, p. 10.

172. Ibid., p. 5.

173. Ibid., p. 6.

174. Ellis 1880, p. 312. The higher pitch was actually later measured to be 438.9.

175. Ibid., pp. 312–313.

176. Ibid., p. 313.

177. As quoted in Ellis 1880. Italics in the original.

178. Ibid.

179. Gesellschaft der Musikfreunde in Wien 1885, p. 2.

180. Ibid.

181. Ibid., p. 3.

182. K. K. Ministerium 1885, p. 11.

183. Anon. 1884–1885b, p. 364. This article was reprinted in K. K. Ministerium (1884–1885) 5, pp. 154–156.

184. Anon. 1884–1885b, p. 364.

185. Anon. 1884–1885a, pp. 427–430.

186. Anon. 1884–1885b, pp. 363–366.

187. Ibid., p. 366.

188. Kalkbrenner 1884–1885, pp. 205–207.

189. *Zeitschrift für Instrumentenbau* 5 (1884–1885), pp. 35–36, here p. 36.

190. Ibid.

191. K. K. Minsiterium 1885, pp. 3–4, and Sächsiches Hauptstaatsarchiv, in Dresden, "Herbeiführung eines internationalen Stimmtons," Ministerium des Innern, 1885–91, A. 847, Nr. 17537, folio 30.

192. K. K. Ministerium 1885, p. 10.

193. Ibid., p. 11.

194. Ibid. See also Grassi-Landi 1885.

195. Letter from Rudolph Koenig to London, 7 November 1888. University of Toronto Archives. London Papers. I would like to thank David Pantalony for bringing this letter to my attention.

196. See, e.g., Meerens 1864. Meerens was drawing upon the work of Delezenne 1857; see also Delezenne circa 1850s.

197. Meerens 1864, p. 11.

198. If a′ is 432, then a is 216, and A is 103.

199. Meerens 1895, p. 14.

200. K. K. Ministerium 1885, pp. 12, 15.

201. Meerens 1895, p. 22.

202. Ibid.

203. Meerens 1894, p. 10, and Meerens 1892, p. 56.

204. Meerens 1892, p. 56.

205. Ibid., p. 59.

206. Meerens 1864, p. 11.

207. Meerens 1894, p. 10.

208. K. K. Ministerium 1885, pp. 13–14.

209. Ibid., p. 14.

210. Ibid.

211. Ibid.

212. Meerens lauded Blaserna's efforts. See Meerens 1892, p. 59.

213. K. K. Ministerium 1885, p. 15.

214. Ibid.

215. Ibid., p. 16.

216. Ibid.

217. Ibid.

218. Ibid., p. 18.

219. Ibid.

220. Ibid., pp. 19–20.

221. Ibid., p. 20.

222. Ibid., p. 24.

223. Ibid., p. 25.

224. Ibid., p. 29.

225. Ibid.

226. Böhm 1970–1980: 13, pp. 10–11.

227. K. K. Ministerium 1885, p. 37.

228. Ibid., p. 38.

229. Ibid.

230. Ibid., p. 39.

231. Ibid., p. 40.

232. Ibid., pp. 20–24, 28, 30.

233. Loewenherz 1888, p. 261.

234. Ibid.

235. Ibid., p. 263.

236. Ibid.

237. Ibid., pp. 263–264.

238. Ibid., p. 265.

239. Ibid., pp. 265–266.

240. Ibid., p. 266.

241. Helmholtz 1889, pp. 65–67.

242. Reichel 1883, pp. 47–51.

243. Leman 1890, pp. 77–87, 170–183, 197–202.

Chapter 8

1. Irving 2002, p. 193.

2. See, e.g., Metzner 1998, pp. 113–159.

3. Kramer 2002, pp. 68–99.

4. William Weber 1975, p. 10.

5. Ibid., p. 22.

6. Ibid., p. 20.

7. Ibid.

8. Ibid., p. 34.

9. *Musical Review*, 11 March 1845, p. 2, as cited in William Weber 1975, p. 20.

10. *Wiener Zuschauer*, 10 June 1846, pp. 733–734, as cited in William Weber 1975, p. 20.

11. William Weber 1975, pp. 19–21.

12. As cited in William Weber 1975, p. 22.

13. William Weber 1975, p. 34.

14. For the seminal study on regimes of instruction in the early nineteenth century, see Foucault 1977. For an outstanding account of the role of pedagogy in physics, see Warwick 2003.

15. See Charlton and Musgrave 2001, pp. 75–77.

16. Logier circa 1818.

17. "J. B. Logier" in Schilling 1835–1838: 4 (1837), pp. 434–437, here p. 434; and E. Weber 1834, p. 237.

18. A London Musician 1824, col. 409.

19. Logier circa 1818, p. 1.

20. Ibid., p. 7.

21. Ibid.

22. Ibid., p. 5.

23. Ibid.

24. Ibid., p. 8.

25. Ibid. Italics in the original.

26. Ibid., p. 6.

27. Ibid.

28. Ibid., p. 7.

29. Ibid., p. 4.

30. Ibid., p. i.

31. Ibid., p. ii.

32. One account challenged this success, claiming that by 1818 Logier's method was enjoying only limited success, with instructors in London, Dublin, Edinburgh,

Manchester, Halifax, Cork, Newry, Cashel, Chester, Liverpool, Newcastle-upon-Tyne, Preston, Sheffield, and Rochester teaching a total of only fifty students. C. F. Ma. 1818, col. 896. Another article contradicted this view, arguing that Logier had over 400 students by 1816 in two London schools alone. See A London Musician 1824, col. 411.

33. Logier 1824, p. 35.

34. Logier 1827a, p. xii, n.

35. Ibid., p. 9. See also E. Weber 1834, p. 237.

36. Logier circa 1818, p. ii.

37. Charlton and Musgrave 2001, p. 76.

38. As quoted in Eigeldinger 1986, p. 96. Original: Kalkbrenner 1831, p. 4.

39. C. F. Ma. 1818, col. 896.

40. Parker 1824, p. 209.

41. C. F. Ma. 1818, col. 896.

42. Logier 1818, p. 10.

43. As cited in Logier 1824, pp. 37–38. See also C. F. Ma. 1818, cols. 893–898.

44. As cited in Logier circa 1818, p. 9.

45. Anon. 1817a, p. 24.

46. Ibid., pp. 25–26.

47. Sievers 1819, cols. 597–598.

48. Ibid., col. 599.

49. Anon. 1817a, p. 29.

50. Ibid., p. 38.

51. Kollmann 1828 (originally 1822), p. 9. Although the date of publication does not appear in the work, one can date it by an advertisement; see *Intelligenz-Blatt zur allgemeinen musikalischen Zeitung* 30 (1828), col. 52. Kollmann originally published his article in *Allgemeine musikalische Zeitung* 24 (1822), cols. 769–774, 785–794, 801–814.

52. As cited in anon. 1817a, p. 25.

53. Müller 1828, p. 41. E. Weber, music teacher in Statgard, Pommerania, who employed Logier's chiroplast and teaching method, denied this charge, arguing that the chiroplast was not designed to teach twelve students simultaneously. He also denied that the contraption harmed his students' fingers. E. Weber 1834, p. 238.

54. Kollmann 1828 (originally 1822), p. 9.

55. Müller 1828, p. 45.

56. Ibid.

57. Ibid., p. 47.

58. Ibid., pp. 50–51.

59. Kollmann 1828 (originally 1822), p. 24.

60. Ibid., p. 25.

61. Ibid.

62. Charlton and Musgrave 2001, p. 76.

63. E. Weber 1834, p. 238.

64. Schünemann 1932, p. 38.

65. Ibid., p. 39.

66. Ibid., p. 40.

67. Stöpel 1821, cols. 553–557. Stöpel was somewhat surprised to find that in Britain, only female pupils were instructed in this fashion. See also cols. 553–554 n.

68. Ibid., cols. 553–554.

69. Ibid., col. 555.

70. Ibid.

71. Logier 1824, pp. 15–18. See also Stöpel 1826, pamphlet attached to p. 52, fol. 1. The pages of this pamphlet are not numbered; hence, I shall refer to them as folio numbers.

72. Stöpel claimed the Institute was opened on 3 October. Stöpel 1826, fol. 2.

73. Logier 1824, p. 22.

74. Ibid.

75. Loesser 1954.

76. Stöpel 1826, fol. 2, n.

77. Ibid., fol. 2–3.

78. Logier 1824, p. 23.

79. Ibid., p. 24.

80. Ibid., pp. 25–27.

81. Ibid., p. 30.

82. Ibid., pp. 33–34.

83. "J. B. Logier," in Schilling 1835–1838: 4 (1837), p. 435.

84. Logier 1827b. Note that Franz Stöpel claimed that he translated this work and that it appeared in 1821. I have yet to find that edition, or another mention of it. See also Logier 1827c.

85. Loewe 1825, pp. 41–42, 57–58.

86. Marx 1825, p. 60. Marx repeated his Anglophobia in a review of Logier 1827 and 1829 five years later, arguing that England was a "country left behind that has never accomplished anything in music save to purchase it as if it were a fashionable article." See Marx 1830, p. 413.

87. Anon. 1834, p. 5.

88. Ibid.

89. Ibid., p. 6.

90. Ibid.

91. Breitung 1841, p. 18.

92. Ibid., p. 19.

93. Ibid.

94. Brendel 1867, p. 43.

95. Czerny 1830 (1961), p. 16.

96. Brendel 1867, pp. 39–62.

97. Ibid., p. 42.

98. Kollmann argued that the chiroplast could be harmful to some children's hands and fingers, which might be too small for the finger-guide, rubbing against its corners, thereby causing injury to the soft bones of the fingers. Indeed, short fingers could not bend sufficiently to pass through the apparatus resulting in the children playing with their fingers straight, which is contrary to the rules of sound piano playing. See Kollmann 1828 (originally 1822), pp. 25–26. Rather famously, Robert Schumann needed to give up his piano-playing career as a result of damage to his fingers caused by a chiroplast.

99. Walther 1732, p. 638. See also Marpurg 1749–1750, p. 7, Beer 1719, p. 176, and Schleuning 1984, p. 64.

100. Brossard 1705, p. 221.

101. Anon. 1749, p. 418; Scheibe 1745, pp. 252–253; and Schleuning 1984, p. 64.

102. Schleuning 1984, p. 64.

103. Scheibe 1745, pp. 253–254.

104. Chladni 1821a, p. 34.

105. Ibid., pp. 34–35.

106. See the review of the performance of Ms. Kirchgessner, the blind harmonica virtuoso, in the Berlin court and Deutscher Theater, Kunzen and Reinhardt 1793, part I, p. 25. Such commercialism was seen as reducing the art of music to "*Handwerk.*" Genres of music that enjoyed a larger public appeal, such as folk songs, were often particularly accused of becoming "nachahmender Handarbeit für gangbaren Marktpreis." Ibid., part II, p. 36.

107. Benda 1819, cols. 817–818n.

108. Pederson 1994, pp. 87–107.

109. Marx 1828a, pp. 23–24, as translated in Pederson 1994, p. 104.

110. Müller 1830, part I, p. x.

111. Ibid.

112. Ibid., p. 309.

113. Ibid., p. 310.

114. "Seit einigen Jahren hat man angefangen Sonaten fürs Clavier mit gutem Beifall zu setzen: bisher haben sie noch die rechte Gestalt nicht, und wollen mehr Gerühet werden, als rühen, d.i. sie zielen mehr auf die Bewegung der Finger, als der Herzen. Doch ist die Verwunderung über eine ungewöhnliche Fertigkeit auch eine Art der Gemüths-Bewegung, die nicht selten den Neid gebieret. . . . Die Franzsosen werden nun auch in diesem Sonaten-Handel, wowie in ihrem neuern Cantaten, zu lauter Italienern; es läufft aber meist auf ein Flickwerk, auf lauter zusammengestoppelte Cläusulgen hinaus, und ist nicht natürlich." Mattheson 1739, pp. 233–234, as reprinted in Schleuning 1984, p. 348. Johann Sebastien Bach's son, Carl Philip Emmanuel, concurred: "Es ist unstreitig ein Vorurtheil, als wenn die Stärcke eines Clavieristen in der blossen Geschwindigkeit bestände. Man kan die fertigsten Finger, einfache und doppelte Triller haben, die Applicatur verstehen, vom Blatte treffen, es mögen so viele Schlüssel im Lauffe des Stückes vorkommen als sie wollen, alles ohne viele Mühe aus dem Stegereif transponiren, Decimen, Duodecimen greiffen, Läufer und Kreutzsprünge von allerley Arten machen können, und was dergleichen mehr ist; und man kan bey dem allen doch nicht ein deutlicher, ein gefälliger, ein rührender Clavieriste seyn. Die Erfahrung lehret es mehr als zu oft, wie die Treffer und geschwinden Spieler von Profeßion nichts weniger als diese Eigenschaften

besitzen, wie sie zwar durch die Finger das Gesicht in Verwunderung setzen [Augenmusik, or music for the eyes], der empfindliche Seele eines Zuhörers aber gar nichts zu thun geben. Sie überraschen das Ohr, ohne es zu vergnügen, und betäuben den Verstand, ohne ihm Genug zu thun." Bach 1753, part I, p. 115. See also Schleuning 1984, p. 348.

115. Steuber 1810, col. 322.

116. Ibid., col. 321.

117. R. 1807, col. 214.

118. Müller 1830, part II, p. 340.

119. Ibid., p. 341.

120. Ibid.

121. Gathy 1835, p. 317.

122. Ibid.: "die Kraft, die Vollendung der Ausführung."

123. Burkhard 1832, p. 357.

124. Schilling 1835–1838: 6 (1838), pp. 780–784 (which is actually a misprint in the text, as it should be 780–782), here p. 780.

125. Ibid., p. 781.

126. Ibid.

127. Ibid.

128. Ibid.

129. Ibid., p. 784 (it is actually p. 782, but misprinted as 784).

130. Ibid.

131. Walker 1983–1996: 1, p. 168.

132. Ibid., p. 169.

133. Schütz circa 1830, p. 34.

134. Ibid., p. 36.

135. Laphalèque 1830, p. 61.

136. Louis Spohr to W. Speyer, 5 June 1830. As quoted in Courcy 1957: 1, p. 392, and Borer 1995, p. 6.

137. Schütz circa 1830, p. 50.

138. Ibid.

139. Houghton, Jr., 1943, p. 55.

140. Maréchal 1904, pp. 111–112. As translated in Walker 1983–1996: 1, p. 308.

141. Borer 1995, p. 1.

142. Ibid., pp. 1–2.

143. Ibid., p. 12.

144. Ibid., pp. 12–13: "They all hurt; the last one kills."

145. Guhr 1829a.

146. Guhr 1829b, p. 78. See also Guhr 1829a, p. 1.

147. Gottfried Weber 1829b, p. 76.

148. Ibid., p. 77.

149. Ibid.

150. Guhr 1829a, p. 2 (italics my own).

151. Ibid., p. 1.

152. Stowell 1985, p. 2. Other influential teaching texts used throughout the German territories included Spohr 1832, Blumenthal 1818, and Kieninger circa 1825. And the German cellist and founder of the Dresden School of Cellists, Justus Johann Friedrich Dotzauer wrote his renowned *Violincellschule* op. 165 in 1832 and his *Violincellschule für den ersten Unterricht nebst 40 Übungsstücken* op. 126 in 1836, excerpts of which are still used by young cellists throughout the world. These works drew upon his *Méthode de Violinschule* of 1825. See Seyfried 1825, pp. 249–262.

153. Pestiscus 1807, cols. 161–162.

154. Ibid., cols. 177–178.

155. Stowell 1985, p. 4.

156. Rink 2002, p. 80.

157. Stowell 1985, p. 4.

158. Rink 2002, pp. 79–80.

159. Ibid., and Guhr 1829a, pp. 6, 10. See also Metzner 1998, p. 132.

160. Guhr 1829a, p. 5.

161. Ibid., pp. 3–4.

162. Ibid., p. 10.

163. Ibid., p. 4.

164. Bennati 1831, p. 53.

165. Ibid.

166. Ibid.

167. Ibid., pp. 53–57.

168. Ibid., pp. 57–58.

169. Ibid., p. 58.

170. Ibid.

171. Ibid.

172. Ibid., p. 59.

173. Ibid.

174. Goethe to Carl Friedrich Zelter, 9 June 1831, in Goethe 1887–1919, part IV, 48: pp. 224–225.

175. Müller 1830, part II, p. 325.

176. Ibid., pp. 287–288. Müller maintains that Guhr's hope "became ridiculous" (p. 287). Although Guhr's school was not successful, Müller's comment was published only one year after Guhr's work. Hence, Müller was not the most objective critic.

177. Müller 1830, part II, p. 287.

178. Eigeldinger 1986, p. 16.

179. Ibid.

180. Ibid.

181. Walker 1983–1996: 1, p. 161.

182. Ibid.

183. Ibid., pp. 161–162.

184. Heine 1925: 9, p. 275, as translated in Walker 1983–1996: 1, p. 163.

185. See, e.g., Plantinga 1984, pp. 180–203.

186. Ibid., p. 184.

187. Metzner 1998, pp. 113–159, and Kramer 2002, pp. 68–99.

188. Walker 1983–1996: 1, pp. 302–303.

189. Ibid., pp. 285, 315–316.

190. F.-J. Fétis, "Nouvelles de Paris: Concerts spirituels," in *La revue musicale* 3 (1828), pp. 253–254, as translated in Walker 1983–1996: 1, p. 133. See also Metzner 1998, p. 139.

191. Franz Liszt, "Lettres d'un bachelier ès musique" (first published as a series of twelve articles in *La Revue et gazette musicale de Paris* in 1837–1839), in Chantavoine 1912, p. 108. As translated by Metzner 1998, p. 144.

192. As quoted in Walker 1983–1996: 1, p. 177.

193. Ibid.

194. Ramann 1880–1883: 2, pp. 127–128, as translated in Walker 1983–1996: 1, p. 130. Munito was a performing dog with an amazing range of skills, constantly impressing scientific communities with his physical and mental feats.

195. "Genie," in Schilling 1835–1838: 3 (1836), pp. 190–193, here p. 191.

196. Ibid.

197. Steuber 1810, col. 321.

198. Ibid., col. 323.

199. Ibid., col. 324.

200. Hoffmann 1810, col. 633.

201. Eigeldinger 1986, p. 16.

202. Ibid., p. 13. See also Rosen 1995, p. 368.

203. Rosen 1995, p. 368.

204. As cited in Eigeldinger 1986, p. 27. See also Walker 1983–1996: 1, p. 297.

205. As cited in Eigeldinger 1986, p. 27. Chopin also disapproved of Liszt's penchant for showmanship. Walker 1983–1996: 1, p. 184. For the importance of athletics in disciplining the mind and body, see Warwick 1998, pp. 288–326, and Warwick 2003.

206. As cited in Eigeldinger 1986, p. 27.

207. As cited in Eigeldinger 1986, p. 96. Italics in the original.

208. As cited in Eigeldinger 1986, p. 29.

209. As quoted in Eigeldinger 1986, pp. 40–41. Italics in the original.

210. Kollmann 1828 (originally 1822), p. 23.

211. Rosen 1995, p. 382.

212. Ibid., p. 364.

213. Ibid.

214. G. Weber 1827, p. 183.

215. Ibid., pp. 183–184.

216. Ibid., p. 184.

217. Ibid., p. 185.

218. Ibid.

219. Ibid. Those terms refer to J. S. Bach's style of keyboard playing.

220. Ibid., p. 186.

221. Ibid., p. 187.

222. Ibid.

223. Schilling defined the touch of a pianist as the manner in which the keys of a keyboard instrument are set into motion. It required "a tenderness of feeling" that "resided in the fingertips." Echoing Logier's main contention, Schilling stressed that only the fingers, and not the hands or arms, should provide the force when playing. Schilling 1835–1838: 1 (1835), p. 224.

224. G. Weber 1827, p. 186.

225. Ibid.

226. Ibid., pp. 187–188.

227. Ibid., pp. 185, 187–188.

228. Chladni 1827c, pp. 188–192.

229. He recommended that the reader consult Ernst Heinrich and Wilhelm Eduard Weber 1825.

230. Chladni 1802, pp. 70, 201–202.

231. Ibid., p. 201.

232. A function follows the law of continuity if (1) $f(x_0)$ is defined, so that x_0 is in the domain of f, (2) $\lim x \to x_0 f(x)$ exists for x in the domain of f, and (3) $\lim x \to x_0 f(x) = f(x_0)$.

233. Chladni 1802, p. 70.

234. Ibid., pp. 70–71.

235. Chladni 1827c, p. 191.

236. Ibid., pp. 191–192.

237. Schwinning 1835, pp. 240–254. G. Weber 1827 was reprinted in *Caecilia* 17 (1835), pp. 233–240.

238. Schwinning 1835, p. 242.

239. Ibid., pp. 244–245.

240. Ibid., pp. 245–246.

241. Ibid., pp. 246–247.

242. Ibid., p. 247.

243. Ibid.

244. Ibid., p. 248.

245. Ibid.

246. Ibid., p. 248–249.

247. Ibid., p. 251.

248. Ibid., p. 252. As the ethnomethodologist David Sudnow realized, and Schwinning himself argued, one critically learns how to anticipate and connect the touch of a pleasant tone with the ensuing tones. See Sudnow 2002.

249. Helmholtz 1863.

250. Helmholtz 1954.

251. Ibid., p. 75.

252. Ibid., pp. 75–76.

253. Ibid., p. 77.

254. Ibid., p. 80.

255. Ibid., p. 79.

256. See Hiebert 2003, p. 438.

257. Helmholtz 1954, p. 80.

258. Ibid., p. 19.

259. Ibid., pp. 80–81. If more magnification was required, the tube and eyepiece of a microscope could be placed behind the metal plate.

260. The end of the other prong is thickened in order to counterbalance the weight of the doublet. The iron loop B attached to the upper prong moves horizontally. If it is moved toward to fork's yoke, the pitch will be increased (or sharpened) slightly. By moving the loop toward L, the pitch lowers (or flattens). E is an electromagnet causing the fork to vibrate with a constant, uniform motion. Ibid., p. 81 n.

261. Ibid., p. 82.

262. Ibid., pp. 83–84.

263. Ibid., p. 387.

264. Ibid., p. 85.

265. Ibid.

266. Ibid., p. 86.

267. Hiebert 2003, p. 436.

268. Ibid., pp. 436–439.

269. As quoted in Hiebert 2003, p. 437. See also Riemann 1888, Germer 1878, Riemann 1911, and Schmitt 1894.

270. For an outstanding article on the physiology of virtuosity in the late nineteenth and early twentieth centuries, see Dierig 2001, pp. 413–440.

271. As quoted in Kiener 1952, p. 72, n. 1.

Glossary

adiabatic process A process in which heat is neither liberated nor absorbed.

aeoline (aeolodicon) A keyboard reed instrument powered by a bellows, related to the harmonium.

beat A pulse resulting from the superposition of two waves generated by two different pitches sounding together.

Bildungsbürgertum The German educated-upper-middle class.

calorimeter An insulated container with a thermometer used to measure the amount of heat given off during a reaction.

chiroplast A mechanical device, invented by Johann Bernhard Logier, which was used to strengthen and train the fingers of children, particularly girls, learning to play piano.

comma A very small musical interval. The interval of a Pythagorean comma is approximately 1.0136:1. A syntonic comma (also called a Ptolemaic comma) has an interval of 81:80.

combination tone A pitch that is generated when two pitches are sounded simultaneously. The pitch of the combination tone is lower than both initial pitches.

diapason normal The French standard tuning fork, established in 1859, equivalent to 435 vibrations per second (Hertz).

equal temperament A system of tuning in which the octave is divided up into twelve equal increments; each semitone is the twelfth root of 2 (or 1.059) times greater than the preceding semitone.

fifth (perfect) A musical interval where the ratio of the two pitches is 2:3 in just intonation.

fourth (perfect) A musical interval where the ratio of the two pitches is 4:3 in just intonation.

γ (Greek "gamma") Expression used by physicists to denote the ratio of specific heat at constant pressure (c_p) to the specific heat at constant volume (c_v).

harmonium A pedal-pumped keyboard reed instrument, usually powered by pressure bellows.

isothermal process A process whereby maximum heat transfer occurs between an elastic fluid and its surroundings.

just intonation A system of tuning in which the intervals among all notes are based on small whole numbers.

labial pipe (flue pipe) An organ pipe in which the pitch is determined by the air passing over the lip of the pipe.

longitudinal vibrations Vibrations that oscillate in the same directions in which they propagate. An example is the air column of an organ pipe.

meantone temperament A system of tuning in which the major third is kept pure (i.e., in the ratio of 5:4) and then divided in half to make two whole tones of equal size, and fifths are slightly flattened.

middle C c′ in eighteenth- and nineteenth-century pitch notation, or C4 in present-day scientific pitch notation. It represents the middle C key on the piano. In just intonation, it sounds at 256 vibrations per second.

monochord (tonometer) A musical and physical instrument in which a string is suspended between two pegs. The string's pitch can be calculated mathematically, given its length, mass, and elasticity.

octave A musical interval in which the ratio of the two pitches is 2:1.

pitch The frequency of a tone, usually measured in vibrations per second (vps), or Hertz (Hz).

reed pipe A pipe in which the pitch is determined both by the properties of a vibrating reed and an air column. Pulsating reed pipes have their reeds beat against the end of the an air column, whereas in free-vibrating reeds, the reed fits perfectly into the aperture at one end of the air column.

specific heat The amount of heat needed to raise a particular quantity of gas a certain temperature.

temperament A system of tuning necessary for tuning keyboard instruments.

thirds (perfect) A major third is a musical interval where the ratio of the two pitches is 5:4. The ratio of pitches of a minor third is 6:5.

tonometer See ***monochord***.

upper partials (also known as **harmonic overtones**) Noninteger multiples of the fundamental (root) tone. They give different instruments their unique timbre.

References

Archives

Das Geheime Staatsarchiv Preußischer Kulturbesitz, Berlin: Rep. 120A, Abt. XIV, Fach 2, Nr. 32, Bd. 1; Rep. 120A, Abt. IX, Fach 1, Nr. 2, Bd. 1 (1814–1862); Rep. 76, Ve 9, Abt. XVc, Nr. 8.

Handschriftenabteilung der Gesamthochschul-Bibliothek Kassel-Landesbibliothek und Murhandsche Bibliothek der Stadt Kassel, 4°Ms. Hass, Briefwechsel von Louis Spohr.

Österreiches Staatsarchiv, Wien; Haus-, Hof-, und Staatsarchiv, Hoftiteln der Musikinstrumentenbauer, Obersthofmeisteramt 1845, Rubric 12/33.

Sächsiches Hauptstaatsarchiv, Dresden, Ministerium des Innern, 1885–91, A. 847, Nr. 17537.

Stadtsarchiv Leipzig, II, Sekt. H 1518.

Stadtsarchiv Wien, Hauptregistrar 1826, Departmentbücher M, 16152.

Universitätsarchiv Göttingen Kur. 4Vb 95a.

Books and Articles

Abrogast, Alois Wolfgang. 1992. "Sieben Briefe von Johann Heinrich Scheibler (1777–1837) an Louis Spohr (1784–1859)." *Die Heimat. Krefelder Jahrbuch* 63: 157–168.

Adress-Buch für Berlin mit Einschluss der näheren Umgehend für das Jahr 1835. 1835. Berlin: H. A. W. Logier.

Adreß-Kalender für die Königlichen Haupt- und Residenzstäde Berlin und Potsdam auf das Jahr [1830–1837, 1840–1845, 1847–1866]. 1830–1837, 1840–1845, 1847–1866. Berlin: August Rücker (1830–1836); Rücker und Püchler (1837); J. W. Boike (1840–1845, 47–50); A.W. Hayn (1851–1855) and (1856–1866).

Ahrens, Christian. 1996. "Zur Entwicklung des Harmoniums und seiner Terminologie." In *Das Harmonium in Deutschland. Bau, wirtschaftliche Bedeutung und musikalische Nutzung eines "historischen" Musikinstrumentes,* ed. Christian Ahrens and Gregor Klinke, pp. 20–38. Frankurt am Main: Erwin Bochinsky.

Alder, Ken. 1995. "A Revolution in Measure: The Political Economy of the Metric System in France." In *The Values of Precision,* ed. M. Norton Wise, pp. 39–71. Princeton: Princeton University Press.

Alexander-von-Humboldt Foundation, editor. 2003. *Festliche Musik zum Jubiläum.* CD. Berlin: Stereo Gena.

Anon. 1749. *Kurzgefaßtes Musikalisches Lexicon.* Chemnitz: Johann Christoph and Johann David Stößel.

Anon. 1807. "Chladni in Heidelberg." *Journal des Luxus und der Moden* 22: 395.

Anon. 1813. "Noch ein Wort über schlechten Kirchengesang, nebst einigen Vorschlägen zur Verbesserung desselben." *Allgemeine musikalische Zeitung* 15: cols. 717–724.

Anon. 1814. "Ueber Verbesserung des Orgelspielens beym öffentlichen Gottesdienste." *Allgemeine musikalische Zeitung* 16: cols. 657–662.

Anon. 1817a. *General Observations Upon Music, and Remarks on Mr. Logier's System of Musical Education.* Edinburgh: Duncan Stevenson.

Anon. 1817b. "Mälzels Metronom." *Allgemeine musikalische Zeitung* 19: cols. 417–422.

Anon. 1818. "Recension von Gesangsbildungslehre für den Männerchor, von Pfeiffer u. Nägeli." *Allgemeine musikalische Zeitung* 20: cols. 533–539.

Anon. 1819. "Andeutungen zur Verbesserung der Musik beym evangelischen Gottesdienste." *Allgemeine musikalische Zeitung* 21: cols. 517–523.

Anon. 1821. "Maelzel's Metronome." *Quarterly Musical Magazine and Review* 3: 303.

Anon. 1828. *Die Versammlung der Deutschen Naturforscher und Aerzte in Berlin im Jahre 1828, kritisch beleuchtet.* Leipzig: no publisher listed.

Anon. 1829. "Ueber die heutigen Geschmack in der Musik." *Allgemeine musikalische Zeitung* 31: cols. 285–294.

Anon. 1834. "Beiträge zum Studium des Pianofortespiels." *Neue Leipziger Zeitschrift für Musik* 1: 5–6, 9–10.

Anon. 1835. "Ein Wort gegen zu hohe Stimmung." *Allgemeine musikalische Zeitung* 37: cols. 205–207.

Anon. 1836. *Kurze Abhandlung über den Metronomen von Mälzl und dessen Anwendung als Tempobezeichnung sowohl als bei dem Unterricht in der Musik.* Mainz: B. Schott's Sons.

Anon. 1842. "Humorvolle Erinnerungen eines Unbekannten an die 19. Versammlung in Braunschweig 1841." In *Die 19. Versammlung deutscher Naturforscher und Ärzte zu Braunschweig, im September 1841, und deren Charaktere, Situationen und Forschungen. Ein humoristisches Album für die Mitglieder, Theilnehmer, Freunde und Freundinnen der Versammlung.* Leipzig: Kollmann.

Anon. 1847 (1976). "Kunstnachrichten: Mechanische Musik aus 'Deutsche Illustrirte Zeitung' 1847." Reproduced (1976) in *Das mechanische Musikinstrument* 2: 4–6.

Anon. 1859. "The Normal Diapason." *Journal of the Society of Arts* (June 3): 492–498.

Anon. 1884–1885a. "Petition zur Einführung einer Normalstimmung." *Zeitschrift für Instrumentenbau* 4: 427–430.

Anon. 1884–1845b. "Aufruf zur Einführung einer allgemeinen Normalstimmung in Deutschland." *Zeitschrift für Instrumentenbau* 4: 363–366. Reprinted (1885) in *Zeitschrift für Instrumentenbau* 5: 154–156.

Anon. 1887. "Ueber die Musik mechanischer Musikwerke und ihr Verhältniss zur lebenden Musik. Mit besonderer Berücksichtigung der im 'Akustischen Cabinet' von Kaufmann & Sohn in Dresden aufgestellten Musikkunstwerke." *Zeitschrift für Instrumentenbau* 7: 121–123, 133–137.

Anon. 1851 (1976). "Das Orchestrion von Fr. Th. Kaufmann. Aus 'Deutsche Illustrirte Zeitung' 1851." Reprinted (1976) in *Das mechanische Musikinstrument* 2: 7–9. Original: "Deutsche Illustrirte Zeitung," 1851: 379–381.

Anon. 1998. "Handwerke und Künste." *Das mechanische Musikinstrument* 24: 27–39.

Applegate, Celia. 1998. "How German Is It? Nationalism and the Origins of Serious Music in Early Nineteenth-Century Germany." *19th Century Music* 21: 274–296.

Applegate, Celia, and Pamela Potter, editors. 2002. *Music and German National Identity.* Chicago: The University of Chicago Press.

Ashworth, William J. 1996. "Memory, Efficiency, and Symbolic Analysis: Charles Babbage, John Herschel, and the Industrial Mind." *Isis* 87: 629–653.

Audsley, George Ashdown. 1965. *The Art of Organ-building: A Comprehensive Historical, Theoretical, and Practical Treatise on the Tonal Appointment and Mechanical Construction of Concert-room, Church, and Chamber Organs*, 2 vols. New York: Dover Publications.

Bach, C. P. E. 1753. *Versuch über die wahre Art das Clavier zu spielen. Zwei Theile.* Berlin: Christian Friedrich Henning.

Baker, Nancy Kovaleff, and Thomas Christiansen, editors. 1995. *Aesthetics and the Art of Musical Composition in the German Enlightenment. Selected Writings of Johann Georg Sulzer and Heinrich Christoph Koch.* Cambridge: Cambridge University Press.

Barbour, J. Murray. 1951. *Tuning and Temperament. A Historical Survey.* East Lansing: Michigan State College Press.

Bartels, J. H., and J. C. G. Fricke, editors. 1831. *Amtlicher Berichte über die Versammlung deutscher Naturforscher und Aerzte in Hamburg im September 1830.* Hamburg: Perthes and Besser.

Beer, Johann. 1719. *Musicalische Discurse durch die Principia der Philosophie deducirt: und in gewisse Capitel eingetheilt, deren Innhalt nach der Vorrede zu finden: nebst einem Anhang von eben diesem Autore, genannt der musicalische Krieg zwischen der Composition und der Harmonie.* Nuremberg: Monath.

Bell, James F. 1991. "The Late-Twentieth-Century Resolution of a Mid-Nineteenth-Century Dilemma Generated by the Eighteenth-Century Experiments of Ernst Chladni on the Dynamics of Rods." *Archive for History of Exact Sciences* 43: 251–273.

Benda, Carl. 1819. "Bemerkungen über Spiel und Vortrag des Adagio, für Dilettanten, und Dilettantinnen des Klavierspiels." *Allgemeine musikalische Zeitung* 21: cols. 817–823.

Benelli, Anton. 1818. "Einige Bemerkungen über Lehrer der Singkunst." *Allgemeine musikalische Zeitung* 20: cols. 697–702.

Bennati, Francesco. 1831. "Notice Physiologique: Sur Paganini." *Revue de Paris* 26: 52–60.

Berg, Maxine. 1990. *The Machinery Question and the Making of Political Economy 1815–1848.* Cambridge: Cambridge University Press.

Bernhardt, W. 1856. *Dr. Ernst Chladni der Akustiker. Eine Biographie und geschichtliche Darstellung seiner Entdeckungen zur Erinnerung an seinen hundertjährigen Geburtstag den 30. November 1856.* Wittenberg: Franz Mohr's Buchhandlung.

Billroth, G. 1830. "Andeutungen zur Geschichte der protestantischen Kirchenmusik." *Berliner allgemeine musikalische Zeitung* 7: cols. 405–413.

Bindseil, Heinrich Ernst. 1839. *Akustik mit sorgfältiger Berücksichtigung der neuern Forschungen.* Potsdam: Verlag der Horvath'schen Buchhandlung, J. E. Witte.

Biot, Jean-Baptiste. 1802. "Sur la théorie du son." *Journal de physique* 55: 173–182.

Biot, Jean-Baptiste. 1810 [1809]. "Versuche über die Fortpflanzung des Schalles durch feste Körper und durch sehr lange Röhre." *Annalen der Physik* 35 (5 of the new series): 407–432. Original, 1809: "Expériences sur la propagation du son à travers les corps solides et à ravers l'air, dans des tuyaux très-allongs." *Mémoires de physique et de chimie de la Societé d'Arcueil* 2 (1809): 405–423.

Biot, Jean-Baptiste. 1817. *Précis élémentaire de physique expérimentale,* 2 parts. Paris: Deterville.

Bleiler, E. F., editor. 1967. *The Best Tales of Hoffmann.* New York: Dover.

Blein, Ange-François-Alexandre, baron. 1827. *Exposé de quelques principes nouveux sur l'acoustique et la théorie des vibrations, et leur l'application a plusieurs phénomènes de la physique.* Paris: no publisher listed.

Blessing, Kurt. 1991."Die Schwarzwälder Orchestrionindustrie und ihr Umfeld im 19. Jahrhundert." *Das mechanische Musikinstrument* 15: 5–26.

Blumenthal, J. von. 1818. *Abhandlung über die Eigenthümlichkeit des Flageolet als Anleitung zur praktischen Ausübung desselben auf der Violine, Op. 43.* Vienna.

Blumner, Martin. 1891. *Geschichte der Sing-Akademie zu Berlin. Eine Festgabe zur Säcularfeier am 24. Mai 1891.* Berlin: Horn and Raasch.

Böhm, Walter. 1970–1980. "Josef Stefan." In *Dictionary of Scientific Biography,* ed. Charles Coulston Gillispie, 16 vols., 13: 10–11. New York: Scribner.

Borer, Philippe. 1995. *The Twenty-Four Caprices of Niccolò Paganini and Their Significance for the History of Violin Playing and the Music of the Romantic Era.* Genoa: Civico Istituto di Studi Paganiniani.

Bourdieu, Pierre. 1977. *Outline of a Theory of Practice.* Trans. Richard Nice. Cambridge: Cambridge University Press.

Bowers, Q. David. 1972. *Encyclopedia of Automatic Musical Instruments.* New York: Vestal Press.

Bowie, Andrew. 2002. "Music and the Rise of Aesthetics." In *The Cambridge History of Nineteenth-Century Music,* ed. Jim Samson, pp. 29–54. Cambridge: Cambridge University Press.

Brauns, Heinrich. 1906. *Der Übergang von der Handweberei zum Fabrikbetrieb in der Niederrheinischen Samt- und Seiden-Industrie und die Lage der Arbeiter in dieser Periode.* Leipzig: Verlag von Duncker and Humblot.

Breitung, Carl. 1841. *Der erste Clavier-Lehrer, eine methodisch katechetische Anleitung, den ersten Clavier-Unterricht schon mit Kindern von 4 bis 6 Jahren zu beginnen und auf eine gründliche bildende und anziehende Weise zu betreiben,* 2nd edition. Berlin: Wilhelm Hermes.

Brendel, Franz. 1867. *Geist und Technik im Clavier-Unterricht. Andeutungen zur methodischen Gestaltung desselben unter technischen, pädagogischen und künstlichen Gesichtspuncten. Für Lehrer und Lernende, Eltern und Erzieher.* Leipzig: C. F. Siegel.

Brewster, David. 1819. *A Treatise on the Kaleidoscope.* Edinburgh/London: Archibald Constable/Longman, Hurst, Rees, Orme, and Brown, and Hurst, Robinson.

Brockhaus, F. A., editor. 1824–1837. *Allgemeine deutsche Real-Encyclopädie für die gebildete Stände- ein Conversations-Lexikon,* 7th edition, 12 vols. Leipzig: Brockhaus.

Brossard, Sébastien de. 1705. *Dictionaire de musique,* 2nd edition. Paris: Christophe Ballard.

Bucciarelli, Louis, and Nancy Dworksy. 1980. *Sophie Germain. An Essay in the History of the Theory of Elasticity.* Dordrecht, Holland: R. Reidel.

Buchner, Alexander. 1992. *Mechanische Musikinstrumente.* Hanau am Main: Werner Dausien.

Bürja, Abel. 1790. *Beschreibung eines musikalischen Zeitmessers.* Berlin.

Burkhard, Johann Andreas Christian. 1832. *Neues vollständiges musikalisches Wörterbuch, enthaltend die Erklärung aller in der Musik vorkommenden Ausdrück für Musiker und Musikfreunde.* Ulm: J. Ebnerschen Buchhandlung.

Buschmann, Gustav Adolf. 1988. "Hundert Jahre des Harmoniumbaues und anderer Zungeninstrumente, 1810–1910." *Das mechanische Musikinstrument* 12: 8–9.

Busse, Friedrich Gottlieb von. 1793. "Neue Bemerkungen über die Vogeltöne auf Geigen und Harfen." In Kunzen and Reinhardt 1793, part II, pp. 177–178, 185–187.

Butt, John. 2001. "Choral Music." In *The Cambridge History of Nineteenth-Century Music,* ed. Jim Samson, pp. 213–236. Cambridge: Cambridge University Press.

Cahan, David. 1993. "Helmholtz and the Civilizing Power of Science." In *Hermann von Helmholtz and the Foundations of Nineteenth-Century Science,* ed. David Cahan, pp. 559–601. Berkeley and Los Angeles: The University of California Press.

Cahan, David, editor. 1993. *Hermann von Helmholtz and the Foundations of Nineteenth-Century Science.* Berkeley and Los Angeles: The University of California Press.

Carnot, Sadi. 1824. *Réflexions sur la puissance motrice du feu.* Paris: Bachelier.

Cavaillé-Coll, Aristide. 1859. *De la détermination du ton normal ou du diapason pour l'accord des instruments de musique.* Paris: de Soye et Bouchet.

Cavaillé-Coll, Aristide. 1979. *Complete Theoretical Works of A. Cavaillé-Coll.* A facsimile edition with introduction and notes by Gilbert Huybens. Buren, the Netherlands: Fritz Knuf.

Cavaillé-Coll, Cécile, and Emmanuel Cavaillé-Coll. 1982. *Aristide Cavaillé-Coll. Ses orginines—sa vie—ses oeuvres.* LePoirè-sur Vie: Archevé d'imprimer sur les presses de l'Imprimerie graphique de l'Ouest.

C. E. H. 1820. "Ueber Singvereine." *Allgemeine musikalische Zeitung* 31: 37–41.

C. F. Ma. 1818. "Einige Nachricht über den Chiroplasten und die neue musikalische Unterweisung des Herrn Logier in England." *Allgemeine musikalische Zeitung* 20: cols. 893–898.

Chantavoine, Jean, editor. 1912. *Pages romantiques.* Paris/Leipzig: Alcan/Breitkopf and Härtel.

Charlton, David, and Michael Musgrave. 2001. "Johann Bernhard Logier." In *The New Grove Dictionary of Music and Musicians,* ed. Stanley Sadie, 15: 75–77. London: Macmillan.

Chladni, E. F. F. 1787. *Entdeckungen über die Theorie des Klanges.* Leipzig: Weidmanns Erben und Reich.

Chladni, E. F. F. 1790a. "Von dem Euphon, einem neuerfundenen musikalischen Instrumenten." *Journal des Luxus und der Moden* 5: 539–543.

Chladni, E. F. F. 1790b. "Von dem Euphon, einem neuerfundenen musikalischen Instrumenten." *Journal von und für Deutschland* 7: 200–202.

Chladni, E. F. F. 1791. "Erklärung, die Erfindung des Euphons betreffend." *Intelligenz-Blatt des Journals des Luxus und der Moden* 6: xxiii.

Chladni, E. F. F. 1793a. "Neue Bemerkungen über die Vogeltöne auf Geigen und Harfen." In Kunzen and Reinhardt 1793, part II, pp. 177–181, 185–187.

Chladni, E. F. F. 1793b. "Über das Euphon, von dessem Erfinder, dem Doktor Chladni." In Kunzen and Reinhardt 1793, part II, pp. 77–78.

Chladni, E. F. F. 1793c. "Ueber die Längentöne einer Saite." In Kunzen and Reinhardt 1793, part I, pp. 33–35.

Chladni, E. F. F. 1795. "Musik: Nachricht von einigen neuen Vervollkommnungen des Euphons, von dessen Erfinder." *Journal des Luxus und der Moden* 10: 309–313.

Chladni, E. F. F. 1796. *Über die Longitundinalschwingungen der Saiten und Stäbe.* Erfurt: Georg Adam Keyser.

Chladni, E. F. F. 1798a. "Bemerkungen über die Töne einer Pfeife in verschiedener Gasarten." *Magazin für die neuesten Zustand der Naturkunde mit Rücksicht auf die dazu gehörigen Hülfswissenschaften* 1: 65–79.

Chladni, [E.] C. [*sic*] F. F. 1798b. "Observations on the Tones Produced by an Organ-Pipe in Different Kinds of Gas." *The Philosophical Magazine; Comprehending the Various Branches of Science, the Liberal and Fine Arts, Geology, Agriculture, Manufactures and Commerce* 4: 275–282.

Chladni, E. F. F. 1798c. "Auszug aus der Schrift: Über Longitudinalschwingungen der Saiten und Stäbe. Nebst beygefügten Bemerkungen über die Fortleitung des Schalles durch feste Körper." *Magazin für den neuesten Zustand der Naturkunde* 1: 7–17.

Chladni, E. F. F. 1800a. "Eine neue Art, die Geschwindigkeit der Schwingungen bei einem jeden Tone durch den Augenschein zu bestimmen, nebst einem Vorschlage zu einer festen Tonhöhe." *Annalen der Physik* 5: 1–9.

Chladni, E. F. F. 1800b. "Nachricht von dem Clavicylinder, einem neuerfundenen Instrumente, enthaltend Bemerkungen über einige etwas damit verwandter Tastaturinstrumente." *Allgemeine musikalische Zeitung* 2: cols. 305–313.

Chladni, E. F. F. 1801a. "Musik." *Journal des Luxus and der Moden* 16: 310–313.

Chladni, E. F. F. 1801b. "Zweyte Nachricht von dem Clavicylinder." *Allgemeine musikalische Zeitung* 3: cols. 386–387.

Chladni, E. F. F. 1802. *Die Akustik.* Leipzig: Breitkopf and Härtel.

Chladni, E. F. F. 1803. "Dr. Chladni's Klavicylinder." *Journal des Luxus und der Moden* 18: 136–139.

Chladni, E. F. F. 1809. *Traité d'acoustique.* Paris: Chez Coucier.

Chladni, E. F. F. 1815. "Einige akustische Notizen, aus einem Schreiben des Hrn. Dr. Chladni." *Allgemeine musikalische Zeitung* 17: cols. 14–15.

Chladni, E. F. F. 1817. *Neue Beyträge zur Akustik.* Leipzig: Breitkopf and Härtel.

Chladni, E. F. F. 1821a. *Beyträge zur praktischen Akustik und zur Lehre vom Instrumentenbau, enthaltend die Theorie und Anleitung zum Bau des Clavicylinders und damit verwandter Instrumente.* Leipzig: Breitkopf and Härtel.

Chladni, E. F. F. 1821b. "Weitere Nachtrichten von dem neulich in der musikalischen Zeitung erwähnten chinesischen Blasinstrumente Tscheng oder Tschiang." *Allgemeine musikalische Zeitung* 23 (1821): cols. 369–374.

Chladni, E. F. F. 1821c. "Nachrichten von neuern die Theorie des Schalles und Klanges betreffend Aufsätze, nebst einigen Bemerkungen." *Allgemeine musikalische Zeitung* 23: cols. 593–602.

Chladni, E. F. F. 1821–1826. "Nachrichten von einigen theils wirklichen (theils vielleicht nur angeblichen) neueren Erfindungen und Verbesserungen musikalischer Instrumente." *Allgemeine musikalische Zeitung* 23 (1821): cols. 393–398; 26 (1824): cols. 809–814; 27 (1825): cols. 725–730; and 28 (1826): cols. 693–696.

Chladni, E. F. F. 1822. "Musikalisch-literarische Nachrichten, nebst einigen Bemerkungen mitgetheilt." *Allgemeine musikalische Zeitung* 24: cols. 178–179.

Chladni, E. F. F. 1826a. "Einige Bemerkungen über die Wellenlehre von Ernst Heinrich Weber, Professor in Leipzig, und Wilhelm Weber in Halle." *Allgemeine musikalische Zeitung* 28: cols. 17–23.

Chladni, E. F. F. 1826b. "Über seine Aufnahme bey Napoléon und sonst in Paris." *Caecilia* 5: 137–144.

Chladni, E. F. F. 1826c. "Wellenlehre auf Experimente gegründet, oder über die Wellen tropfbarer Flüssigkeiten, mit Anwendung auf die Schall- und Lichtwellen. Von den Brüdern Ernst Heinrich Weber, Professor in Leipzig und Wilhelm Weber in Halle. Angezeigt in besonderer Beziehung auf die Tonlehre und mit einigen Bemerkungen." *Caecilia* 4: 189–212.

Chladni, E. F. F. 1826d. "Nachricht von einer neuen Art von Blasinstrument, nebst einigen Bemerkungen mitgetheilt." *Allgemeine musikalische Zeitung* 28: cols. 40–41.

Chladni, E. F. F. 1827a. *Kurze Uebersicht der Schall- und Klanglehre, nebst einem Abhange, die Entwickelung und Anordnung der Tonverhältnisse betreffend.* Mainz: B. Schott's Sons.

Chladni, E. F. F. 1827b. "Ueber eine neue sehr lehrreiche Abhandlung des Hrn. Doctor Wilhelm Weber in Halle, die Gesetze der Zungenpfeifen betreffend." *Allgemeine musikalische Zeitung* 29: cols. 281–284.

Chladni, E. F. F. 1827c. "Über die verschiedene Beschaffenheit des Klanges eines Instruments, nachdem es vom verschiedenen Spielern behandelt wird, auch über das Zerschlagen der Claviersaiten." *Caecilia* 6: 188–192.

Choquel, Henri-Louis. 1762. *La musique rendue sensible par la méchanique, ou nouveau systeme pour apprendre facilement la musique soi-même.* Paris: C. Ballard.

Chung, Hae-Bon. 1974. "Das Krefelder Seidengewerbe im 19. Jahrhundert (ca. 1815–1880)." Ph.D. dissertation, The University of Bonn.

Cohen, H. F. 1984. *Quantifying Music: The Science of Music at the First Stage of the Scientific Revolution, 1580–1650.* Dordrecht: D. Reidel.

Collins, H. M. 1985. *Changing Order.* London: SAGE Press.

Courcy, Geraldine I. C. de. 1957. *Paganini, the Genose,* 2 vols. Norman: The University of Oklahoma Press.

Crosland, Maurice. 1978. *Gay-Lussac: Scientist and Bourgeois.* Cambridge: Cambridge University Press.

Cutler, W. H., and the Editor. 1824. "The Metronome." *Quarterly Musical Magazine and Review* 6: 31–33.

Czerny, Carl. 1830 (1961). *Briefe über den Unterricht auf dem Pianoforte vom Anfang bis zur Ausbildung als Anhang zu jeder Clavierschule.* Stuttgart: Antiquariat-Verlag Zimmermann, Diabelli and Co. Facsimile.

Darrigol, Olivier. 1998. "From Organ Pipes to Atmospheric Motions: Helmholtz on Fluid Mechanics." *Historical Studies in the Physical and Biological Sciences* 29: 1–51.

Dearling, Robert. 1996. *The Encyclopedia of Musical Instruments.* New York: Smithmark Publishers.

Degen, Heinz. 1956. "Die Naturforscherversammlung zu Berlin im Jahre 1828 und ihre Bedeutung für die deutsche Geitesgeschichte." *Naturwissenschaftliche Rundschau* 9: 333–340.

Delezenne, Charles. 1854. "Sur le ton des orchestras et des orgues." In *Mémoire de la Société des Sciences à Lille, 1826–1837.* Lille: L. Danel.

Delezenne, Charles. 1857. *Mémoire sur les valeurs numériques des notes de la gamme.* Lille: L. Danel.

Delezenne, Charles. Circa 1850s. *Note sur le ton des orchestres et des orgues.* Lille: L. Danel.

De Morgan, Augustus. 1864. "On the Beats of Imperfect Consonances." *Transactions of the Cambridge Philosophical Society* 10: 129–145.

Dettelbach, Michael. 1996. "Global Physics and Aesthetic Empire: Humboldt's Physical Portrait of the Tropics." In *Visions of Empire: Voyages, Botany, and Representations of Nature,* ed. David Philip Miller and Peter Hanns Reill, pp. 258–301. Cambridge: Cambridge University Press.

Dierig, Sven. 2001. "Con Sordino for Piano and Brain—Bohemian Neuroscience in a 1900 Cultural Metropolis." *Configurations—A Journal of Literature, Science and Technology* 9: 413–440.

Dove, Heinrich Wilhelm. 1828. *Ueber Maass und Messen, oder, Darstellung der bei Zeit-, Raum- und Gewichts-Bestimmungen üblichen Maasse, Messinstrumente und Messmethoden, nebst Reductionstafeln,* 2nd edition. Berlin: Sandersche Buchhandlung.

Dreist, K. A. 1811. "Zweytes Wort über die Gesangbildungs-Lehre nach Pestalozzi's Grundsätzen von M. J. Pfeiffer and H. G. Nägeli." *Allgemeine musikalische Zeitung* 13: cols. 833–842, 858–859, 876–878.

Dryden, John. 1979. *Alexander's Feast or the Power of Music. An Ode, written in Honour of St. Cecilia.* Set to Music by George Frederic Handel. New York: Edwin F. Kalmus.

Dulong, Pierre Louis. 1829a. "Recherches sur la chaleur spécifique des fluids élastiques." *Annales de chimie et de physique* 41: 113–159.

Dulong, Pierre Louis. 1829b. "Untersuchungen über die specifische Wärme der elastischen Flüssigkeiten." *Annalen der Physik und Chemie* 16 of the new series (92 of the original): 438–479.

Ecker, Alexander. 1883. *Lorenz Oken.* Trans. Alfred Tulk. London: Kegan Paul.

Edler, Arnfried. 1983. "The Social Status of Organists in Lutheran Germany from the 16th through the 19th Century." In *Social Status of the Professional Musician from the Middle Ages to the 19th Century*, ed. Walter Salmen, trans. Herbert Kaufman and Barbara Reisner, pp. 61–93. New York: Pendragon.

Eigeldinger, Jean-Jacques. 1986. *Chopin: Pianist and Teacher as Seen by His Pupils.* Trans. Naomi Shohet with Krysia Osostowicz and Roy Howat, and ed. Roy Howat. Cambridge: Cambridge University Press.

Ellis, Alexander J. 1880. "On the History of Musical Pitch." *Journal of the Society of Arts* 28: 293–336, 401–403.

Elste, Martin. 1987. "Von der Flötenuhr zum Leierkasten. Automatische Musikinstrumente in Berlin." In *Handwerk im Dienste der Musik. 300 Jahre Berliner Musikinstrumentenbau*, ed. Dagmar Droysen-Reber, Martin Elste, and Gesine Haase, pp. 91–99. Berlin: Staatliches Institut für Musikforschung Preußischer Kulturbesitz.

Elvers, Rudolf, editor. 1986. *Felix Mendelssohn: A Life in Letters.* Trans. Craig Tomlinson. New York: Fromm International.

Engel, Hans. 1960. *Musik und Gesellschaft: Baustein zu einer Musiksoziologie.* Berlin: M. Hesse.

Fétis, François-Joseph. 1828. "Nouvelles de Paris: Concerts spirituels." *La revue musicale* 3: 253–254.

Fétis, François-Joseph. 1830. *La musique mise a la portée de tout le monde, exposé succinct de tout ce qui est nécessaire pour juger de cet art, et pour en parler, sans l'avoir étudié.* Paris: A. Mesnier.

Finn, Bernard S. 1964. "Laplace and the Speed of Sound." *Isis* 55: 7–19.

Fischer, Ernst Gottfried. 1825. "Versuche über die Schwingung Saiten, besonders zur Bestimmung eines sichern Maasstabes für die Stimmung." Reprinted with remarks by E. F. F. Chladni in *Allgemeine musikalische Zeitung* 27 (1825): cols. 501–511. Originally appeared in *Abhandlungen der Königlichen Preussichen Akademie der Wissenschaften*, Berlin, 1824.

Fischer, Walther. 1952. "Zur Bedeutung der Gesellschaft Deutscher Naturforscher und Ärzte." *Naturwissenschaftliche Rundschau* 5: 349–352.

Fischer-Dieskau, Dietrich. 1997. *Carl Friedrich Zelter und das Berliner Musikleben seiner Zeit. Eine Biographie.* Berlin: Nicolai.

Fleck, Stefan. 2001. "Das 'Belloneon' von J. G. Kaufmann. Eine lange Weg einer schwierigen Restaurierung." *Das mechanische Musikinstrument* 27: 7–18.

Flehr, Helmut. 1983. "Geschichte des Ärztlichen Vereins in Frankfurt am Main und sein standpolitisches Wirken." Ph.D. dissertation, The University of Mainz.

Forkel, J. N. 1788–1801. *Allgemeine geschichtliche Musik*, 2 vols. Leipzig: Im Schwickertschen Verlag.

Foucault, Michel. 1977. *Discipline and Punish: The Birth of the Prison*. Trans. A. Sheridan. New York: Pantheon.

Fox, Robert. 1971. *The Caloric Theory of Gases from Lavoisier to Regnault*. Oxford: Clarendon Press.

Fox, Robert. 1997. "The Velocity of Sound." In *Pierre-Simon Laplace, 1749–1827. A Life in Exact Science*, Charles C. Gillispie with the collaboration of Robert Fox and Ivor Gratton-Guinness, pp. 199–202. Princeton: Princeton University Press.

Frängsmyr, Tore, J. L. Heilbron, and Robin E. Rider, editors. 1990. *The Quantifying Spirit in the Eighteenth Century*. Berkeley and Los Angeles: The University of California Press.

Frevert, Ute, editor. 1988. *Bürgerinnen und Bürger: Geschlechterverhältnisse im 19. Jahrhundert: Zwölf Beiträge mit einem Vorwort von Jürgen Kocka*. Göttingen: Vandenhoeck and Ruprecht.

Frevert, Ute. 1991. *Ehrenmänner: das Duell in der bürgerlichen Gesellschaft*. Munich: C. H. Beck.

Fröhlich. 1820. "Ueber die musikalische Feyer des katholischen Gottesdienstes überhaupt, und die Art einer dem Zeitbedürfnisse gemässen Einrichtung und Verbesserung derselben." *Allgemeine musikalische Zeitung* 22: cols. 369–380, 389–396, 405–413, 421–430.

Gabory. 1770. *Manuel utile et curieux sur la mesure du temps*. Angers: Chez Parisot.

Galison, Peter. 2003. *Einstein's Clocks, Poincaré's Maps: Empires of Time*. London: Norton.

Gathy, August. 1835. *Musikalisches Conversations-Lexicon, Encyclopädie der gesammten Musik-Wissenschaft für Künstler, Kunstfreunde und Gebildete*. Leipzig, Hamburg, and Itzehoe: Schuberth and Niemeyer.

Gay-Lussac, Joseph-Louis, and Jean-Joseph Welter. 1822. "Sur la dilatation de l'air." *Annales de chimie et de physique* 19: 436–437.

Germer, Heinrich. 1878. *Die Technik des Clavierspiels nach den verschiedenen Materien methodisch geordnet und in progressiver Folge. Für den Studiengebrauch bearbeitet*. Leipzig: C. F. Leede.

Gesellschaft der Musikfreund in Wien, editor. 1885. *Exposé zu der Eingabe der Gesellschaft der Musikfreunde und Genossen in Wien an das k. k. Cultus- und Unterrichtsministerium, betreffend die Herbeiführung einer einheitlichen musikalischen Normalstimmung.* Vienna: J. B. Wallishausser.

Girardin, J., and Charles Laurens. 1854. *Dulong de Rouen. Sa vie et ses ouvrages.* Rouen: H. Rivoire.

Gleichmann. 1820. "Ueber die Erfindung der Aeoline oder des Aeolodikon." *Allgemeine musikalische Zeitung* 22: cols. 505–508.

Goethe, Johann Wolfgang von. 1887–1919. *Goethes Werke. Herausgegeben im Auftrage von der Grossherzogin Sophie von Sachsen.* Weimar edition (abbreviated WA), 143 vols. Weimar: Verlag Hermann Böhlaus.

Gooday, Graeme J. N. 2004. *The Morals of Measurement. Accuracy, Irony, and Trust in Late Victorian Electrical Practice.* Cambridge: Cambridge University Press.

Gouk, Penelope. 1999. *Music, Science, and Natural Magic in Seventeenth-Century England.* New Haven, Conn.: Yale University Press.

Grassi-Landi, Bartolomeo. 1885. *L'armonia dei suoni: col vero corista o diapason normale, considerazioni.* Rome: Tipografia Vaticana.

Guhr, Carl. 1829a. *Ueber Paganini's Kunst die Violine zu spielen: ein Anhang zu jeder bis jetzt erschienen Violinschule nebst einer Abhandlung über das Flageoletspiel in einfachen und Doppeltönen, den Heroen der Violine Rode, Kreutzer, Baillot, Spohr.* Mainz: B. Schott's Sons.

Guhr, Carl. 1829b. "Paganinis Kunst die Violine zu spielen." *Caecilia* 11: 78–88.

Guillemin, Amédée. 1881. *Le monde physique.* Paris: Libraire Hachette et Cie.

Guthmann, F. 1806. "Ein neuer Taktmesser, welcher aber erst erfunden werden soll." *Allgemeine musikalische Zeitung* 8: cols. 117–119.

G. W. F. 1837. "Nachrichten aus London über die neue Stimmungsart des Herrn Heinrich Scheibler, durch Ausführung des Herrn Wortmann." *Allgemeine musikalische Zeitung* 39: cols. 240–242.

Haas, Friedrich. 1886. *Anleitung über Scheibler's musikalische und physikalische Tonmessung nach gemachten Beobachtungen und geschrieben für den Gebrauch der Orgelstimmung.* Erfurt: E. Weingart.

Haase, Rudolf. 1957. "Johann Heinrich Scheibler und seine Bedeutung für die Akustik." In *Beiträge zur Musik im Rhein-Mass-Raum*, ed. Carl Maria Brand and Karl Gustav Fellerer, Heft 19, pp. 57–63. Cologne: Arno Volk-Verlag.

Hacking, Ian. 1983. *Representing and Intervening: Introductory Topics in the Philosophy of Natural Science.* Cambridge: Cambridge University Press.

Hällström, Gustav Gabriel. 1832. "Von den Combinationstönen." *Annalen der Physik und Chemie* 24 of the new series (100 of the original): 438–466.

Hamburger, Michael, editor and translator. 1951. *Beethoven: Letters, Journals, and Conversations.* London: Thames and Hudson.

Harding, Rosamond E. M. 1983. *The Metronome and it's* [sic] *Precursors.* Henley-on-Thames: Gresham Books.

Harleß, Emil. 1853. "Stimme." In *Handwörterbuch der Physiologie mit Rücksicht auf physiologische Pathologie,* ed. Rudolf Wagner, pp. 505–706. Brunswick: Friedrich Vieweg und Sohn.

Harrison, John. 1775. *Description Concerning such a Mechanism as Will Afford a Nice, or True Mensuration of Time; Together with some Account of the Attempts for the Discovery of the Longitude by the Moon: as also an Account of the Discovery of the Scale of Musick.* London: printed for the author.

H[äser], A[ugust] F[erdinand]. 1827. "Wie muss die Begleitung des Kirchengesangs mit der Orgel in protestantischer Kirche beschaffen seyn?" *Allgemeine musikalische Zeitung* 29: cols. 723–727.

Häser, A[ugust]. F[erdinand]. 1829. "Über die wissenschaftliche Begründung der Musik durch Akustik." *Allgemeine musikalische Zeitung* 31: cols. 53–57, 69–71, 91–94, 110–114, 143–147.

Haspels, J. J. L. 1987. "Automatic-Musical Instruments: Their Mechanics and Their Music, 1580–1820." Ph.D. dissertation, University of Utrecht.

Hatfield, Gary. 1993. "Helmholtz and Classicism: The Science of Aesthetics and the Aesthetics of Science." In *Hermann von Helmholtz and the Foundations of Nineteenth-Century Science,* ed. David Cahan, pp. 522–558. Berkeley and Los Angeles: The University of California Press.

Haupt, Günther. 1927. "J. N. Mälzels Briefe an Breitkopf und Härtel." *Der Baer. Jahrbuch von Breitkopf und Härtel* 3: 122–145.

Haupt, Helga. 1960. "Wiener Instrumentenbauer von 1791 bis 1815." *Studien zur Musikwissenschaften* 24: 120–184.

Haynes, Bruce. 2002. *A History of Performing Pitch. The Story of "A."* Lanham, Maryland and Oxford: The Scarecrow Press.

Hdt. 1847. "Die Notwendigkeit einer allgemeinen gleichmässigen deutschen Stimmhöhe." *Allgemeine musikalische Zeitung* 49: cols. 801–805.

Hegel, G. W. F. 1974. *Aesthetics: Lectures on Fine Art.* Trans. T. M. Knox. Oxford: Oxford University Press.

Heilbron, John L. 1993. *Weighing Imponderables and Other Quantitative Science.* Berkeley and Los Angeles: The University of California Press.

Heine, Heinrich. 1925. *Sämtliche Werke,* 10 vols. Munich: Georg Müller.

Heine, Heinrich. 1970–. *Heinrich Heines Werke. Säkularausgabe. Briefe. Lebenszeugnisse.* Berlin/Paris: Akademie Verlag/Editions de CNRS.

Helmholtz, Hermann von. 1856. "Ueber Combinationstöne." *Annalen der Physik und Chemie* 99 of the new series (175 of the original): 497–540.

Helmholtz, Hermann von. 1859. "Theorie der Luftschwingungen in Röhren mit offenen Enden." *Journal für reine und angewandte Mathematik* 57: 1–72.

Helmholtz, Hermann von. 1861. "Zur Theorie der Zungenpfeifen." *Annalen der Physik und Chemie* 24 of the new series (114 of the original): 321–327.

Helmholtz, Hermann von. 1863. *Die Lehre von den Tonempfindungen als physiologische Grundlage für die Theorie der Musik.* Brunswick: Friedrich Vieweg.

Helmholtz, Hermann von. 1882–1895. *Wissenschaftliche Abhandlungen,* 3 vols. Leipzig: J. A. Barth.

Helmholtz, Hermann von. 1885. *On the Sensations of Tone as a Physiological Basis for the Theory of Music.* Trans. Alexander J. Ellis. London: Longmans, Green.

Helmholtz, Hermann von. 1889. "Kleinere (Original-) Mittheilungen. Bestimmungen über die Prüfung und Beglaubigung von Stimmgabeln." *Zeitschrift für Instrumentenkunde* 9: 65–67.

Helmholtz, Hermann von. 1954. *On the Sensations of Tone as a Physiological Basis for the Theory of Music.* Trans. (1885) Alexander J. Ellis, with a new introduction by Henry Margenau. New York: Dover Publications.

Henkel, A. 1829. "Ueber Volksgesang, besonders den deutschen." *Berliner allgemeine musikalische Zeitung* 6: 413–416.

Hensel, Sebastian, editor. *Die Familie Mendelssohn: 1729–1847. Nach Briefen und Tagebüchern.* Frankfurt am Main: Insel.

Hermann, C. F. 1822. "Ueber die Anwendung der Orgel bey der Kirchenmusik." *Allgemeine musikalische Zeitung* 24: cols. 678–682.

Heyde, Herbert. 1980. *Katalog zu den Sammlungen des Händel-Hauses in Halle. Part 7. Blasinstrumente, Orgeln und Harmoniums.* Halle: Händel-Haus.

Heyde, Herbert. 1994. *Musikinstrumentenbau in Preussen.* Tutzing: Hans Schneider.

Hiebert, Elfrieda Franz. 2003. "Helmholtz's Musical Acoustics: Incentive for Practical Techniques in Pedaling and Touch at the Piano." In *Proceedings of the*

International Musicological Society, Intercongressional Symposium (IMS), Budapest, August 2000, ed. Liszt Ferenc Academy of Music, pp. 425–443. Budapest: International Musicological Society.

Hobsbawm, Eric, and Terence O. Ranger, editors. 1983. *The Invention of Tradition.* Cambridge: Cambridge University Press.

Hoffmann, E. T. A. 1810. "Recension." *Allgemeine musikalische Zeitung* 12: cols. 630–642, 652–659.

Hoffmann-Donner, Heinrich, editor. 1867. *Ein Liederbuch für Naturforscher und Aerzte. Als Festgabe für die Mitglieder der 41. Versammlung in Frankfurt a. M.* Frankfurt am Main: J. D. Sauerländer's Verlag.

Horstig. 1806. "Nach einer erfolgten Aufforderung, etwas von den Wirkungen der Tonkunst, und des Gesanges insbesondere, auf gesellige Verhältnisse in moralischer Hinsicht zu sagen." *Allgemeine musikalische Zeitung* 9: cols. 129–132.

Houghton, Jr., Walter E. 1943. "The English Virtuoso in the Seventeenth Century." *Journal of the History of Ideas* 3: 51–73, 190–218.

Humboldt, Alexander von. 1989. *Ideen zu einer Geographie der Pflanzen, nebst einem Naturgemälde der Tropenländer.* In his *Schriften zur Geographie der Pflanzen*, ed. Hanno Beck. Darmstadt: Wissenschaftliche Buchgesellschaft.

Humboldt, Alexander von, and Hinrich Lichtenstein, editors. 1829. *Amtlicher Bericht über die Versammlung deutscher Naturforscher und Ärzte zu Berlin im September 1828, erstattet von den damaligen geschäftsführen A. v. Humboldt und H. Lichtenstein.* Berlin: T. Trautwein.

Hummel, Johann Nepomuk. 1828. *Ausführlich theoretisch-practische Anweisung zum Piano-Fortespiel.* Vienna: T. Haslinger.

Irving, John. 2002. "The Invention of Tradition." In *The Cambridge History of Nineteenth-Century Music*, ed. Jim Samson, pp. 178–212. Cambridge: Cambridge University Press.

Jackson, Myles W. 1994. "Natural and Artifical Budgets: Accounting for Goethe's Economy of Nature." *Science in Context* 7: 409–431.

Jackson, Myles W. 2000. *Spectrum of Belief: Joseph von Fraunhofer and the Craft of Precision Optics.* Cambridge, Mass.: MIT Press.

Jackson, Myles W. 2001. "Music and Science during the Scientific Revolution." *Perspectives on Science* 9: 106–115.

Jackson, Myles W. 2003. "Harmonious Investigators of Nature: Music and the Persona of the German *Naturforscher* in the Nineteenth Century." *Science in Context* 16: 121–145.

Jackson, Myles W. 2004. "Physics, Machines, and Musical Pedagogy in Nineteenth-Centry Germany." *History of Science* 42: 371–418.

J. C. H. 1819. "Einige Worte über die musikalische Bildung jetziger Zeit." *Allgemeine musikalische Zeitung* 21: cols. 565–576, 581–586.

Johnston, James F. W. 1831. "Meeting of the Cultivators of Natural Science and Medicine at Hamburgh in September." *Edinburgh Journal of Science*, new series, 4: 190–244.

Jorgensen, Owen H. 1991. *Tuning. Containing the Perfection of Eighteenth-Century Temperament, the Lost Art of Nineteenth-Century Temperament and the Science of Equal Temperament.* East Lansing: Michigan State University Press.

Jousse, Jean. 1832. *An Essay on Temperament.* London: printed for the author.

Jurkowitz, Edward. 2002. "Helmholtz and the Liberal Unification of Science." *Historical Studies in the Physical Sciences* 32: 291–317.

Kaestner, J. G. 1830. "Ueber das Handwerk in der Kunst." *Berliner allgemeine musikalische Zeitung* 7: 297–300.

Kalkbrenner, A. 1884–1885. "Die Normalstimmung und die Consequenz für die Militär-Musik." *Zeitschrift für Instrumentenbau* 5: 205–207.

Kalkbrenner, Frédéric. 1831. *Méthode pour apprendre le piano-forte a l'aide du guide-mains.* Paris: Ignaz Pleyel.

Kaufmann, Friedrich. "Ueber Crescendo- und Diminuendo-Züge an Orgeln." *Allgemeine musikalische Zeitung* 25 (1823): cols. 119–120.

Kempelen, Wolfgang von. 1791. *Mechanismus der menschlichen Sprache nebst einer Beschreibung seiner sprechenden Maschine.* Vienna: J. V. Degen.

Kielmeyer, C. von, and G. Jäger, editors. 1835. *Amtlicher Bericht über die Versammlung deutscher Naturforscher und Ärzte zu Stuttgart im September 1834.* Stuttgart: J. M. Metzler.

Kiener, Hélène. 1952. *Marie Jaëll, 1846–1925; problèmes d'esthétique et de pédagogie musicales.* Préf. d'André Siegfried. Paris: Flammarion.

Kieninger, J. M. Circa 1825. *Theoretische und praktische Anleitung für angehende Violinspieler, nach den besten Methoden eingerichtet.* Graz.

Kipnis, Nahim. 1991. *History of the Principle of Interference of Light.* Berlin: Birkhäuser Verlag.

Kirnberger, J. P. 1771–1779. *Die Kunst des reinen Satzes in der Musik,* 2 vols. Berlin: Decker and Hamburg.

Kivy, Peter. 2001. *The Possessor and the Possessed: Handel, Mozart, Beethoven, and the Idea of Musical Genius.* New Haven, Conn.: Yale University Press.

K. K. Ministerium für Cultus und Unterricht, editor. 1885. *Beschlüsse und Protokolle der Internationalen Stimmton-Conferenz in Wien 1885.* Vienna: K. K. Schulbücher Verlag.

Klein, Heinrich. 1799. "Nachtricht: Beschreibung meiner Tastenharmonika." *Allgemeine musikalische Zeitung* 1: cols. 675–679.

Klusen, Ernst, Hermann Stoffels, and Theo Zart. 1979–1980. *Das Musikleben der Stadt Krefeld, 1780–1945*, 2 vols. Cologne: Arno Volk-Verlag.

Knecht. 1803. "Ueber die Stimmung der musikalischen Instrumente überhaupt, und der Orgel insbesondere." *Allgemeine musikalische Zeitung* 5: cols. 529–535.

Koch, Heinrich Christoph. 1782–1793. *Versuch einer Anleitung zur Composition*, 2 vols. Rudolstadt and Leipzig: Adam F. Böhme.

Koch, Heinrich Christoph. 1802. *Musikalisches Lexikon welches die theoretische und praktische Tonkunst, encyclopädisch bearbeitet, alle alten und neuen Kunstwörter erklärt, und die alten und neuen Instrumente beschrieben*. Frankfurt am Main: August Hermann dem Jüngern.

Kollmann, A. F. 1828. "Bemerkungen über Hrn. J. B. Logier's sogenanntes 'Neues System des Musikunterrichts.'" In A. F. B. Kollmann and C. F. Müller, *Ueber Logier's Musikunterrichts-System*. Munich: K. B. Hof-Musikalischen- und Instrumenten-Handlung von Falter und Sohn, pp. 5–37. Originally published in *Allgemeine musikalische Zeitung* 24 (1822): 769–774, 785–794, 801–814.

Kramer, Lawrence. 2002. *Musical Meaning: Toward a Critical History*. Berkeley and Los Angeles: The University of California Press.

Kraushaar, Otto. 1838. *Construction der gleichschwebenden Temperatur ohne Scheibler'sche Stimmgabeln auf mathematischen Instrumenten. Mit Rücksicht auf die Scheibler'sche Erfindung*. Cassel: Theodor Fischer bei Krieger'sche Buchhandlung.

Kruta, Vladislav. 1973. "Eindrücke aus der Berliner Naturforscher-Versammlung (1828) in Briefen eines Teilnehmers." *Sudhoffs Archiv. Zeitschrift für Wissenschaftsgeschichte* 57: 152–170.

Kümmel, Werner Friedrich. 1987. *Heinrich Hoffmann: Gesammelte Gedichte, Zeichnungen und Karikaturen*. Frankfurt am Main.

Kunzen, F. A., and J. F. Reinhardt, editors. 1793. *Studien für Tonkünstler und Musikfreunde. Eine historisch-kritische Zeitschrift mit neun und dreissig Musikstücken von verschiedenen Meistern fürs Jahr 1792 in zwei Theile*. Berlin: Verlag der neuen Musikhandlung.

Kützing, Carl (also sp. Karl). 1836. *Theoretisch-praktischer Handbuch der Orgelbaukunst*. Leipzig: J. P. Dalp.

Kützing, Carl (also sp. Karl). 1838. *Beiträge zur praktischen Akustik als Nachtrag für Fortepiano- und Orgelbaukunst.* Bern, Chur and Leipzig: J. F. I. Dalp.

l'Affillard, Michel. 1705. *Principes tres-faciles pour bien apprendre la musique, qui conduiront promptement ceux qui ont du naturel pour le Chant; jusqu'au point de chanter toute sorte de musique proprement, & à livre ouvert.* Paris: Chrisophe Ballard.

Lagrange, Joseph-Louis. 1761–1773. *Mélanges de philosophie et de mathématique de la Société royale de Turin.* 4 vols. Turin: Société royale de Turin.

Laphalèque, G. Imbert de. 1830. *Notice sur le célèbre violiniste Nicolo Paganini.* Paris: E. Guyot.

Laplace, Pierre-Simon. 1817 [1816]. "Ueber die Länge des Secunden-Pendels, und die Geschwindigkeit des Schalles in verschiedenen Mitteln." *Annalen der Physik* 57 (27 of the new series): 225–243. Original: 1816, "Sur la vitesse du sons dans l'air et dans l'eau," *Annales de chimie et de physique* 3: 238–241.

Laplace, Pierre-Simon. 1822. "Note sure la vitesse du son." *Annales de chimie et physique* 20: 266–269.

Lecomte, Auguste François. 1857. "Mémoire explicatif de l'invention de Scheibler. Pour introduire une exactitude inconnue avant lui, dans l'accord des instruments de musique." In *Mémoire de la Société Imperiale des Sciences, de l'Agriculture et des Arts, de Lille,* pp. 47–136. Lille: L. Danel.

Lehmann, Johann Traugott. 1827. *Gründliches vollständiges und leichtfassliches Stimmsystem; oder Anweisung wie jeder Pianoforte oder Klavierinstrumente auf die beste und leichteste Art, rein und richtig in kurzer Zeitstimmen lernen kann.* Leipzig: C. E. Kollmann.

Lehmann, Johann Traugott. 1830. *Anleitung die Orgel rein und richtig stimmen zu lernen und in guter Stimmung zu erhalten. Nebst einer Beschreibung über den Bau der Orgel.* Leipzig: Breitkopf and Härtel.

Leipp, A. E., and M. Castellango. 1977. "Du diapason et de sa relativité." *La revue musicale* 294: 5–39.

Leipziger Adreßkalender auf das Schult-Jahr 1824 [1826]. 1824/1826. Leipzig: Wilhelm Staritz.

Leman. 1890. "Ueber die Normalstimmgabel der Physikalisch-Technische Reichstanstalt und die absolute Zählung ihrer Schwingungen." *Zeitschrift für Instrumentenkunde* 10: 77–87, 170–183, 197–202.

Lenoir, Timothy. 1997. "The Politics of Vision: Optics, Painting, and Ideology in Germany 1845–95." In his *Instituting Science: The Cultural Production of Scientific Disciplines,* pp. 131–178. Stanford: Stanford University Press.

Leonhardt, Henrike. 1990. *Der Taktmesser. Johann Nepomuk Mälzel- Ein lückenhafter Lebenslauf.* Hamburg: Kellner.

Lindley, Mark. 2001a. "Temperaments." In *The New Grove Dictionary of Music and Musicians,* ed. Stanley Sadie, vol. 25, pp. 248–268. London: Macmillan.

Lindley, Mark. 2001b. "Pythagorean Intonation." In *The New Grove Dictionary of Music and Musicians,* ed. Stanley Sadie, vol. 20, pp. 643–645. London: Macmillan.

Lindley, Mark. 2001c. "Just [pure] Intonation." In *The New Grove Dictionary of Music and Musicians,* ed. Stanley Sadie, vol. 13, pp. 290–295. London: Macmillan.

Lindner, M. Fr. W. 1811. "Was ist für die Gesangs-Bildung geschehn?" *Allgemeine musikalische Zeitung* 13: cols. 3–8, 17–23, 33–43, 49–59.

Lindsay, R. Bruce. 1966. "The Story of Acoustics." *Journal of Acoustical Society of America* 39: 629–644.

Lindsay, R. Bruce. 1973. *Acoustics: Historical and Philosophical Development.* Stroudsburg, Penn.: Dowden, Hutchinson, and Ross.

Lissajous, Jules Antoine. 1857. "Mémoire sur l'étude optique des mouvements vibratoires." *Annales de chimie* (3rd series) 51: 147–231.

Lissajous, Jules Antoine. 1873. "Sur le phonoptomètre, instrument propre à l'étude optique des mouvements périodique ou continues." *Comptes rendus hebdomadaires des séances de l'Académie des sciences* 76: 878–890.

Lissajous, Jules Antoine. 1885a. "Note sur un moyen nouveau de mettre en évidence le mouvement vibratoire des corps." *Comptes rendus hebdomadaires des séances de l'Académie des science* 41: 93–95.

Lissajous, Jules Antoine. 1885b. "Note sur une méthode nouvelle applicable à l'étude des mouvements vibratoires." *Comptes rendus hebdomadaires des séances de l'Académie des science* 41: 814–817.

Liszt, Franz. 1912. "Lettres d'un bachelier ès musique" (first published as a series of twelve articles in *La revue et gazette musicale de Paris* in 1837–1839). In *Pages romantiques,* ed. Jean Chantavoine. Paris/Leipzig: Alcan/Breitkopf and Härtel.

Lloyd, Llewlyn S., and Hugh Boyle. 1963. *Intervals, Scales, and Temperaments.* London: Macdonald.

Loehr, Johann Joseph. 1836. *Ueber die Scheibler'sche Erfindung überhaupt und dessen Pianoforte- und Orgel-Stimmung insbesondere.* Krefeld: C. M. Schüller.

Loesser, Arthur. 1954. *Men, Women, and Pianos: A Social History.* New York: Simon and Schuster.

Loewe, K. 1825. "Ueber Logiers Musik-System." *Berliner allgemeine musikalische Zeitung* 2: 41–42, 57–58.

Loewenherz, L[eopold]. 1888. "Ueber die Herstellung von Stimmgabeln." *Zeitschrift für Instrumentenkunde* 8: 261–267.

Logier, J[ohann]. B[ernhard]. 1818. *An Authentic Account of the Examination of Pupils, Instructed in the New System of Musical Education; before Certain Members of the Philharmonic Society, and Others.* London: R. Hunter.

Logier, J[ohann]. B[ernhard]. Circa 1818. *The First Companion to the Royal Patent Chiroplast, or Hand Director. A New Invented Apparatus for Facilitating the Attainment of a Proper Execution on the Piano-Forte.* London: I. Green.

Logier, J[ohann]. B[ernhard]. 1824. *A Short Account of the Progress of J. B. Logier's System of Musical Education in Berlin, and Its Subsequent Introduction by Orders of the Prussian Government into the Public Seminaries, for Its General Promulgation Through the Prussian States with a Brief Sketch of the Present State of Music in Berlin.* London: R. Hunter.

Logier, J[ohann]. B[ernhard]. 1827a. *A System of the Science of Music and Practical Composition; Incidentally Comprising What Is Usually Understood by the Term Thorough Bass.* London: J. Green.

Logier, Johann Bernhard. 1827b. *System der Musikwissenschaft und der praktischen Composition.* Berlin: Heinrich Adolf Wilhelm Logier.

Logier, Johann Bernhard. 1827c. *Lehrbuch der musikalischen Composition. Auszug aus J. B. Logier's System der Musik-Wissenschaft, herausgegeben auf Veranstaltung des Verfassers. Zum Gebrauch für Schulen.* Berlin: Heinrich Adolf Wilhelm Logier.

Logier, Johann Bernward. 1827/1829. *Anweisung zum Unterricht im Klavierspiel und der musikalischen Komposition.* Berlin: Logier.

A London Musician. 1824. "Entstehung und Verbreitung der Logier'schen Lehrart." *Allgemeine musikalische Zeitung* 26: cols. 409–412.

Loulie, Étienne. 1696. *Éléments ou principes de musique, mis dans un nouvel ordre.* Paris: C. Ballard.

Maley, Jr., V. Carlton. 1990. *The Theory of Beats and Combination Tones 1700–1863.* New York: Garland Publishing.

Maréchal, Henri. 1904. *Rome: Souvenirs d'un musicien.* Paris: Hachette.

Marpurg, Friedrich Wilhelm, editor. 1749–1750. *Der critische Musicus an der Spree.* 50 Stücke. Berlin: A. Haude and J. C. Spener.

Marpurg, Friedrich Wilhelm. 1756. *Principes du clavecin.* Berlin: Haude et Spener.

Marpurg, Friedrich Wilhelm. 1776. *Versuch über die musikalische Temperatur.* Breslau: J. F. Korn.

Marx, A[dolf]. B[ernhard]. 1825. "Zusatz aus andrer Feder." *Berliner allgemeine musikalische Zeitung* 2: 58–60, 65–67, 73–75.

Marx, A[dolf]. B[ernhard]. 1827. "Einfluß der Musik auf Sittlichkeit." *Berliner allgemeine musikalische Zeitung* 4: 395–397.

Marx, A[dolf]. B[ernhard]. 1828a. "Mösers und Rombergs Quartete." *Berliner allgemeine musikalische Zeitung* 5: 23–24.

Marx, A[dolf]. B[ernhard]. 1828b. "Ueber Choral-Gesang als Gesangsbildungsmittel in Schulen." *Berliner allgemeine musikalische Zeitung* 5: 211–214.

Marx, A[dolf]. B[ernhard]. 1830. "Beurtheilungen." *Berliner allgemeine musikalische Zeitung* 7: 413–414.

Marx, Karl. 1857–1858 (1973). *Grundrisse der Kritik der politischen Ökonomie*. Trans. Martin Nicolaus. New York: Randon House.

Mattheson, Johann. 1739. *Der vollkommenene Capellmeister*. Hamburg: Christian Gerold.

Meerens, Charles. 1857a. *Table de logarithmes acoustiques, depuis 1 jusqu'a 1200, précédée du'une instruction élémentaire*. Lille: L. Danel.

Meerens, Charles. 1857b. *Mémoire sur les valeurs numériques des notes de la gamme*. Lille: L. Danel.

Meerens, Charles. Circa 1850s. *Note sur le ton des orchestres et des orgues*. Lille.

Meerens, Charles. 1864. *Instruction élémentaire du calcul musical et philosophie de la musique*. Brussels: Schott.

Meerens, Charles. 1892. *Acoustique musicale*. Brussels: J.-B. Katto.

Meerens, Charles. 1894. *L'avenir de la science musicale*. Brussels: J.-B. Katto.

Meerens, Charles. 1895. *Le tonometre: d'aprés l'invention de Scheibler. Nouvelle démonstration a la portée de tout le monde*. Paris: Chez Colombier/Brussels: J.-B. Katto.

Melde, Franz. 1866/1888. *Ueber Chladni's Leben und Wirken, nebst einem chronologischen Verzeichnis seiner literärischen Arbeiten*. Marburg: C. L. Pfeil. Second edition: 1888, Marburg: N. G. Elwert'sche Verlagsbuchhandlung.

Mémoires de la classe des sciences mathématiques et physiques de l'Institut [National, Impérial] de France. 1811. Paris.

Mendel, Arthur. 1968. "On the Pitches in Use in Bach's Time, Parts I and II." In *Studies in the History of Musical Pitch. Monographs by Alexander J. Ellis and Arthur Mendel*, pp. 187–224. New York: DeCapo. Original: *Musical Quarterly* 41 (1955): 332–354, 466–480.

Mendel, Arthur. 1978. "Pitch in Western Music since 1500: A Re-Examination." *Acta musciologica* 50: 1–93.

Mendel, Hermann, editor. 1870–1877. *Musikalisches Conversations-Lexikon. Eine Encycklopädie der gesammten musikalischen Wissenschaften für Gebildete aller Stände, unter Mitwirkung der literarischen Commission der Berliner Tonkünstlervereins.* 11 vols. Berlin: Robert Oppenheim.

Metzner, Paul. 1998. *Crescendo of the Virtuoso: Spectacle, Skill, and Self-Promotion in Paris during the Age of Revolution.* Berkeley and Los Angeles: The University of California Press.

Michaelis, C. F. 1795. *Ueber den Geist der Tonkunst, Erster Versuch.* Leipzig: In der Schäferische Buchhandlung.

Michaelis, C. F. 1800a. *Mittheilungen zur Beförderung der Humanität und des guten Geschmacks.* Leipzig: G. Benjamin Meißner.

Michaelis, C. F. 1800b. *Moralische Vorlesungen.* Weissenburg in Franken: Verlag des oberdeutschen Addresse- und Industrie-Komptoirs.

Michaelis, C. F., and J. G. Schicht. 1814. "Aufforderung zur Festsetzung und gemeinschaftlichen Annahme eines gleichen Grundtones der Stimmung der Orchesters." *Allgemeine musikalische Zeitung* 16: cols. 772–776.

Moltke, Helmuth von. 1892. *Gesammelte Schriften und Denkwürdigkeiten des General-Feldmarschalls Grafen Helmuth von Moltke*, vol. 6, Briefe 3. Sammlung. Berlin: E. S. Mittler.

Montal, Claude. 1836. *L'art d'accorder soi-même son piano.* Geneva: Minkoff. Reprint, 1976.

Morrell, Jack, and Arnold Thackray, editors. 1984. *Gentlemen of Science: Early Correspondence of the BAAS.* London: Offices of the Royal Historical Society.

Morus, Iwan Rhys. 1998. *Frankenstein's Children: Electricity, Exhibition, and Experiment in Early-Nineteenth-Century London.* Princeton: Princeton University Press.

Mosse, George L. 1996. *The Image of Man: The Creation of Modern Masculinity.* Oxford University Press.

Mueller, Barbara, editor. 1952. *Goethe's Botanical Writings.* Honolulu: University of Hawaii Press.

Müller, C. F. 1828. "Betrachtung des Logier'schen Musikunterrichts-Systems." In A. F. B. Kollmann and C. F. Müller, *Ueber Logier's Musikunterrichts-System*, pp. 38–56. Munich: K. B. Hof-Musikalischen- und Instrumenten-Handlung von Falter und Sohn.

Müller, Johann Christian. 1788. *Anleitung zum Selbstunterricht auf der Harmonika* Leipzig: Siegfried Lebrecht Crusius.

Müller, Johannes. 1838–1840. *Handbuch der Physiologie des Menschen für Vorlesungen.* 2 vols. Coblenz: Hölscher.

Müller, Wilhelm Christian. 1830. *Aesthetisch-historische Einleitungen in die Wissenschaft der Tonkunst. Zwei Theile. Theil I: Versuch einer Aesthetik der Tonkunst in Zusammenhang mit den übrigen Schönen Künsten nach geschichtlicher Entwickelung. Theil II: Übersicht einer Chronologie der Tonkunst mit Anwendungen allgemeiner Civilisation und Kultur-Entwickelung.* Leipzig: Breitkopf and Härtel.

Muncke, G. W. 1833. "Bemerkungen über die Versuche des Hern. Lenz im Betreff der Drehungen des Coulombschen Wagebalkens, und Nachricht von den akustischen Versuchen des Hrn. Scheibler." *Annalen der Physik und Chemie* 29 new series (105 of the original): 381–403.

Näke, Karl. 1862. *Ueber Orchesterstimmung. Den deutschen Kapellmeistern bei ihrer Versammlung in Dresden, den 28. September 1862 gewidmet.* Dresden: Liepsch and Reichardt.

Nipperdey, Thomas. 1972. "Verein als soziale Struktur in Deutschland im späten 18. und frühen 19. Jahrhundert." In *Geschichtswissenschaft und Vereinswesen im 19. Jahrhundert. Beiträge zur Geschichte historischer Forschung in Deutschland,* ed. Hartmut Boockmann et al., pp. 1–44. Göttingen: Vanderhoeck and Ruprecht.

Noble, David F. 1984. *Title Forces of Production: A Social History of Industrial Automation.* New York: Knopf.

Nöding, Kaspar. 1823. *Johann Heinrich Völler's, Hof-Instrumentenbauer und Mechanikus in Cassel, Lebensbeschreibung.* Marpurg: Kreiger'sche Schriften.

Oertling, August. 1843. "Ueber die Prüfung plan-paralleler Gläser und Beschreibung das dabei in Anwendung gebrachten Instruments." In *Verhandlungen des Vereins zur Beförderung des Gewerbefleisses in Preußen,* pp. 60–72. Berlin: Petsch.

Ohm, Georg Simon. 1839. "Bemerkungen über Combinationstöne und Stösse." *Annalen der Physik und Chemie* 47 of the new series (123 of the original): 463–466.

Oken, Lorenz. 1823. "Versammlung der deutschen Naturforscher und Aerzte zu Leipzig am 18 September 1822." *Isis* 1: cols. 553–559.

Oken, Lorenz. 1827. "Versammlung der deutschen Naturforscher und Aerzte zu Dresden, vom 18ten bis 23sten September 1826." *Isis* 20: cols. 296–407.

Oken, Lorenz. 1828. "Versammlung der Naturforscher und Aerzte im September zu München." *Isis* 21: cols. 417–596.

Oken, Lorenz. 1829. "Versammlung der Naturforscher und Aerzte zu Berlin, im September 1828." *Isis* 22: cols. 217–450.

Olesko, Kathryn M. 1995. "The Meaning of Precision: The Exact Sensibility in Early Nineteenth-Century Germany." In *The Values of Precision*, ed. M. Norton Wise, pp. 103–134. Princeton: Princeton Univesity Press.

Olesko, Kathryn M. 1996. "Precision, Tolerance, and Consensus: Local Cultures in German and British Resistance Standards." *Archimedes: New Studies in the History and Philosophy of Science and Technology* 1: 117–156 (ed. Jed. Z. Buchwald).

Ord-Hume, Arthur W. J. G. 1986. *Harmonium: The History of the Reed Organ and Its Makers*. London: David and Charles Newton Abbot.

Ottner, Helmut. 1977. *Der Wiener Instrumententenbau, 1815–1833*. Tutzing: Hans Schneider.

Outram, Derinda 1984. *Georges Cuvier. Vocation, Science, and Autonomy in Post-Revolutionary France.* Manchester: Manchester University Press.

Owen, Barbara. 2001. "Reed organ." In *The New Grove Dictionary of Music and Musicians*, ed. Stanley Sadie and John Tyrrell, vol. 21, pp. 64–71. London and New York: Macmillan.

Paddison, Max. 2002. "Music as Ideal: The Aesthetics of Autonomy." In *The Cambridge History of Nineteenth-Century Music*, ed. Jim Samson, pp. 318–342. Cambridge: Cambridge University Press.

Pajot, Louis-Léon. 1735. "Description et usage d'un Métromètre ou Machine pour battre les Mesures et les Temps de toutes sortes d'Airs." In *Histoire de l'Academie Royale des Sciences, année 1732*, pp. 182–195. Paris.

Palisca, Claude V. 1961. "Scientific Empiricism in Musical Thought." In *Seventeenth Century Science and the Arts*, ed. Hedley Howell Rhys, pp. 91–137. Princeton: Princeton University Press.

Pantalony, David. 2005. "Rudolph Koenig's Workshop of Sound: Instruments, Theories, and the Debate over Combination Tones." *Annals of Science* 62: 57–82.

Parker, John R. 1824. *A Musical Biography: or Sketches of the Lives and Writings of Eminent Musical Characters. Interspersed with an Epitome of Interesting Musical Matter.* Boston: Stone and Fovell.

Pederson, Sanna. 1994. "A. B. Marx, Berlin Concert Life, and German National Identity." *19th Century Music* 18: 87–107.

Pellisov. 1834. "Ueber die Kirchenmusik des katholischen Cultus." *Allgemeine musikalische Zeitung* 36: cols. 721–744.

Perne, François-Louis. 1821. "Nachricht von der im Uebungs-Saale des königlichen Conservatoriums der Musik in Paris aufgestellten, crescendo und decrescendo spielbaren Orgel (*Orgue expressif*)." *Allgemeine musikalische Zeitung* 23: cols. 133–139, 149–156.

Pestiscus. 1807. "Ueber musikalische Lehrbücher und die neuesten unter derselben." *Allgemeine musikalische Zeitung* 9: 161–166, 177–183.

Pfannenstiel, Max. 1958. *Kleines Quellenbuch zur Geschichte der Gesellschaft Deutscher Naturforscher und Ärzte. Gedächtnisschrifte für die hundertste Tagung der Gesellschaft.* Berlin: Springer Verlag.

Pfeiffer, Michael Traugott, and Hans-Georg Nägeli. 1810. *Gesangsbildungs-Lehre nach Pestalozzischen Grundsätzen.* Zurich: H. G. Nägeli.

Plantinga, Leon. 1984. *Romantic Music: A History of Musical Style in Nineteenth-Century Europe.* London: Norton.

Poggendorff, J. C. 1834. "Zusatz des Herausgebers." *Annalen der Physik und Chemie* 32 new series (108 of the original): 520–528.

Poisson, Siméon-Denis. 1808. "Mémoire sur la theorie su son." *Journal de l'École Polytechnique* 7: 319–392.

Poisson, Siméon-Denis. 1835–1836. *Lehrbuch der Mechanik. Nach der zweiten sehr vermehrten Ausgabe übersetzt von Moriz A. Stern,* 2 parts. Berlin: G. Riemer. Original: 1811, *Traité de mécanique.* Paris: Courcier.

Polanyi, Michael. 1994. *Personal Knowledge: Towards a Post-Critical Philosophy.* Chicago: University of Chicago Press.

Poppe, Johann Heinrich Moritz. 1810. *Praktisches Handbuch für Uhrmacher, Uhrenhändler, und für Uhrenbesizzer,* 2nd edition. Leipzig: Commersche Buchhandlung.

Powell, Ardal. 1791. *The Virtuoso Flute-Player.* Cambridge: Cambridge University Press.

Purcell, Henry. 1696. *A Choice of Collection of Lessons for the Harpsichord or Spinnet Composed by Ye Late Mr Henry Purcell.* London: Henry Playford.

Quandt, C. F. 1791. "Antwort auf Hern. Dr. Chladni's Erklärung, die Erfindung des Euphons betreffend." *Intelligenz-Blatt des Journals des Luxus und der Moden* 6: xxiv.

Quandt, C. F. 1800. "Die Harmonika aus der 17ten Jahrhundert." *Allgemeine musikalische Zeitung* 3: cols. 151–152.

R. 1807. "Ueber blinde Musiker." *Allgemeine musikalische Zeitung* 9: cols. 209–219.

Ramann, Lina, editor. 1880–1883. *Franz Liszts Gesammelte Schriften.* 6 vols. Leipzig: Breitkopf and Härtel.

Rasch, Rudolf. 2002. "Tuning and Temperament." In *The Cambridge History of Western Music Theory,* ed. Thomas Christensen, pp. 193–222. Cambridge: Cambridge University Press.

Rave, Paul Ortwin. 1962. *Karl Friedrich Schinkel. Berlin. Dritter Teil. Bauten für Wissenschaft, Verwaltung, Heer, Wohnbau und Denkmäler*. Berlin: Deutscher Kunstverlag.

Ravetz, Jerome R. 1972. *Scientific Knowledge and Its Social Problems*. Oxford: Clarendon Press.

Rbt. 1931. "Johann Heinrich Schiebler-Heydweiler (1777–1837)." *Die Heimat. Mittheilungen der Vereine für Heimatkunde in Krefeld-Uerdingen am Rhein* 10: 170–182.

Reichardt, Johann Friedrich. 1776. *Ueber die Pflichten des Ripien-Violinisten*. Berlin and Leipzig: G. J. Decker.

Reichel, C. 1883. "Ueber die Justirung der Stimmgabeln auf genau vorgeschriebene Schwingungszahlen." *Zeitschrift für Instrumentenkunde* 3: 47–51.

Reichmeister, J. C. 1828. *Die Orgel in einem guten Zustande und Stimmung zu erhalten. Ein unentbehrliches Handbuch für angehende Organisten und Schullehrer, nebst vorausgeschichter Beschreibung der Orgel nach allen ihren Teilen*. Leipzig: Fest.

Reichmeister, J. C. 1832. *Unentbehrliches Hilfsbuch beim Orgelbau. Ein treuer Ratgeber für Kommunen, Kirchenbeamten, Organisten und alle, die, welche bei dem Neubau oder der Hauptreparatur einer Orgel wesentliche Obliegenheiten zu erfüllen haben*. Leipzig: A. Fest'schen Verlags-Buchhandlung.

Ribock, Justus Johannes Heinrich. 1782. *Bemerkungen über die Flöte, und Versuch einer kurzen Anleitung zur besseren Einrichtung und Behandlung derselben*. Stendahl: Franzen und Grosse.

Riehn, Rainer. 1979. "Beethovens Verhältnis zum Metronom." In *Musik Konzepte* 8: *Beethoven: das Problem der Interpretation*, ed. Heinz-Klaus Metzger and Rainer Riehn, pp. 70–84. Munich: Edition Text und Kritik.

Riemann, Hugo. 1888. *Kathechismus des Klavierspiels*. Leipzig: M. Hesse.

Riemann, Ludwig. 1911. *Das Wesen des Klavierklanges und seine Beziehung zum Anschlag. Eine akustisch-ästhetische Untersuchung für Unterricht und Haus*. Leipzig: Breitkopf and Härtel.

Riemer, Friedrich Wilhelm, editor. 1833–1834. *Briefwechsel zwischen Goethe und Zelter in der Jahren 1796 bis 1832*. Berlin: Duncker und Humblot.

Rink, John. 2002. "The Profession of Music." In *The Cambridge History of Nineteenth-Century Music*, ed. Jim Samson, pp. 55–86. Cambridge: Cambridge University Press.

Riskin, Jessica. 2003. "The Defecating Duck, Or, The Ambiguous Origins of Artificial Life." *Critical Inquiry* 20: 599–633.

Ritter, Hermann. 1920. *Alte rheinische Fabrikantenfamilien und ihre Industrien*. Cologne: Heinrich Z. Gonski.

Rochlitz, Friedrich. 1818. "Zur Geschichte des musikalischen Metronomen." *Allgemeine musikalische Zeitung* 20: cols. 468–472.

Roeber, August. 1834. "Untersuchungen des Hern. Scheibler in Crefeld über die sogenannten Schläge, Schwebungen oder Stösse." *Annalen der Physik und Chemie* 31 of the new series (108 of the original): 333–362 and 492–520.

Roeber, August. 1839. "Akustik." In *Repertorium der Physik. Enthaltend eine vollständige Zusammenstellung der neuern Fortschritte dieser Wissenschaft*, ed. Heinrich Wilhelm Dove, 4 vols., 1838–1841, 3 (1839), pp. 1–141. Berlin: Verlag von Veit and Comp.

Röllig, Karl L. Circa 1786. *Zwölf kleine Tonstücke für die Harmonika*. Leipzig: Breitkopf.

Röllig, Karl L. 1787. *Über die Harmonika*. Leipzig: Breitkopf and Härtel.

Rosen, Charles. 1995. *The Romantic Generation*. Cambridge, Mass.: Harvard University Press.

Rösen, Heinrich. 1965. "Sicherheitswache." *Die Heimat. Krefelder Jahrbuch* 36: 38–39.

Rudorff, Ernst, editor. 1900. *Briefe von Carl Maria von Weber an Hinrich Lichtenstein*. Brunswick: George Westermann.

Sachs, Curt. 1912–1913. "Zur Frühgeschichte der durchschlagenden Zunge." *Zeitschrift für Instrumentenbau* 33: 1323–1325.

Sachs, Curt. 1913. *Real-Lexikon der Musikinstrumente, zugleich ein Polyglossar für das gesamte Instrumentengebiet*. Berlin: J. Bard.

Sachs, Curt. 1920. *Handbuch der Musikinstrumenenkunde*. Leipzig: Breitkopf and Härtel.

Sachs, Curt. 1940. *A History of Musical Instruments*. New York: W. W. Norton.

Sauer, Leopold. 1813. "Notizen." *Allgemeine musikalische Zeitung* 15: cols. 117–120.

Sauer, Leopold. 1824. "Ueber das Crescendo in des Abt Voglers Orchestrion." *Allgemeine musikalische Zeitung* 26: cols. 370–375.

Sauveur, Joseph. 1701. *Principes d'acoustique et de musique: ou, Système general des intervalles des sons, et de son application à tous les systêmes et à tous instruments de musique*. Paris: s. n.

Schaffer, Simon. 1990. "Genius in Romantic Natural Philosophy." In *Romanticism and the Sciences*, ed. Andrew Cunningham and Nicholas Jardine, pp. 82–98. Cambridge: Cambridge University Press.

Schaffer, Simon. 1992. "Late Victorian Metrology and Its Instrumentation: A Manufactory of Ohms." In *Invisible Connections: Instruments, Institutions, and Science*, ed. Robert Bud and Susan E. Cozzens, pp. 23–56. Bellingham, Wash.: SPIE.

Schaffer, Simon. 1994. "Babbage's Intelligence: Calculating Engines and the Factory System." *Critical Inquiry* 21: 203–277.

Schaffer, Simon. 1995. "Accurate Measurement Is an English Science." In *The Value of Precision*, ed. M. Norton Wise, pp. 135–172. Princeton: Princeton Univesity Press.

Schaffer, Simon. 1997. "Experimenters' Techniques, Dyers' Hands, and the Electric Planetarium." *Isis* 88: 456–483.

Scheibe, Johann Adolf. 1745. *Critischer Musikus. Neue vermehrte und verbesserte Auflage.* Leipzig: Bernhard Christian Breitkopf.

Scheibler, [Johann] H[einrich]. 1816. "Die Aura." *Allgemeine musikalische Zeitung* 18: cols. 505–511.

Scheibler, [Johann] Heinrich. 1834a. *Der physikalische und musikalische Tonmesser, welcher durch den Pendel, dem Auge sichtbar, die absoluten Vibrationen der Töne, der Haupt-Gattungen von Combination-Tönen, so wie die schärfste Genauigkeit gleichschwebender und mathematischer Accorde beweist.* Essen: G. D. Bädeker.

Scheibler, [Johann] Heinrich. 1834b. *Mittheilungen über das Wesentliche des musikalischen und physikalischen Tonmessers.* Krefeld: C. M. Schüller.

Scheibler, [Johann] Heinrich. 1836. *Anleitung, die Orgel unter Beibehaltung ihrer momentanen Höhe oder nach einiem bekannten* a *vermittels des Metronoms nach Stössen gleichschweben zu stimmen.* Krefeld: C. M. Schüller.

Scheibler, [Johann] Heinrich. 1837a. *Mittheilungen an die Versammlung der deutschen Naturforscher zu Bonn im Jahr 1835 über das wesentliche des physikalischer und musikalischer Tonmessers.* Crefeld: no publisher listed.

Scheibler, [Johann] Heinrich. 1837b. *Ueber mathematische Stimmung, Temperaturen und Orgelstimmung nach Vibrations-Differenzen oder Stössen.* Krefeld: no publisher listed.

Scheibler, [Johann] Heinrich. 1838. *Schriften über musikalische und physikalische Tonmessung und deren Anwendung auf Pianoforte- und Orgelstimmung.* Krefeld: C. M. Schüller.

Schicht, J. G. 1812. *Grundregeln der Harmonie, nach dem Verwechslungs-System entworfen und mit Beispielen erläutert.* Leipzig: Breitkopf and Härtel.

Schiller, Friedrich. 1967. *On the Aesthetic Education of Man.* Trans. Elizabeth M. Wilkinson and L. A. Willoughby. Oxford: Clarendon Press.

Schilling, Gustav, editor. 1835–1838. *Encyclopädie der gesammten musikalischen Wissenschaften, oder das Universal-Lexicon der Tonkunst,* 6 vols. Stuttgart: Franz Heinrich Köhler.

Schindler, Anton. 1840. *Biographie von Ludwig van Beethoven.* Münster: Aschendorff.

Schindler, Anton. 1855. "Die gegenwärtige hohe Orchester-Stimmung und ihr Ausgang." *Niederrheinische Musik-zeitung für Kunstfreunde und Künstler* 3: 66–68.

Schipperges, Heinrich, editor. 1968. *Die Versammlung Deutscher Naturforscher und Ärzte im 19. Jahrhundert.* Stuttgart: Gentner.

Schipperges, Heinrich. 1972. "Einführung." In Hans Querner and Heinrich Schipperges, editors, *Wege der Naturforschung 1822–1972 im Spiegel der Versammlungen Deutscher Naturforscher und Ärzte.* Berlin: Springer Verlag.

Schleuning, Peter. 1984. *Geschichte der Musik in Deutschland. Das 18. Jahrhundert: Der Bürger erhebt sich.* Hamburg: Rowohlt.

Schmidt, Alois. 1909. *Johann Heinrich Schreibler-Heydweiller (1777–1837).* Bonn: Carl Georgi.

Schmidt, Bernhard. 2002. *Lied-Kirchenmusik-Predigt im Festgottesdienst Friedrich Schleiermachers. Zur Rekonstruction seiner liturgischen Praxis.* New York: Walter de Gruyter.

Schmitt, Hans. 1894. *Ueber die Kunst des Anschlags: Vorlesung, gehalten im Wiener wissenschaftlichen Club.* Vienna: Doblinger.

Schmitz, Hans W. 1983. "Die Familie Kaufmann. 150 Jahre Entwickelung der selbstspielenden Musikinstrumente." *Das mechanische Muiskinstrument* 8: 10–34.

Schneider, Wilhelm. 1834. *Historisch-technische Beschreibung der musicalishen Instrumente, ihres Alters, Tonumfangs und Baues, ihrer Erfinder, Verbesserer, Virtuosen und Schulen, nebst einer fasslichen Anweisung zur gründlichen Kenntniss und Behandlung derselben.* Neisse and Leipzig: Theodor Henning.

Schneider, Wilhelm. 1835. *Die Orgelregister, deren Enstehung, Name, Behandlung, Benutzung und Mischung.* Leipzig: Friese.

Schröder, Cornelia. 1959. *Carl Friedrich Zelter und die Akademie: Dokumente und Briefe zur Entstehung der Musik-Sektion in der Preußischer Akademie der Künste.* Berlin: Akademie der Künste.

Schünemann, Georg. 1932. *Carl Friedrich Zelter, der Begründer der Preussischen Musikpflege.* Berlin: Max Hesses Verlag.

Schünemann, Georg. 1937. *Carl Friedrich Zelter. Der Mensch und sein Werk.* Berlin: Berliner Bibliophilen-Abend.

Schütz, F. G. J. Circa. 1830. *Leben, Charakter und Kunst des Ritters Nicolo Paganini.* Ilmenau: Bernhard Friedrich Voigt.

Schulze, Johann Friedrich. 1829. "Verbesserungen in Orgelbau." *Allgemeine musikalische Zeitung* 31: cols. 189–191.

Schuster, Carolyn. 1991. "Les Orgues Cavaillé-Coll au salon, au théatre et au concert." Ph.D. dissertation, University of Tours.

Schwägrichten, Friedrich, and Gustav Kunze. 1823. "Statuten der Gesellschaft der deutschen Naturforscher und Aerzte." *Isis* 1: cols. 1–3.

Schweigger, J. S. C. 1823. "Eine rückwärts und vorwärts gerichtete Betrachtung, als Ergebnis der ersten Tagung der Gesellschaft Deutscher Naturforscher und Aerzte." *Journal für Chemie und Physik* 37: 455–458.

Schwiening. 1837. "Ueber den Unterschied der bisherigen und der Scheibler'schen Stimm-Methode und über die Wichtigkeit der letztern." *Caecilia* 19: 217–241.

Schwinning. 1835. "Ueber den Klang der Saiteninstrument." *Caecilia* 17: 240–254.

Seibt, Ilsabe. 1998. *Friedrich Schleiermacher und das Berliner Gesangbuch von 1829.* Göttingen: Vandenhoeck and Ruprecht.

Sembdner, Helmut, editor. 1970. *Heinrich von Kleist. Sämtliche Werke und Briefe*, 5th edition, 2 vols. Munich: Carl Hanser Verlag.

Seyfried, Ignaz von. 1825. "Recension von *Méthode de Violinschule*, J. J. F. Dotzauer. *Violinzell-Schule* von J. F. F. Dotzauer (Mainz: B. Schott und Söhne)." *Caecilia* 3: 249–262.

Shapin, Steven. 1994. *A Social History of Truth: Civility and Science in Seventeeth-Century England.* Chicago: University of Chicago Press.

Siefert, Helmut. 1969. *Das naturwissenschaftliche und medizinische Vereinswesen im deutschen Sprachgebiet: 1750–1850.* Frankfurt am Main: self-published.

Sievers, G. L. P. 1817. "Ueber den jetztigen Zustand der Musik in Frankreich, besonders in Paris." *Allgemeine musikalische Zeitung* 19: cols. 265–277, 281–292, 297–304.

Sievers, G. L. P. 1819. "Pariser musikalische Allerley." *Allgemeine musikalische Zeitung* 21: cols. 588–593, 597–603.

Simpson, Christopher. 1665. *The Principles of Practical Musick Microform: Delivered in a Compendious, Easie, and New Method, for the Instruction of Beginners, either in Singing or Playing upon Instruments: To Which Are Added, some Short and Easie Ayres Designed for Learners.* London: William Godbid for Henry Brome.

Smith, Crosbie W., and M. Norton Wise. 1989. *Energy and Empire: A Biographical Study of Lord Kelvin.* Cambridge: Cambridge University Press.

Spohr, Louis. 1832. *Violinschule, mit erlaeuternden Kupfertafeln.* Vienna: T. Haslinger.

Sprengel, P. N. 1771. *Handwerke und Künste in Tabellen*, 8th collection. Berlin: Buchhandlung der Realschule.

Stadlen, Peter. 1979. "Beethoven und das Metronom." In *Musik Konzepte* 8: *Beethoven: das Problem der Interpretation*, ed. Heinz-Klaus Metzger and Rainer Riehn, pp. 12–33. Munich: Edition Text und Kritik.

Staudinger, Hans. 1913. *Individuum und Gemeinschaft in der Kulturorganisation des Vereins. Zwei Teile. Teil I: Formen und Schichten, dargestellt am Werdegang der musikalisch-geselligen Organisation.* Jena: Eugen Diederich.

Stephenson, Bruce. 1994. *The Music of the Heavens. Kepler's Harmonic Astronomy.* Princeton: Princeton University Press.

Steuber. 1810. "Ueber die ästhetische Bildung des componirenden Tonkünstlers." *Allgemeine musikalische Zeitung* 12: cols. 321–324, 793–799.

Stöckel, G. E. 1800. "Ueber die Wichtigkeit der richtigen Zeitbewegung eines Tonstücks, nebst einer Beschreibung eines musikalischen Chronometers und dessen Anwendung für Komponisten, Ausführer, Lehrer und Lernende der Tonkunst." *Allgemeine musikalische Zeitung* 3: cols. 657–666, 673–679.

Stöpel, Franz. 1821. "Auszug aus einem Berichte an das königlichen Preussischen Ministeriums des Cultus, das Logierische Lehr- und Unterrichts-System in der Musik betreffend." *Allgemeine musikalische Zeitung* 23: cols. 553–557.

Stöpel, Franz. 1826. "Zur Beantwortung der 'Erklärung' des Musiklehrers J. B. Logier in No. 69 des Frankfurter Journals d. J." *Intelligenzblatt zur Caecilia* 4, pamphlet attached to p. 52.

Stowell, Robin. 1985. *Violin Technique and Performance Practice in the Late Eighteenth and Early Nineteenth Centuries.* Cambridge: Cambridge University Press.

Strohmann. 1810. "Notizen." *Allgemeine musikalische Zeitung* 12: col. 968.

Strohmann. 1811. "Verbesserung der Rohrwerke in der Orgel." *Allgemeine musikalische Zeitung* 13: cols. 153–157.

Strombeck, F. K. von, and David Mansfeld, editors. 1842. *Amtlicher Bericht über die neunzehnten Versammlung der Gesellschaft deutscher Naturforscher und Aerzte zu Braunschweig.* Brunswick: Friedrich Vieweg and Son.

Sudhoff, Karl. 1922. *Hundert Jahre Deutscher Naturforscher-Versammlungen. Gedächtnisschrift zur Jahrhundert-Tagung der Gesellschaft Deutscher Naturforscher und Ärzte, Leipzig, im September 1922.* Leipzig: F. C. W. Vogel.

Sudnow, David. 2002. *Ways of the Hand*, 2nd edition. Cambridge, Mass.: MIT Press.

Sulzer, Johann Georg. 1771–1774. *Allgemeine Theorie der schönen Künste.* 3 vols. Leipzig: M. G. Weidmanns Erben und Reich.

Sweet, Paul R. 1980. *Wilhelm von Humboldt.* 2 vols. Columbus: Ohio State University Press.

Tacke, Hans-Christian. 2002. *Johann Gottlob Töpfer (1791–1870). Leben-Werk-Wirksamskeit.* Kassel: Bärenreiter.

Tans'ur, William. 1756. *New Musical Grammar or the Harmonical Spectator.* London: J. Hodges.

te Heese, Anke. 2001. "Vom naturgeschichtlichen Investor zum Staatsdiener: Sammler und Sammlungen der Gesellschaft Naturforschender Freunde zu Berlin um 1800." In *Sammeln als Wissen: Das Sammeln und seine wissenschaftliche Bedeutung,* ed. Anke te Heesen and E. C. Spary, pp. 62–84. Göttingen: Wallstein Verlag.

Thayer, Alexander, et al., editors. 1901–1911. *Ludwig van Beethoven's Leben.* 5 vols. Berlin: W. Weber.

Thompson, Emily. 2002. *The Soundscape of Modernity: Architectural Acoustics and the Culture of Listening in America, 1900–1933.* Cambridge, Mass.: MIT Press.

Thun, Alphons. 1879. *Die Industrie am Niederrhein und ihre Arbeiter. Erster Theil. Die linksrheinische Textilindustrie.* Leipzig: Verlag von Duncker & Humblot.

Tiggelen, Philippe John van. 1987. *Componium: The Mechanical Musical Improvisor.* Louvain-la-Neuve: Institut Supérieur d'Archèologie et d'Histoire de l'Art, Collège Érasme.

Töpfer, Johann Gottlob. 1833. *Die Orgelbau-Kunst, nach einer neuen Theorie dargestellt und auf mathematische und physicalische Grundsätze gestützt.* Weimar: Wilhelm Hoffmann.

Töpfer, Johann Gottlob. 1834. *Erster Nachtrag zu Orgelbau-Kunst, welcher die vollständigung der Mensuren zu den Labialstimmen und die Theorie der Zungenstimmen mit den dazu gehörigen Mensur-Tabellen, nebst einer Anweisung zur Verfertigung derselben.* Weimar: Wilhelm Hoffmann.

Töpfer, Johann Gottlob. 1842. *Die Scheibler'sche Stimm-Methode.* Erfurt: Verlag der Kunst- und Musikalien-Handlung von Wilhelm Körner.

Töpfer, Johann Gottlob. 1843. *Die Orgel, Zweck und Beschaffenheit ihrer Theile, Gesetze ihrer Construction, und Wahl der dazu gehörigen Materialen.* Erfurt: G. Wilhelm Körner.

Töpfer, Johann Gottlob. 1855. *Lehrbuch der Orgelbaukunst, nach den besten Methoden älterer und neurer, in ihrem Fach ausgezeichneter Orgelbaumeister und begründet auf mathematische und physicalische Gesetze.* Weimar: Bernhard Friedrich Voigt.

Turner, R. Steven. 1977. "The Ohm-Seebeck Dispute, Hermann von Helmholtz, and the Origins of Physiological Acoustics." *British Journal for the History of Science* 10: 1–24.

Tyndall, John. 1867. *Sound.* New York: Appleton.

Ullmann, Dieter. 1996. *Chladni und die Entwicklung der Akustik von 1750–1860.* Berlin: Birkhäuser.

Uthe. 1811. "Berichtigung einiger Ausserungen des Herrn Strohmann in seinem Aufsatze in No. 9 d. mus. Zeit. v. diesem Jahre." *Allgemeine musikalische Zeitung* 13: cols. 429–431.

Vieth, [Gerhard Ulrich Anton]. 1805. "Über Combinationstöne, in Beziehung auf einige Streitsschriften über die zweier englischer Physiker, Th. Young und J. Gougg." *Annalen der Physik* 21: 265–314.

Vogler, Georg Joseph [Abt], 1801. "Data zur Akustik." *Allgemeine musikalische Zeitung* 3: cols. 517–525, 533–540, 519–554, 565–571.

"Vorbemerkung der Redaktion." 1835. *Allgemeine musikalische Zeitung* 37: cols. 485–488.

Walker, Alan. 1983–1996. *Franz Liszt.* 3 vols. New York: Alfred A. Knopf.

Walker, D. P. 1967. "Kepler's Celestial Music." *Journal of the Warburg and Courtauld Institutes* 30: 228–250.

Walther, Johann Gottfried. 1732. *Musikalisches Lexikon oder musikalische Bibliothec.* Leipzig: Wolffgang Deer.

Wapnick, Joel, and Peter Friedman. 1980. "Effects on Dark–Bright Timbral Variation on the Perception of Flatness and Sharpness." *Journal of Research in Music Education* 28: 176–184.

Warwick, Andrew C. 1998. "Exercising the Student Body: Mathematics and Athleticism in Victorian Cambridge." In *Science Incarnate: Historical Embodiments of Natural Knowledge,* ed. Christopher Lawrence and Steven Shapin, pp. 288–326. Chicago: University of Chicago Press.

Warwick, Andrew C. 2003. *Masters of Theory: Cambridge and the Rise of Mathematical Physics.* Chicago: University of Chicago Press.

Weber, Carl Maria von. 1812. "Der Trompeter, eine Maschine von der Erfindung des Mechanicus Hern. Friedrich Kaufmann, in Dresden." *Allgemeine musikalische Zeitung* 14: cols. 633–636.

Weber, E. 1834. "Ueber den Werth des Chiroplasten in Bezug auf den in No. 2 der Leipziger Zeitschrift für Musik unter der Rubik; Ueber den Handleiter enthaltenen Aufsatz." *Neue Leipziger Zeitschrift für Musik* 1: 237–239.

Weber, Ernst Heinrich, and Wilhelm Eduard Weber. 1825. *Die Wellenlehre auf Experimente gegründet oder über die Wellen tropfbarer Flüssigkeiten mit Anwendung auf die Schall- und Lichtwellen.* Leipzig: Gerhard Fleischer.

Weber, Ernst Heinrich, and Wilhelm Eduard Weber. 1826. "Allgemein fassliche Darstellung des Vorganges, durch welchen Saiten und Pfeifen dazu gebracht werden, einfache Töne und Flageolettöne hervorbringen, nebst Erörterungen der

Verschiedenheit des Zustandes, in dem sich schalleitende, das Selbsttönen erregende, selbsttönende und resonirende Körper befinden." *Allgemeine musikalische Zeitung* 28: cols. 185–200, 205–213, 221–236.

Weber, Gottfried. 1813. "Noch einmal ein Wort über den musikalischen Chronometer oder Taktmesser." *Allgemeine musikalische Zeitung* 15: cols. 441–447.

Weber, Gottfried. 1814. "Ueber die jetzt bevorstehende wirkliche Einführung des Taktmessers." *Allgemeine musikalische Zeitung* 16: cols. 445–449, 461–464.

Weber, Gottfried. 1824. *Versuch einer geordneten Theorie der Tonsetzkunst*, 2 vols. Mainz: B. Schott's Söhne.

Weber, Gottfried. 1827. "Vorwort" to E. F. F. Chladni, "Über die verschiedene Beschaffenheit des Klanges eines Instruments, nachdem es vom verschiedenen Spielern behandelt wird, auch über das Zerschlagen der Claviersaiten." *Caecilia* 6: 183–188. Reprinted in *Caecilia* 17 (1835): 233–240.

Weber, Gottfried. 1829a. "Vorwort" to Wilhelm Weber, "Compensation der Orgelpfeifen." *Caecilia* 11: 184–186.

Weber, Gottfried. 1829b. "Vorwort" to Carl Guhr, "Paganinis Kunst die Violine zu spielen." *Caecilia* 11: 76–78.

Weber, Gottfried. 1833. "Vortwort" to Wilhelm Weber, "Ueber die Erzeugung der Aliquottöne auf Zungenpfeifen und auf den Clarinetten." *Caecilia* 12: 1–2.

Weber, Rainer. 1992. "Was sagen die Holzblasinstrumente zur Mozarts Kammerton?" *Tibia* 4: 291–298.

Weber, Wilhelm. 1827. *Leges oscillationis oriundae, si duo corpora diversa celeritate oscillantia, ita conjunguntur ut oscillare non possint nisi simul et synchronice*. Halle: Henrich Eduard Floss.

Weber, Wilhelm. 1828a. "Compensation der Orgelpfeifen, ein Vortrag des Prof. Wilhelm Weber zu Halle, bei der Versammlung der deutschen Naturforscher zu Berlin, den 19. September 1828." *Annalen der Physik und Chemie* 14 of the new series (90 of the original): 397–408.

Weber, Wilhelm. 1828b. "Zur Aërodynamik und Akustik. II. Etwas über resonirende Luftsäule und Lufträume von Wheatstone." In *Jahrbuch der Chemie und Physik* 23 of the new series: 327–333 (ed. J. S. C. Schweigger and Fr. W. Schweigger-Seidel).

Weber, Wilhelm. 1829a. "Compensation der Orgelpfeifen [with an introduction by Gottfried Weber]." *Caecilia* 11: 186–204.

Weber, Wilhelm. 1829b. "Ueber die Construction und den Gebrauch der Zungenpfeifen." *Annalen der Physik und Chemie* 16 of the new series (92 of the original): 193–206.

Weber, Wilhelm. 1829c. "Versuche mit Zungenpfeifen." *Annalen der Physik und Chemie* 16 of the new series (92 of the original): 415–438.

Weber, Wilhelm. 1829d. "Theorie der Zungenpfeifen." *Annalen der Physik und Chemie* 17 of the new series (93 of the original): 192–246.

Weber, Wilhelm. 1829e. "Ueber die Tartinischen Töne." *Annalen der Physik und Chemie* 15 of the new series (91 of the original): 216–222.

Weber, Wilhelm, 1829f. "Ueber die zweckmässige Einrichtung eines Monochords oder Tonmessers und den Gebrauch desselben, zum Nutzen der Physik und Musik." *Annalen der Physik und Chemie* 15 of the new series (91 of the original): 1–19.

Weber, Wilhelm. 1830a. "Lebensbild E. F. F. Chladni." In *Allgemeine Encyclopädie der Wissenschaften und Künste* 21: 177–190 (ed. J. S. Ersch und J. G. Gruber).

Weber, Wilhelm. 1830b. "Ueber die specifische Wärme fester Körper insbesondere der Metalle." *Annalen der Physik und Chemie* 20 of the new series (96 of the original): 177–213.

Weber, Wilhelm. 1830c. "Ueber die Erzeugung der Aliquottöne auf Zungenpfeifen und auf der Klarinette." *Caecilia* 12: 3–26.

Weber, Wilhelm. 1833. "Vergleichung der Theorie der Saiten und Blas-Instrumente." *Annalen der Physik und Chemie* 28 of new series (104 of the original): 1–17.

Weber, Wilhelm. 1892–1894. *Wilhelm Weber's Werke*. Hrsg. von der königlichen Gesellschaft der Wissenschaften zu Göttingen, 6 vols. Berlin: Julius Springer.

Weber, William. 1975. *Music and the Middle Class: The Social Structure of Concert Life in London, Paris, and Vienna*. London: Groom Helm.

Wehmeyer, Grete. 1993. *Prestißißimo. Die Wiederentdeckung der Langsamkeit in der Musik*. Hamburg: Rowohlt.

Wehran, August Heinrich. 1853. *An Essay on the Theory and Practice of Tuning in General, and on Scheibler's Invention of Tuning Pianofortes and Organs by the Metronome in Particular, for the Use of Musicians, Tuners, and Organ Builders*. London: Robert Cocks.

Weiske, J. G. 1790. *Zwölf geistliche prosaiche Gesänge, mit Beschreibung eines Taktmessers*. Leipzig: Breitkopf and Härtel.

Weißflog, Egon. 1999. "Der Trompeterautomat von Seyffert und Heinrich aus den Jahren 1816/1818." *Das mechanische Musikinstrument* 24: 1–4.

Wendt, Am. 1811. "Betrachtungen über Musik, und insbesondere über den Gesang, als Bildungsmittel in der Erziehung." *Allgemeine musikalische Zeitung* 13: cols. 281–287, 297–303, 316–324, 333–341.

Wenk, August Heinrich. 1798. *Beschreibung eines Chronometers oder musikalischen Taktmessers und seines vortheilhaften Gebrauchs für das musikliebende Publikum.* Magdeburg: G. C. Keil.

Werber, Karl. 1976. *Wilhelm Weber: Biographen hervorragender Naturwissenschaftler, Techniker und Mediziner* 22. Leipzig: B. G. Teubner.

Werner, Johann Gottlob. 1823. *Lehrbuch, das Orgelwerk nach allen seinen Theilen kennen, erhalten, beurtheilen und verbessern zu lernen.* Merseburg: F. Kobitzsch.

Wertheim, Guillaume. 1848. "Mémoire sur la vitesse du sons dans les liquids." *Annales de chimie et physique* 23: 434–475.

Wheatstone, Charles. 1828. "On the Resonances, or Reciprocated Vibrations of Columns of Air." *Quarterly Journal of Science, Literature, and Art* 5 of the new series: 175–183.

Wiederkehr, Karl Heinrich. 1960. "Wilhelm Webers Stellung in der Entwicklung der Elektrizitätslehre." Ph.D. dissertation, University of Hamburg.

Wiederkehr, Karl Heinrich. 1967. *Wilhelm Eduard Weber. Erforscher der Wellenbewegung und der Elektrizität, 1804–1891.* Stuttgart: Wissenschaftliche Verlagsgesellschaft.

Wilke, Friedrich. 1821. "Warum findet man so viele schlechte Orgeln, und wie möchte diesem Uebel abzuhelfen seyn?" *Allgemeine musikalische Zeitung* 23: cols. 625–630, 641–646.

Wilke, Friedrich. 1823. "Ueber Crescendo- und Diminuendo-Züge an Orgeln." *Allgemeine musikalische Zeitung* 25: cols. 113–119, 149–155.

Wilke, Friedrich. 1824. "Ueber das Wirken des Abts und Geh. Raths Vogler im Orgelbaufache." *Allgemeine musikalische Zeitung* 26: cols. 673–677, 689–694.

Wilke, Friedrich. 1825. "Ueber die von Hrn. Kaufmann in der Leipziger musikalischen Zeitung Jahrgang 25, No. 8, Seite 118 und 119 gegebene Bemerkung, den Nachtheil eines gleichstarken Orgelwindes zu den Labial- und Zungenstimmen betreffend." *Allgemeine musikalische Zeitung* 27: cols. 541–544.

Wilke, Friedrich. 1834. "Ueber compensirte Zungenpfeifen." *Caecilia* 16: 64.

Wilke, Friedrich, and Friedrich Kaufmann. 1823. "Ueber die Erfindung der Rohrwerke mit durchschlagenden Zungen." *Allgemeine musikalische Zeitung* 25: cols. 149–155.

Wise, M. Norton. 1999. "Architectures for Steam." In *The Architecture of Science*, ed. Peter Galison and Emily Thompson, pp. 107–140. Cambridge, Mass.: MIT Press.

Wolfram, Johann Christian. 1815. *Anleitung zur Kenntnis, Beurteilung und Erhaltung der Orgeln; für Orgelspieler und alle diejenigen, welche bei Erbauung, Reparatur und Erhaltung dieser Instrumente interessiert sind.* Gotha: C. Steudel.

Wonneberger, Lothar. 1989. "Eine interessante Variante eines Trompeterautomaten." *Das mechanische Muiskinstrument* 13: 9–10.

Woolhouse, Wesley Stoker Barker. 1835. *Essay on Musical Intervals, Harmonics, and the Temperament of the Musical Scale.* London: J. Souter.

Wright, Thomas. 1795. *A Concerto for the Harpsichord or Piano Forté, with Accompaniments for Two Violins, Two Oboes, Two Horns, a Tenore, and Bass, etc.* London: Preston and Son.

Young, Thomas. 1802. "In Reply to Mr. Gough's Letters." *Nicholson's Journal* 3: 145–146.

Young, Thomas. 1855. *Miscellaneous Works of the Late Thomas Young.* 3 vols. Ed. George Peacock and John Leitch. London: J. Murray.

Zang, Johann Heinrich. 1804. *Der vollkommene Orgelmacher, oder Lehre von der Orgel und Windprobe, der Reparatur und Stimmung der Orgeln und anderer Tasteninstrumente.* Nuremberg: A. G. Schneider and Weige.

Index

Printed in the United States
By Bookmasters